Moments, Positive Polynomials and Their Applications

Imperial College Press Optimization Series

ISSN 2041-1677

Series Editor: Jean Bernard Lasserre *(LAAS-CNRS and Institute of Mathematics, University of Toulouse, France)*

Vol. 1: Moments, Positive Polynomials and Their Applications
by Jean Bernard Lasserre

Vol. 2: Examples in Markov Decision Processes
by A. B. Piunovskiy

Imperial College Press Optimization Series Vol. 1

Moments, Positive Polynomials and Their Applications

Jean Bernard Lasserre

LAAS-CNRS and Institute of Mathematics
University of Toulouse, France

Imperial College Press

Published by

Imperial College Press
57 Shelton Street
Covent Garden
London WC2H 9HE

Distributed by

World Scientific Publishing Co. Pte. Ltd.
5 Toh Tuck Link, Singapore 596224
USA office: 27 Warren Street, Suite 401-402, Hackensack, NJ 07601
UK office: 57 Shelton Street, Covent Garden, London WC2H 9HE

British Library Cataloguing-in-Publication Data
A catalogue record for this book is available from the British Library.

First published 2010 (Hardcover)
Reprinted 2016 (in paperback edition)
ISBN 978-1-911299-73-8

MOMENTS, POSITIVE POLYNOMIALS AND THEIR APPLICATIONS
Imperial College Press Optimization Series — Vol. 1

ISBN-13 978-1-84816-445-1
ISBN-10 1-84816-445-9

To my daughter Julia and to Carole

Preface

Consider the following list of problems:

- Finding the global minimum of a function on a subset of $\mathbb{R}^n$,
- Solving equations,
- Computing the Lebesgue volume of a subset $\mathbf{S} \subset \mathbb{R}^n$,
- Computing an upper bound on $\mu(\mathbf{S})$ over all measures μ satisfying some moment conditions,
- Pricing exotic options in Mathematical Finance,
- Computing the optimal value of an optimal control problem,
- Evaluating an ergodic criterion associated with a Markov chain,
- Evaluating a class of multivariate integrals,
- Computing Nash equilibria,
- With $\widehat{f}$ the convex envelope of a function f, evaluate $\widehat{f}(\mathbf{x})$ at some given point $\mathbf{x}$.

The above seemingly different and unrelated problems share in fact a very important property: they all can be viewed as a particular instance of the *Generalized Moment Problem* (GMP in short)! And of course the above list is not exhaustive!

It is known that the GMP has great modeling power with impact in several branches of Mathematics and also with important applications in various fields. However, and as illustrated by the above list, in its full generality the GMP cannot be solved numerically. According to Diaconis (1987), "the theory [of moment problems] is not up to the demands of applications".

One invoked reason is the high complexity of the problem: "numerical determination ... is feasible for a small number of moments, but appears to be quite difficult in general cases", whereas Kemperman (1987) points out the lack of a general algorithmic approach. Indeed quoting Kemperman:

"...a deep study of algorithms has been rare so far in the theory of moments, except for certain very specific practical applications, for instance, to crystallography, chemistry and tomography. No doubt, there is a considerable need for developing reasonably good numerical procedures for handling the great variety of moment problems which do arise in pure and applied mathematics and in the sciences in general...".

The main purpose of this book is to show that the situation becomes much nicer for the GMP with polynomial data (and sometimes one may even consider GMPs with some piecewise polynomial data or even rational functions). Indeed, results from real algebraic geometry and functional analysis have provided new characterizations of polynomials positive on a basic semi-algebraic set $\mathbb{K} \subset \mathbb{R}^n$, and *dual* results on moment sequences that can be represented by finite Borel measures supported on $\mathbb{K}$. This beautiful duality is nicely captured by standard duality in convex analysis, applied to some appropriate convex cones. Moreover, these characterizations are not only very elegant and simple to state, but more importantly, it also turns out that they are amenable to practical computation as they can be checked by semidefinite programming (and sometimes linear programming), two well known powerful techniques of convex optimization.

Conjunction of the above theoretical breakthrough with the development of semidefinite programming has allowed to define a numerical scheme based on semidefinite programming to approximate, and sometimes solve exactly, the original GMP. It merely consists of a hierarchy of semidefinite relaxations of the GMP where each semidefinite relaxation is a convex optimization problem for which efficient public softwares are available. As shown in the book, the beautiful duality between moments and positive polynomials is perfectly expressed by standard duality in convex optimization, when applied to these semidefinite relaxations.

This book is an attempt to convince the reader that one may now consider solving (or at least approximating as closely as desired) some difficult problems that were thought to be out of reach a few years ago, or for which only heuristics were available. Of course, since such problems remain hard, this methodology has some practical limitations mainly due to the size of the original problem, especially in view of the present status of semidefinite programming solvers (used as a black box subroutine to solve each semidefinite relaxation). However, we also indicate how sparsity or symmetries (when present) can be exploited so as to handle problems of larger size. Hence much remains to be done but we hope that this methodology will open the door to a more systematic and efficient treatment of such problems.

We introduce this methodology in a unified manner. To do so we first provide a general numerical scheme for solving (or approximately solving) the abstract GMP with polynomial data. Convergence and several other properties are proved in a very general context. Next, we illustrate in detail the above general methodology when applied to several instances of the GMP in various and very different applications in Global Optimization, Algebra, Probability and Markov Chains, Optimal Control, Mathematical Finance, Multivariate Integration, etc..., possibly with *ad hoc* adjustments if and when necessary. Depending on the problem on hand, additional insights are also provided.

The book is divided into two main parts, and here is a brief chapter-by-chapter description of its content. Part I is devoted to the theoretical basis that supports the proposed methodology to solve or approximate the abstract GMP. Part II is devoted to illustrate (and sometimes complement) the methodology for specific instances of the GMP in various domains.

Part I

Chapter 1 describes the abstract basic GMP and its dual with a few examples. We also provide some general results concerning the GMP with polynomial data and its dual, and show why the theory of moments and its dual theory of positive polynomials can be useful to solve the GMP.
Chapter 2 reviews basic results of real algebraic geometry on the representation of positive polynomials, among which the fundamental Positivstellensatze of Krivine, Stengle, Schmüdgen, Putinar and Jacobi and Prestel. We also provide some additional representation results that take account of several specific cases like for instance, finite varieties, convex semi-algebraic sets, representations that preserve sparsity, etc.
Chapter 3 is the *dual* of Chapter 2 as most results are the *dual* analogues of those described in Chapter 2. Indeed the problem of representing polynomials that are positive on a set $\mathbb{K}$ has a dual facet which is the problem of characterizing sequences of reals that are moment sequences of some finite Borel measure supported on $\mathbb{K}$. Moreover, as we shall see, this beautiful duality is nicely captured by standard duality in convex analysis, applied to some appropriate convex cones. We review basic results in the moment problem and also particularize to some specific important cases like those

in Chapter 2.
Chapter 4, now with the appropriate theoretical tools on hand, describes the general methodology to solve the abstract GMP. That is, we provide a hierarchy of semidefinite (or sometimes linear) relaxations of the basic GMP, whose associated sequence of optimal values converges to the optimal value of the GMP. Variants of this methodology are also provided to handle additional features such as countably many moment constraints, several measures, or GMP with sparsity properties. This chapter should allow the reader to see what is the basic idea underlying the methodology that permits to solve (or approximate) a problem formulated as a particular instance of the GMP.

Part II

Part II of the book is devoted to convince the reader about the power of the moment approach for solving the generalized moment problem with polynomial data. Each of the next chapters illustrates (and sometimes complements) the above methodology on some particular important instances of the GMP. For each chapter we detail the semidefinite relaxations of Chapter 4 in the specific context of the chapter. Moreover, depending on the specificity of the problem on hand, additional insights are also provided.
Chapter 5 is about global optimization, probably the simplest instance of the generalized moment problem. We detail the semidefinite relaxations of Chapter 4 for minimizing a polynomial on $\mathbb{R}^n$ and on a compact basic semi-algebraic set $\mathbb{K} \subset \mathbb{R}^n$. We also discuss the linear relaxations and several particular cases, e.g. when $\mathbb{K}$ is a polytope or a finite variety. In particular, this latter case encompasses all $0-1$ discrete optimization problems.
Chapter 6 is about solving systems of polynomial equations. Of course, if the goal is to search for just one solution, one may minimize any polynomial criterion and see the problem as a particular case of Chapter 5. But we also consider the case where one searches for *all* complex or *all* real solutions and show that the moment approach is well-suited to solve this problem and provides the first algorithm to compute all real solutions without computing all the complex solutions, in contrast with the usual algebraic approaches based on Gröbner bases or homotopy.
Chapter 7 covers some applications in probability. We first consider the problem of computing an upper bound on $\mu(\mathbf{S})$ for some subset $\mathbf{S} \subset \mathbb{R}^n$, over all measures μ that satisfy certain moment conditions. We then consider

the difficult problem of computing (or at least approximating) the volume of a compact basic semi-algebraic set. We end up with the mass-transfer (or Monge-Kantorovich) problem.

Chapter 8 is about Markov chains and invariant probabilities. We first address the problem of computing an upper bound on $\mu(\mathbf{S})$ for some $\mathbf{S} \subset \mathbb{R}^n$, over all invariant probability measures μ of a given Markov chain on $\mathbb{R}^n$. We then consider the problem of approximating the value of an ergodic criterion, as an alternative to simulation which only provides a random estimate.

Chapter 9 considers an important application in mathematical finance, namely the pricing of exotic options under a no-arbitrage assumption, first with only knowledge of some moments of the distribution of the underlying asset price, and then when one assumes that the asset price obeys some Ito stochastic differential equation.

Chapter 10 considers an application in control. We apply the moment approach to the so-called weak formulation of optimal control problems in which the initial problem is viewed as an infinite linear-programming model over suitable *occupation measures*, an instance of the generalized moment problem.

Chapter 11 considers the following problem: Given a rational function f on a basic semi-algebraic set $\mathbb{K}$, evaluate $\widehat{f}(\mathbf{x})$ at a particular point $\mathbf{x}$ in the domain of the convex envelope $\widehat{f}$ of f. We then consider semidefinite representations for the convex hull co($\mathbb{K}$) of $\mathbb{K}$. That is, finding a set defined by linear matrix inequalities in a lifted space such that co($\mathbb{K}$) is a suitable projection of that set.

Chapter 12 is about approximating the multivariate integral of a rational function or an exponential of a multivariate polynomial. We then consider the moment approach as a tool for evaluating gradients or Hesssians in the maximum entropy approach for estimating an unknown density based on knowledge of some of its moments.

Chapter 13 first considers the problem of minimizing the supremum of finitely many rational functions on a basic semi-algebraic set. Then this is used to compute (or approximate) the value of Nash equlibria for N-player finite games. We end up with applying the moment approach to 2-player zero-sum polynomial games.

Chapter 14, our final chapter, applies the moment approach to provide bounds on functionals of the solution of linear partial differential equations with boundary conditions and polynomial coefficients.

• The applications of the GMP described in this book come from various areas, e.g., Optimal Control, Optimization, Mathematical Finance, Probability and Operations Research. Therefore, in the Appendices at the end of the book, we have provided some brief basic background on results from optimization and probability that are used in some of the chapters.

• At some places in the book, some theorems, lemmas or propositions are framed into a box to emphasize their importance, at least to the author's taste.

• Sometimes, proofs of theorems, lemmas and propositions are postponed for the sake of clarity of exposition and for the reader to avoid from being lost in technical details in the middle of a chapter. Sometimes, if short and/or important, a proof is provided just after the corresponding theorem, lemma or proposition. Finally, if it is too long, or too technical, or not crucial, a proof is simply omitted but a reference is provided in the *Notes and Sources* section at the end of the corresponding chapter.

Acknowledgments

This book has benefited from several stimulating discussions with (among others) M. Anjos, R.E Curto, E. de Klerk, L. Elghaoui, L.A. Fialkow, K. Fukuda, J.W. Helton, M. Kojima, S. Kuhlmann, M. Laurent, M. Marshall, Y. Nesterov, P. Parrilo, D. Pasechnich, V. Powers, A. Prestel, M. Putinar, J. Renegar, B. Reznick, M-F. Roy, K. Schmüdgen, M. Schweighofer, F. Sottile, B. Sturmfels, J. Sun, T. Theobald, K. Toh, L. Tuncel, V. Vinnikov, and H. Wolkowicz, during several visits and/or workshops in the period 2000-2008 at the *Mathematical Sciences Research Institute* (MSRI, Berkeley), The American Institute of Mathematics (AIM, Palo Alto), the *Fields Institute for Research in Mathematical Sciences* (Toronto), the *Institute for Mathematics and its Applications* (IMA, Minneapolis), the *Oberwolfach Institute*, and the *Institute for Mathematical Sciences* (IMS, Singapore). I gratefully acknowledge financial support from the AMS and the above institutes which made these visits possible.

I also want to thank D. Bertsimas, B. Helton, R. Laraki, M. Laurent, T. Netzer, C. Prieur, M. Putinar, P. Rostalski, T. Priéto-Rumeau, E. Trélat, M. Zervos, Xuan Vinh Doan, as some of the results presented in this book have been obtained in collaboration with them, and in particular, I wish to thank my colleague Didier Henrion at LAAS (especially for our collaboration on the software GloptiPoly which he made user friendly with free access). Finally, special thanks are due to J-B. Hiriart-Urruty, J. Renegar and S. Sorin.

For part of this research, financial support from the (French) ANR under grant ANR-NT05-3-41612 is gratefully acknowledged. At last, I want to thank the CNRS institution for providing a very nice and pleasant working environment at the LAAS-CNRS laboratory in Toulouse.

Toulouse, May 2009 *Jean B. Lasserre*

Contents

Preface vii

Acknowledgments xiii

Part I Moments and Positive Polynomials 1

1. The Generalized Moment Problem 3
   - 1.1 Formulations 4
   - 1.2 Duality Theory 7
   - 1.3 Computational Complexity 10
   - 1.4 Summary 12
   - 1.5 Exercises 13
   - 1.6 Notes and Sources 13

2. Positive Polynomials 15
   - 2.1 Sum of Squares Representations and Semi-definite Optimization 16
   - 2.2 Nonnegative Versus s.o.s. Polynomials 20
   - 2.3 Representation Theorems: Univariate Case 22
   - 2.4 Representation Theorems: Mutivariate Case 24
   - 2.5 Polynomials Positive on a Compact Basic Semi-algebraic Set 28
     - 2.5.1 Representations via sums of squares 28
     - 2.5.2 A matrix version of Putinar's Positivstellensatz 33
     - 2.5.3 An alternative representation 34
   - 2.6 Polynomials Nonnegative on Real Varieties 37

2.7 Representations with Sparsity Properties 39
2.8 Representation of Convex Polynomials 42
2.9 Summary 47
2.10 Exercises 48
2.11 Notes and Sources 49

3. Moments 51

3.1 The One-dimensional Moment Problem 53
3.1.1 The full moment problem 54
3.1.2 The truncated moment problem 56
3.2 The Multi-dimensional Moment Problem 57
3.2.1 Moment and localizing matrix 58
3.2.2 Positive and flat extensions of moment matrices 61
3.3 The $\mathbb{K}$-moment Problem 62
3.4 Moment Conditions for Bounded Density 66
3.4.1 The compact case 67
3.4.2 The non compact case 68
3.5 Summary 70
3.6 Exercises 71
3.7 Notes and Sources 71

4. Algorithms for Moment Problems 73

4.1 The Overall Approach 73
4.2 Semidefinite Relaxations 75
4.3 Extraction of Solutions 80
4.4 Linear Relaxations 86
4.5 Extensions 87
4.5.1 Extensions to countably many moment constraints 87
4.5.2 Extension to several measures 88
4.6 Exploiting Sparsity 91
4.6.1 Sparse semidefinite relaxations 94
4.6.2 Computational complexity 96
4.7 Summary 97
4.8 Exercises 98
4.9 Notes and Sources 99
4.10 Proofs 99

4.10.1 Proof of Theorem 4.3 99
4.10.2 Proof of Theorem 4.7 102

Part II Applications 107

5. Global Optimization over Polynomials 109

5.1 The Primal and Dual Perspectives 110
5.2 Unconstrained Polynomial Optimization 111
5.3 Constrained Polynomial Optimization: Semidefinite Relaxations 117
5.3.1 Obtaining global minimizers 119
5.3.2 The univariate case 121
5.3.3 Numerical experiments 122
5.3.4 Exploiting sparsity 123
5.4 Linear Programming Relaxations 125
5.4.1 The case of a convex polytope 126
5.4.2 Contrasting LP and semidefinite relaxations. . . 126
5.5 Global Optimality Conditions 127
5.6 Convex Polynomial Programs 130
5.6.1 An extension of Jensen's inequality 131
5.6.2 The s.o.s.-convex case 132
5.6.3 The strictly convex case 133
5.7 Discrete Optimization 134
5.7.1 Boolean optimization 135
5.7.2 Back to unconstrained optimization 137
5.8 Global Minimization of a Rational Function 138
5.9 Exploiting Symmetry 141
5.10 Summary 143
5.11 Exercises 144
5.12 Notes and Sources 144

6. Systems of Polynomial Equations 147

6.1 Introduction 147
6.2 Finding a Real Solution to Systems of Polynomial Equations 148
6.3 Finding All Complex and/or All Real Solutions: A Unified Treatment 152

6.3.1 Basic underlying idea 155
6.3.2 The moment-matrix algorithm 155
6.4 Summary 160
6.5 Exercises 161
6.6 Notes and Sources 161

7. Applications in Probability 163
7.1 Upper Bounds on Measures with Moment Conditions . 163
7.2 Measuring Basic Semi-algebraic Sets 168
7.3 Measures with Given Marginals 175
7.4 Summary 177
7.5 Exercises 177
7.6 Notes and Sources 178

8. Markov Chains Applications 181
8.1 Bounds on Invariant Measures 183
8.1.1 The compact case 183
8.1.2 The non compact case 185
8.2 Evaluation of Ergodic Criteria 187
8.3 Summary 189
8.4 Exercises 190
8.5 Notes and Sources 191

9. Application in Mathematical Finance 193
9.1 Option Pricing with Moment Information 193
9.2 Option Pricing with a Dynamic Model 196
9.2.1 Notation and definitions 197
9.2.2 The martingale approach 198
9.2.3 Semidefinite relaxations 200
9.3 Summary 202
9.4 Notes and Sources 203

10. Application in Control 205
10.1 Introduction 205
10.2 Weak Formulation of Optimal Control Problems 206
10.3 Semidefinite Relaxations for the OCP 210
10.3.1 Examples 212
10.4 Summary 215

10.5 Notes and Sources 216

11. Convex Envelope and Representation of Convex Sets 219

11.1 The Convex Envelope of a Rational Function 219
11.1.1 Convex envelope and the generalized moment problem 220
11.1.2 Semidefinite relaxations 223
11.2 Semidefinite Representation of Convex Sets 225
11.2.1 Semidefinite representation of co($\mathbb{K}$) 226
11.2.2 Semidefinite representation of convex basic semi-algebraic sets 229
11.3 Algebraic Certificates of Convexity 234
11.4 Summary 239
11.5 Exercises 239
11.6 Notes and Sources 240

12. Multivariate Integration 243

12.1 Integration of a Rational Function 243
12.1.1 The multivariable case 245
12.1.2 The univariate case 246
12.2 Integration of Exponentials of Polynomials 250
12.2.1 The moment approach 251
12.2.2 Semidefinite relaxations 253
12.2.3 The univariate case 254
12.3 Maximum Entropy Estimation 256
12.3.1 The entropy approach 257
12.3.2 Gradient and Hessian computation 259
12.4 Summary 260
12.5 Exercises 262
12.6 Notes and Sources 262

13. Min-Max Problems and Nash Equilibria 265

13.1 Robust Polynomial Optimization 265
13.1.1 Robust Linear Programming 267
13.1.2 Robust Semidefinite Programming 269
13.2 Minimizing the Sup of Finitely Many Rational Functions 270
13.3 Application to Nash Equilibria 273

13.3.1 N-player games 273
13.3.2 Two-player zero-sum polynomial games 276
13.3.3 The univariate case 280
13.4 Exercises 281
13.5 Notes and Sources 282

14. Bounds on Linear PDE 285
14.1 Linear Partial Differential Equations 285
14.2 Notes and Sources 288

Final Remarks 291

Appendix A Background from Algebraic Geometry 293
A.1 Fields and Cones 293
A.2 Ideals 295
A.3 Varieties 296
A.4 Preordering 298
A.5 Algebraic and Semi-algebraic Sets over a Real Closed Field 300
A.6 Notes and Sources 302

Appendix B Measures, Weak Convergence and Marginals 305
B.1 Weak Convergence of Measures 305
B.2 Measures with Given Marginals 308
B.3 Notes and Sources 310

Appendix C Some Basic Results in Optimization 311
C.1 Non Linear Programming 311
C.2 Semidefinite Programming 313
C.3 Infinite-dimensional Linear Programming 316
C.4 Proof of Theorem 1.3 318
C.5 Notes and Sources 319

Appendix D The GloptiPoly Software 321
D.1 Presentation 321
D.2 Installation 321
D.3 Getting started 322
D.4 Description 323
D.4.1 Multivariate polynomials (`mpol`) 324

D.4.2 Measures (meas) 325
D.4.3 Moments (mom) 325
D.4.4 Support constraints (supcon) 327
D.4.5 Moment constraints (momcon) 327
D.4.6 Floating point numbers (double) 328
D.5 Solving Moment Problems (msdp) 329
D.5.1 Unconstrained minimization 329
D.5.2 Constrained minimization 331
D.5.3 Several measures 335
D.6 Notes and Sources 337

Glossary 339

Bibliography 341

Index 359

Part I

Moments and Positive Polynomials

Chapter 1

The Generalized Moment Problem

We describe the abstract basic Generalized Moment Problem (GMP) and its dual with a few examples. We also provide some general results concerning the GMP with polynomial data and its dual and show why the theory of moments and its dual theory of positive polynomials can be useful to solve the GMP.

Problems involving moments of measures arise naturally in many areas of applied mathematics, statistics and probability, economics, engineering, physics and operations research. For instance, how do we obtain optimal bounds on the probability that a random variable belongs to a given set, given some of its moments? How do we price derivative securities in a financial economics framework without assuming any model for the underlying price dynamics, given only moments of the price of the underlying asset? But as we will see throughout the book, some (and even many) other problems seemingly different and which *a priori* do not involve any moment of some measure, have equivalent reformulations which involve moments or generalized moments.

In fact, these problems can be seen as particular instances of a linear infinite dimensional optimization problem, called the **Generalized Moment Problem** (in short, GMP).

Let $\mathbb{K}$ be a Borel subset of $\mathbb{R}^n$ and let $\mathscr{M}(\mathbb{K})$ be the space of finite *signed* Borel measures on $\mathbb{K}$, whose positive cone $\mathscr{M}(\mathbb{K})_+$ is the space of finite Borel measures μ on $\mathbb{K}$. Given a set of indices Γ, a set of reals $\{\gamma_j : j \in \Gamma\}$, and functions $f, h_j : \mathbb{K} \to \mathbb{R}$, $j \in \Gamma$, that are integrable with respect to every measure $\mu \in \mathscr{M}(\mathbb{K})_+$, the GMP is defined as follows:

$$\text{GMP}: \quad \begin{array}{rl} \rho_{\text{mom}} = \displaystyle\sup_{\mu \in \mathscr{M}(\mathbb{K})_+} & \displaystyle\int_{\mathbb{K}} f \, d\mu \\ \text{s.t.} & \displaystyle\int_{\mathbb{K}} h_j \, d\mu \leqq \gamma_j, \qquad j \in \Gamma. \end{array} \tag{1.1}$$

(Recall that the symbol "$\leqq$" stands for either an inequality "$\leq$" or an equality "$=$".) Note that we write "sup" instead of "max" to indicate that an optimal solution might not be attained. In this chapter, we present several examples that illustrate the modeling power of problem (1.1), develop a duality theory that forms the basis of future developments and briefly discuss the complexity of problem (1.1).

1.1 Formulations

In this section, our goal is to illustrate the modeling flexibility of formulation (1.1). Towards this goal, we present three examples from rather diverse areas such as probability theory, financial economics and optimization.

Moment problems in probability

Given vectors $\boldsymbol{\alpha} = (\alpha_1, \ldots, \alpha_n)' \in \mathbb{N}^n$ and $\mathbf{x} = (x_1, \ldots, x_n)' \in \mathbb{R}^n$, let

$$\mathbf{x}^{\boldsymbol{\alpha}} := x_1^{\alpha_1} x_2^{\alpha_2} \ldots x_n^{\alpha_n}.$$

Let $\mathbf{S}$, $\mathbb{K} \subset \mathbb{R}^n$ be given Borel sets with $\mathbf{S} \subset \mathbb{K}$. Then, the problem of finding an optimal bound on the probability that a $\mathbb{K}$-valued random variable $\mathbf{X}$ belongs to $\mathbf{S}$, given some of its moments $\gamma_{\boldsymbol{\alpha}}$ for $\boldsymbol{\alpha} \in \Gamma \subset \mathbb{N}^n$, can be formulated as solving the problem:

$$\begin{aligned} \rho_{\text{mom}} = \sup_{\mu \in \mathscr{M}(\mathbb{K})_+} & \; \mu(\mathbf{S}) \\ \text{s.t.} & \int_{\mathbb{K}} \mathbf{x}^{\boldsymbol{\alpha}} d\mu = \gamma_{\boldsymbol{\alpha}}, \qquad \forall \boldsymbol{\alpha} \in \Gamma, \end{aligned} \tag{1.2}$$

a special case of formulation (1.1) with $f(\mathbf{x}) = 1_{\mathbf{S}}(\mathbf{x})$ and $h_{\boldsymbol{\alpha}}(\mathbf{x}) = \mathbf{x}^{\boldsymbol{\alpha}}$, $\boldsymbol{\alpha} \in \Gamma$.

Moment problems in financial economics

A central question in financial economics is to find the price of a so-called derivative security given information on the underlying asset. Let us take the example of the so-called European call option on an underlying security (this is why an option is called derivative security as its value is derived from another) with strike k and maturity T. It gives its holder the option (but not the obligation) to buy the underlying security at time T at price k. Clearly, if the price S_T is more than k, then the holder will exercise the option and make a profit of $S_T - k$, while if it is less than k, he will not exercise and does not make a profit. Thus, the payoff of this option is $\max(S_T - k, 0)$. Clearly, as the payoff of this option is nonnegative, it has some value. A key problem in financial economics is to determine the price of such an option. This is exactly the area of the 1997 Nobel prize in economics awarded to Robert Merton and Myron Scholes (Fisher Black has passed away in 1995). Under the assumption that the price of the underlying asset follows a geometric Brownian motion and using the no-arbitrage assumption, the Black-Scholes formula provides an explicit and insightful answer to this question.

No arbitrage means that one cannot make money deterministically. For example, if a stock trades in two exchanges, it will trade at the same price, since otherwise there is an arbitrage opportunity. It turns out that the assumption of no-arbitrage is equivalent to the existence of a probability measure μ, such that the price of any European call option with strike k is given by $\mathrm{E}_\mu[\max(S_T - k, 0)]$. If we do not assume a particular stochastic process for the price dynamics S_t, but only moments of the price S_T at time $t = T$, and under the no-arbitrage assumption, the problem of finding an optimal upper bound on the price of a European call option with strike k given the first m moments γ_j, $j = 0, 1, \ldots, m$, ($\gamma_0 = 1$) of the price of the underlying asset, is given by:

$$\begin{aligned} \rho_{\mathrm{mom}} = \quad & \sup_{\mu \in \mathscr{M}(\mathbb{R}_+)_+} \int_{\mathbb{R}_+} \max(x - k, 0)\, d\mu \\ & \text{s.t.} \int_{\mathbb{R}_+} x^j\, d\mu = \gamma_j, \quad j = 0, 1, \ldots, m, \end{aligned} \tag{1.3}$$

a special case of formulation (1.1) with $\mathbb{K} = \mathbb{R}_+$, $f(x) = \max(x - k, 0)$ and $h_j(x) = x^j$. As another example, if the prices p_j of European call options of strike k_j, $j = 1, \ldots, m$ are given, the problem of finding an optimal upper bound on the price of a European call option with strike k is a special case of formulation (1.1) with $\mathbb{K} = \mathbb{R}_+$, $f(x) = \max(x - k, 0)$, $h_j(x) = \max(x - k_j, 0)$, and $\gamma_j = p_j$.

Global optimization over polynomials

With $f : \mathbb{R}^n \to \mathbb{R}$ and $\mathbb{K} \subset \mathbb{R}^n$, consider the constrained optimization problem:

$$\begin{aligned} f^* = \sup\ & f(\mathbf{x}) \\ \text{s.t. } & \mathbf{x} \in \mathbb{K}, \end{aligned} \tag{1.4}$$

which we rewrite as

$$\begin{aligned} \rho_{\mathrm{mom}} = \sup_{\mu \in \mathscr{M}(\mathbb{K})_+} & \int_{\mathbb{K}} f\, d\mu \\ \text{s.t. } & \int_{\mathbb{K}} d\mu = 1. \end{aligned} \tag{1.5}$$

Theorem 1.1. *Problems (1.4) and (1.5) are equivalent, that is* $f^* = \rho_{\mathrm{mom}}$.

Proof. If $f^* = +\infty$, let M be arbitrary large, and let $\mathbf{x} \in \mathbb{K}$ be such that $f(\mathbf{x}) \geq M$. Then, with $\mu := \delta_{\mathbf{x}} \in \mathscr{M}(\mathbb{K})_+$ (with $\delta_{\mathbf{x}}$ being the Dirac measure at $\mathbf{x}$), we have $\int f\, d\mu = f(\mathbf{x}) \geq M$, and so $\rho_{\mathrm{mom}} = +\infty$.

We next consider the case $f^* < +\infty$. Since $f(\mathbf{x}) \leq f^*$ for all $\mathbf{x} \in \mathbb{K}$, then $\int_{\mathbb{K}} f d\mu \leq f^*$ and thus $\rho_{\mathrm{mom}} \leq f^*$. Conversely, with every $\mathbf{x} \in \mathbb{K}$, we associate the Dirac measure $\delta_{\mathbf{x}} \in \mathscr{M}(\mathbb{K})_+$ which is a feasible solution of problem (1.5) with value $f(\mathbf{x})$, leading to $\rho_{\mathrm{mom}} \geq f^*$. This combined with $f^* \geq \rho_{\mathrm{mom}}$ leads to $f^* = \rho_{\mathrm{mom}}$, the desired result. Notice also that if $\mathbf{x}^* \in \mathbb{K}$ is a global minimizer of problem (1.4), then the probability measure $\mu^* := \delta_{\mathbf{x}^*}$ is an optimal solution of problem (1.5). □

In other words, we can formulate the general nonlinear optimization problem as a special case of the generalized moment problem (1.1), which underscores the modeling flexibility of problem (1.1). In contrast to problem (1.4), problem (1.5) is **linear**, and thus **convex**. It is, however, infinite-dimensional.

Given the diversity and generality of these examples, it is evident that the generalized moment problem is a problem of remarkable modeling power. In later chapters and in the exercises we explore several other examples. The goal of the book is to understand the complexity of problem (1.1) and its variations, to explore applications in a variety of applied contexts and to develop algorithms for providing bounds, and sometimes solutions.

For the case $f = 0$, $h_{\boldsymbol{\alpha}} = \mathbf{x}^{\boldsymbol{\alpha}}$, $\boldsymbol{\alpha} \in \Gamma \subset \mathbb{N}^n$, problem (1.1) becomes a *feasibility* problem known as the $\mathbb{K}$-moment problem.

Definition 1.1. Given $\gamma_{\boldsymbol{\alpha}}$, $\boldsymbol{\alpha} \in \Gamma \subset \mathbb{N}^n$ and a set $\mathbb{K} \subseteq \mathbb{R}^n$, **the $\mathbb{K}$-moment problem** asks whether there exists a finite Borel measure $\mu \in \mathscr{M}(\mathbb{K})_+$ such that

$$\int_{\mathbb{K}} \mathbf{x}^{\boldsymbol{\alpha}} d\mu = \gamma_{\boldsymbol{\alpha}}, \quad \boldsymbol{\alpha} \in \Gamma.$$

In the next section, we develop a duality theory that forms the basis of algorithms and relaxations that we will utilize in later chapters.

1.2 Duality Theory

Problem (1.1) is linear and its dual problem is given by:

$$\begin{aligned}\rho_{\text{pop}} = \inf_{\boldsymbol{\lambda}} & \sum_{j\in\Gamma} \gamma_j \lambda_j \\ \text{s.t.} & \sum_{j\in\Gamma} \lambda_j h_j(\mathbf{x}) \geq f(\mathbf{x}), \quad \forall\, \mathbf{x} \in \mathbb{K}. \\ & \lambda_j \geq 0, \quad j \in \Gamma_+\end{aligned} \tag{1.6}$$

where $\Gamma_+ \subseteq \Gamma$ stands for the set of indices j for which the generalized moment constraint is the inequality $\int h_j \, d\mu \leq \gamma_j$.

As we will apply general results from conic duality in convex optimization, we write problems (1.1) and (1.6) as conic optimization problems. So we introduce the convex cones:

$$C(\mathbb{K}) = \{(\boldsymbol{\gamma}, \gamma_0)' : \exists\, \mu \in \mathscr{M}(\mathbb{K})_+ \text{ s.t. } \gamma_0 = \textstyle\int_{\mathbb{K}} f\, d\mu, \; \gamma_j \gtreqqless \int_{\mathbb{K}} h_j \, d\mu, \; \forall\, j \in \Gamma\},$$

$$P(\mathbb{K}) = \left\{(\boldsymbol{\lambda}, \lambda_0)' : \lambda_j \geq 0, \; j \in \Gamma_+; \sum_{j\in\Gamma} \lambda_j h_j(\mathbf{x}) + \lambda_0 f(\mathbf{x}) \geq 0, \quad \forall\, \mathbf{x} \in \mathbb{K}\right\},$$

and rewrite problems (1.1) and (1.6) as

$$\rho_{\text{mom}} = \sup_{\gamma_0} \{\gamma_0 \; : \; (\boldsymbol{\gamma}, \gamma_0)' \in \overline{C(\mathbb{K})}\,\}, \tag{1.7}$$

$$\rho_{\text{pop}} = \inf_{\boldsymbol{\lambda}} \{\sum_{j\in\Gamma} \gamma_j \lambda_j \; : \; (\boldsymbol{\lambda}, -1)' \in \overline{P(\mathbb{K})}\,\}, \tag{1.8}$$

where $\overline{R}$ denotes the closure of the set R (whereas $\operatorname{int} R$ denotes its interior).

The *weak duality* property holds for problems (1.7) and (1.8) if for any two feasible solutions γ_0 of (1.7) and $\boldsymbol{\lambda}$ of (1.8), one has $\gamma_0 \leq \sum_{j\in\Gamma} \lambda_j \gamma_j$ (and so $\rho_{\text{mom}} \leq \rho_{\text{pop}}$). If $\rho_{\text{mom}} < \rho_{\text{pop}}$ then one says that there is a *duality gap* for problems (1.7) and (1.8). Finally *strong duality* holds if there is no duality gap, i.e., $\rho_{\text{pop}} = \rho_{\text{mom}}$.

Using general results of conic duality in convex optimization, we obtain the following weak and strong duality.

Theorem 1.2.
(a) (Weak duality) *The optimal values of (1.7) and (1.8) satisfy* $\rho_{\text{mom}} \leq \rho_{\text{pop}}$.

(b) (Strong duality) *If* $(\gamma, \gamma_0) \in C(\mathbb{K})$ *for some* γ_0 *, and there exists* $\boldsymbol{\lambda} \in \mathbb{R}^{|\Gamma|}$ *(with* $\lambda_j \geq 0$ *for all* $j \in \Gamma_+$*) such that* $(\boldsymbol{\lambda}, -1)' \in \operatorname{Int} P(\mathbb{K})$, *then* $\rho_{\text{mom}} = \rho_{\text{pop}}$ *and problem (1.1) has an optimal solution, that is, the* sup *is attained if it is finite.*

Proof. (a) Let γ_0 and $\boldsymbol{\lambda}$ be arbitrary feasible solutions of (1.7) and (1.8) respectively. By definition of the cone $C(\mathbb{K})$ there exists a finite Borel measure $\mu \in \mathscr{M}(\mathbb{K})_+$ such that

$$\gamma_0 = \int_{\mathbb{K}} f d\mu \leq \int_{\mathbb{K}} \left(\sum_{j \in \Gamma} \lambda_j h_j \right) d\mu = \sum_{j \in \Gamma} \lambda_j \int_{\mathbb{K}} h_j d\mu \leq \sum_{j \in \Gamma} \lambda_j \gamma_j,$$

that is, weak duality holds (and so, $\rho_{\text{mom}} \leq \rho_{\text{pop}}$).
(b) Strong duality follows from general results of conic duality in convex optimization that requires, however, that there exists a vector $(\boldsymbol{\lambda}, -1)'$ in the interior of the cone $P(\mathbb{K})$, which is known as a Slater type condition. □

In important special cases we do not need to impose Slater type conditions for strong duality to hold, as the next theorem states.

Theorem 1.3. *Suppose that* $\mathbb{K}$ *is compact,* f *is bounded and upper-semicontinuous on* $\mathbb{K}$, h_j *is continuous on* $\mathbb{K}$ *for every* $j \in \Gamma$, *and there exists* $k \in \Gamma$ *such that* $h_k > 0$ *on* $\mathbb{K}$. *Then:*
(a) $\rho_{\text{mom}} = \rho_{\text{pop}}$ *and if problem (1.1) has a feasible solution then it has an optimal solution, that is, the* sup *is attained.*
(b) *If the* sup *is attained, problem (1.1) has an optimal solution* μ *supported on finitely many points of* $\mathbb{K}$, *i.e.,* μ *is a finite linear combination of Dirac measures.*

A proof which uses results from infinite-dimensional linear optimization, can be found in Appendix C.4 where an appropriate summarized background is also provided.

There are cases where one can relax compactness of the set $\mathbb{K}$ and still maintain strong duality. One such case is when $-f$ is lower-semicontinuous, bounded from below and inf-compact (or a moment function); see Definition B.11. In addition, for each $\boldsymbol{\alpha} \in \Gamma$, the moment constraint is an inequality constraint $\int_{\mathbb{K}} h_{\boldsymbol{\alpha}} d\mu \leq \gamma_{\boldsymbol{\alpha}}$, and $h_{\boldsymbol{\alpha}}$ is a nonnegative lower-semicontinuous function.

Countably many constraints.

We next consider the moment problem (1.1) and extend Theorem 1.3 when there are *countably many* constraints present in problem (1.1).

Corollary 1.4. *Suppose that the assumptions of Theorem 1.3 hold with Γ a countable set. If (1.1) has a feasible solution, then it is solvable (i.e., the sup is attained), and there is no duality gap, i.e., $\rho_{\text{mom}} = \rho_{\text{pop}}$.*

Proof. Let (Γ_m) be a sequence of finite sets such that $\Gamma_m \subset \Gamma$ for all m and $\Gamma_m \uparrow \Gamma$ as $m \to \infty$. We may assume without loss of generality that there exists an index, say $j = 0 \in \Gamma_1$ (hence $0 \in \Gamma_m$ for all m) such that $h_0 > 0$ on $\mathbb{K}$. Suppose that (1.1) has a feasible solution and consider the moment problem (1.1) with finitely many constraints indexed in Γ_m; let ρ^m_{mom} be its optimal value. Similarly let ρ^m_{pop} denote the optimal value of its dual. By Theorem 1.3, $\rho^m_{\text{mom}} = \rho^m_{\text{pop}}$, for all m. In addition, $\rho_{\text{mom}} \leq \rho^m_{\text{mom}}$ for each m and the sequence (ρ^m_{mom}) is monotone nonincreasing as more constraints are added as m increases. Moreover, let $\mu_m \in \mathscr{M}(\mathbb{K})_+$ be an optimal solution of the primal. As $h_0 > 0$ on $\mathbb{K}$ we have $h_0 \geq \delta$ on $\mathbb{K}$ for some $\delta > 0$, and so $\mu_m(\mathbb{K}) \leq \gamma_0/\delta$ for all m. Therefore, there is a subsequence $\{m_i\}$ and a finite Borel measure μ on $\mathbb{K}$ such that $\mu_{m_i} \Rightarrow \mu$[1] as $i \to \infty$. Fix $j \in \Gamma$ arbitrary so that $h_j \in \Gamma_m$ for all m sufficiently large. As h_j is continuous, $\mathbb{K}$ is compact and $\mu_{m_i} \Rightarrow \mu$, we have

$$\int_{\mathbb{K}} h_j \, d\mu = \lim_{i \to \infty} \int_{\mathbb{K}} h_j \, d\mu_{m_i} \leqq \gamma_j,$$

and so, as j was arbitrary, μ is feasible for the moment problem. Moreover, as f is upper semicontinuous, it is bounded above on $\mathbb{K}$, and

[1]The notation $\mu_{n_j} \Rightarrow \mu$ (standard in probability) stands for the weak convergence of measures; see Definition B.1.

$$\rho_{\text{mom}} \leq \limsup_{i\to\infty} \rho_{\text{mom}}^{m_i} = \limsup_{i\to\infty} \int_{\mathbb{K}} f\, d\mu_{m_i} \leq \int_{\mathbb{K}} f\, d\mu$$

where the last inequality on the right follows from Proposition 1.4.18 in Hernández-Lerma and Lasserre (2003). This proves that μ is a primal solution of the moment problem (with Γ countable), and so $\rho_{\text{mom}}^{m_i} \downarrow \rho_{\text{mom}}$, which also implies $\rho_{\text{mom}}^{m} \downarrow \rho_{\text{mom}}$ because the sequence (ρ_{mom}^{m}) is monotone. Finally, as $\rho_{\text{pop}}^{m} = \rho_{\text{mom}}^{m}$, we also get $\rho_{\text{pop}}^{m} \downarrow \rho_{\text{mom}}$. □

1.3 Computational Complexity

In this section, we consider the complexity of a variant of problem (1.1) with only *inequality* moment constraints:

$$\begin{aligned} \rho_{\text{mom}}^{\leq} = & \sup_{\mu \in \mathscr{M}(\mathbb{K})_+} \int_{\mathbb{K}} f d\mu \\ & \text{s.t.} \int_{\mathbb{K}} h_j d\mu \leq \gamma_j \quad j \in \Gamma, \end{aligned} \tag{1.9}$$

(i.e., $\Gamma = \Gamma_+$), and the corresponding dual problem becomes:

$$\begin{aligned} \rho_{\text{pop}}^{\leq} = \inf & \sum_{j\in\Gamma} \gamma_j \lambda_j \\ \text{s.t.} & \sum_{j\in\Gamma} \lambda_j h_j(\mathbf{x}) \geq f(\mathbf{x}), \quad \forall\, \mathbf{x} \in \mathbb{K}, \\ & \lambda_j \geq 0, \quad j \in \Gamma. \end{aligned} \tag{1.10}$$

All the results we presented earlier regarding strong duality continue to hold for the pair of primal and dual problems (1.9) and (1.10). In order to study the complexity of problem (1.9), we consider the separation problem associated with problem (1.10):

Definition 1.2. The **separation problem** for problem (1.10): Given $\lambda_j \geq 0, \quad j \in \Gamma$, check whether

$$\sum_{j\in\Gamma} \lambda_j h_j(\mathbf{x}) \geq f(\mathbf{x}), \quad \forall\, \mathbf{x} \in \mathbb{K},$$

and if not, find a violated inequality.

Given the equivalence of separation and optimization, as well as strong duality, we need to consider conditions under which the separation problem is solvable in polynomial time.

Theorem 1.5.
(a) *If f is concave, h_j is convex for every $j \in \Gamma$, and $\mathbb{K}$ is a convex set, then problem (1.9) is solvable in polynomial time.*
(b) *If f and h_j, $j \in \Gamma$ are quadratic or piecewise linear functions over p polyhedra $\mathbb{K}_i$, $i = 1, \ldots, p$, that form a partition of $\mathbb{K} = \mathbb{R}^n$ (with p being a polynomial in n and $|\Gamma|$), then problem (1.9) is solvable in polynomial time.*

Proof. (a) The separation problem becomes

$$\inf_{\mathbf{x} \in \mathbb{K}} \sum_{j \in \Gamma} \lambda_j h_j(\mathbf{x}) - f(\mathbf{x}),$$

which in this case is a convex optimization problem, solvable efficiently using the ellipsoid method.
(b) We present the case in which $f(\mathbf{x}) = \mathbf{x}'\mathbf{Q}\mathbf{x} + \mathbf{b}'\mathbf{x} + c$ and $h_j(\mathbf{x}) = \mathbf{x}'\mathbf{Q}_j\mathbf{x} + \mathbf{b}_j'\mathbf{x} + c_j$, $j \in \Gamma$. We have

$$\sum_{j \in \Gamma} \lambda_j h_j(\mathbf{x}) - f(\mathbf{x}) = \mathbf{x}'\overline{\mathbf{Q}}\mathbf{x} + \overline{\mathbf{b}}'\mathbf{x} + \overline{c},$$

where

$$\overline{\mathbf{Q}} = \sum_{j \in \Gamma} \lambda_j \mathbf{Q}_j - \mathbf{Q},$$
$$\overline{\mathbf{b}} = \sum_{j \in \Gamma} \lambda_j \mathbf{b}_j - \mathbf{b},$$
$$\overline{c} = \sum_{j \in \Gamma} \lambda_j c_j - c.$$

To solve the separation problem, we check whether $\overline{\mathbf{Q}}$ is positive semidefinite. If it is not, we decompose $\overline{\mathbf{Q}} = \mathbf{U}'\,\boldsymbol{\Theta}\,\mathbf{U}$, where $\boldsymbol{\Theta}$ is the diagonal matrix of the eigenvalues of $\overline{\mathbf{Q}}$. Let $\theta_i < 0$ be a negative eigenvalue of $\overline{\mathbf{Q}}$. Let $\mathbf{u}$ be a vector with $u_j = 0$, $j \neq i$, and u_i large enough so that

$\theta_i u_i^2 + (\mathbf{U}\overline{\mathbf{b}})_i u_i + \overline{c} < 0$. Let $\mathbf{x}_0 = \mathbf{U}'\mathbf{u}$. Then,

$$\begin{aligned}
\sum_{j\in\Gamma} \lambda_j h_j(\mathbf{x}_0) - f(\mathbf{x}_0) &= \mathbf{x}_0'\overline{\mathbf{Q}}\mathbf{x}_0 + \overline{\mathbf{b}}'\mathbf{x}_0 + \overline{c} \\
&= \mathbf{u}'\mathbf{U}\mathbf{U}'\,\boldsymbol{\Theta}\,\mathbf{U}\mathbf{U}'\mathbf{u} + \overline{\mathbf{b}}'\mathbf{U}'\mathbf{u} + \overline{c} \\
&= \mathbf{u}'\boldsymbol{\Theta}\mathbf{u} + \overline{\mathbf{b}}'\mathbf{U}'\mathbf{u} + \overline{c} \\
&= \sum_{j=1}^{n} \theta_j u_j^2 + \sum_{j=1}^{n} (\mathbf{U}\overline{\mathbf{b}})_j u_j + \overline{c} \\
&= \theta_i u_i^2 + (\mathbf{U}\overline{\mathbf{b}})_i u_i + \overline{c} < 0,
\end{aligned}$$

which produces a violated inequality.

If $\overline{\mathbf{Q}}$ is positive semidefinite, then we solve the convex quadratic optimization problem

$$\rho_0 = \min_{\mathbf{x}\in\mathbb{R}^n} \mathbf{x}'\overline{\mathbf{Q}}\mathbf{x} + \overline{\mathbf{b}}'\mathbf{x} + \overline{c},$$

which can be solved efficiently (e.g. as a semidefinite program after some transformation). If $\rho_0 < 0$ and it is attained as some $\mathbf{x}_0$, then $\mathbf{x}_0$ produces a violated inequality. Otherwise, the given dual solution is feasible. The case of piecewise linear functions is addressed in Exercise 1.3. □

1.4 Summary

A key objective in this book is to provide algorithms for solving problem (1.1). In order to make progress we will restrict ourselves to cases where the functions f and h_j, $j \in \Gamma$, are polynomials (or is some cases, rational functions or piecewise polynomials) and the set $\mathbb{K}$ is a basic semi-algebraic set, i.e.,

$$\mathbb{K} = \{\mathbf{x} \in \mathbb{R}^n \;:\; g_i(\mathbf{x}) \geq 0,\; i = 1, \ldots, m\},$$

where $g_i \in \mathbb{R}[\mathbf{x}]$, for all $i = 1, \ldots, m$.

In particular, in the case where f, $\{h_j\}_{j\in\Gamma} \in \mathbb{R}[\mathbf{x}]$, and $\mathbb{K}$ is a basic semi-algebraic set, then problem (1.6) asks for a polynomial to be nonnegative for all $\mathbf{x} \in \mathbb{K}$. This naturally leads us in Chapter 2 to study nonnegative polynomials (and polynomials nonnegative on a basic semi-algebraic set) a topic of central importance in the development of 20th century mathematics. Similarly, when f, $\{h_j\}_{j\in\Gamma} \in \mathbb{R}[\mathbf{x}]$, then in problem (1.1), only the moments of the unknown measure μ are involved and not μ

itself, which naturally leads us in Chapter 3 to the study of the $\mathbb{K}$-moment problem. In fact, we will see that there is a nice and beautiful duality between the theory of moments and the theory of positive polynomials.

1.5 Exercises

Exercise 1.1. Given vectors $\mathbf{x}_i \in \mathbb{K} \subset \mathbb{R}^n$ and scalars α_i, $i = 1, \ldots, m$, β and γ, formulate the problem of finding a polynomial $f \in \mathbb{R}[\mathbf{x}]$ such that $f(\mathbf{x}_i) = \alpha_i$, $i = 1, \ldots, m$, $\beta \leq f(\mathbf{x}) \leq \gamma$ for all $\mathbf{x} \in \mathbb{K}$ as a generalized moment problem (1.1).

Exercise 1.2. When $\mathbb{K}$ is not compact, prove that Theorem 1.3 is also valid if the data satisfy the following: $-f$ is lower-semicontinuous, bounded from below and inf-compact (or a moment function); see Definition B.11. In addition, for each $\boldsymbol{\alpha} \in \Gamma$, when the moment constraint is an inequality $\int h_{\boldsymbol{\alpha}} d\mu \leq \gamma_{\boldsymbol{\alpha}}$, then $h_{\boldsymbol{\alpha}}$ is a nonnegative lower-semicontinuous function, and h_α is continuous otherwise. (Hint: Use Theorem B.9 and Proposition B.10.)

Exercise 1.3. Prove Theorem 1.5(b) for the case that f and h_j, $j \in \Gamma$, are piecewise linear functions over p convex sets $\mathbb{K}_i$ that form a partition of $\mathbb{K} = \mathbb{R}^n$.

1.6 Notes and Sources

1.1. The fact that no-arbitrage is equivalent to the existence of a martingale measure was originally proved by Harrison and Kreps (1979). For a derivation from linear optimization duality see Bertsimas and Tsitsiklis (1997).

1.2. For a survey on duality results for conic linear problems see Shapiro (2001).

1.3. The ellipsoid method was developed by Shor (1970), and Yudin and Nemirovski (1977). The polynomial time complexity of the method was shown by Khachian (1979). For the equivalence of separation and optimization see Grötschel *et al.* (1988). Theorem 1.5 is from Bertsimas and Sethuraman (2000, Theor. 16.4.4).

Chapter 2

Positive Polynomials

We review basic results of real algebraic geometry on the representation of positive polynomials, among which are the fundamental Positivstellensatz of Krivine, Stengle, Schmüdgen, Putinar and Jacobi and Prestel. We also provide some additional representation results that take into account several specific cases like finite varieties, convex semi-algebraic sets, representations that preserve sparsity, etc.

In the previous chapter, we have seen that the dual problem (1.6) of the generalized moment problem in the case where f, $(h_j)_{j\in\Gamma}$ are polynomials and $\mathbb{K}$ is a basic semi-algebraic set, asks for a polynomial to be nonnegative for all $\mathbf{x} \in \mathbb{K}$.

In one dimension, the ring $\mathbb{R}[x]$ of real polynomials of a single variable has the fundamental property (Theorem 2.5) that every nonnegative polynomial $p \in \mathbb{R}[x]$ is a sum of squares of polynomials, that is,

$$p(x) \geq 0,\ \forall x \in \mathbb{R} \quad \Leftrightarrow \quad p(x) = \sum_{i=1}^{k} h_i(x)^2,\ \forall x \in \mathbb{R},$$

for finitely many polynomials (h_i). In multiple dimensions, however, it is possible for a polynomial to be nonnegative without being a sum of squares. In fact, in his famous address in a Paris meeting of mathematicians in 1900, where he posed important problems for mathematics to be addressed in the 20th century, Hilbert in his 17th problem conjectured that every nonnegative polynomial can always be written as a sum of squares of rational functions. This conjecture was later proved to be correct by Emil Artin in 1926, using the Artin-Schreier theory of real closed fields.

An immediate extension is to consider characterizations of polynomials p that are nonnegative on a basic semi-algebraic set $\mathbb{K}$ defined by polynomial inequalities $g_i(\mathbf{x}) \geq 0$, $i = 1, \ldots, m$. By characterization, we mean a representation of p in terms of the g_i's such that the nonnegativity of p on $\mathbb{K}$ follows immediately from the latter representation in the same way that p is obviously nonnegative, when it is a sum of squares. In other words, this representation of p can be seen as a *certificate* of nonnegativity of p on $\mathbb{K}$.

In this chapter, we review the key representation theorems for nonnegative or positive polynomials. These theorems are the fundamendal building blocks for later chapters. This chapter uses some elementary results from algebraic geometry that we review in Appendix A, so that the book is self-contained.

2.1 Sum of Squares Representations and Semi-definite Optimization

In this section we show that if a nonnegative polynomial has a sum of squares representation then one can compute this representation by using semidefinite optimization methods. Given that semidefinite optimization problems are efficiently solved both from a theoretical and a practical point of view, it follows that we can compute a sum of squares decomposition of a nonnegative polynomial, if it exists, efficiently.

Let $\mathbb{R}[\mathbf{x}]$ denote the ring of real polynomials in the variables $\mathbf{x} = (x_1, \ldots, x_n)$. A polynomial $p \in \mathbb{R}[\mathbf{x}]$ is a sum of squares (in short s.o.s.) if p can be written as

$$\mathbf{x} \mapsto p(\mathbf{x}) = \sum_{j \in J} p_j(\mathbf{x})^2, \qquad \mathbf{x} \in \mathbb{R}^n,$$

for some finite family of polynomials $(p_j : j \in J) \subset \mathbb{R}[\mathbf{x}]$. Notice that necessarily the degree of p must be even, and also, the degree of each p_j is bounded by half of that of p.

Denote by $\Sigma[\mathbf{x}] \subset \mathbb{R}[\mathbf{x}]$ the space of s.o.s. polynomials. For any two real symmetric matrices $\mathbf{A}, \mathbf{B}$, recall that $\langle \mathbf{A}, \mathbf{B} \rangle$ stands for trace($\mathbf{AB}$). Finally, for a multi-index $\boldsymbol{\alpha} \in \mathbb{N}^n$, let $|\boldsymbol{\alpha}| := \sum_{i=1}^n \alpha_i$. Consider the vector

$$\begin{aligned} \mathbf{v}_d(\mathbf{x}) &= (\mathbf{x}^{\boldsymbol{\alpha}})_{|\boldsymbol{\alpha}| \leq d} \\ &= (1, x_1, \ldots, x_n, x_1^2, x_1 x_2, \ldots, x_{n-1}x_n, x_n^2, \ldots, x_1^d, \ldots, x_n^d)', \end{aligned}$$

of all the monomials $\mathbf{x}^{\boldsymbol{\alpha}}$ of degree less than or equal to d, which has dimension $s(d) := \binom{n+d}{d}$. Those monomials form the canonical basis of the vector space $\mathbb{R}[\mathbf{x}]_d$ of polynomials of degree at most d.

Proposition 2.1. *A polynomial $g \in \mathbb{R}[\mathbf{x}]_{2d}$ has a sum of squares decomposition (or is s.o.s.) if and only if there exists a real symmetric and positive semidefinite matrix $\mathbf{Q} \in \mathbb{R}^{s(d)\times s(d)}$ such that $g(\mathbf{x}) = \mathbf{v}_d(\mathbf{x})'\mathbf{Q}\mathbf{v}_d(\mathbf{x})$, for all $\mathbf{x} \in \mathbb{R}^n$.*

Proof. Suppose there exists a real symmetric $s(d) \times s(d)$ matrix $\mathbf{Q} \succeq \mathbf{0}$ for which $g(\mathbf{x}) = \mathbf{v}_d(\mathbf{x})'\mathbf{Q}\mathbf{v}_d(\mathbf{x})$, for all $\mathbf{x} \in \mathbb{R}^n$. Then $\mathbf{Q} = \mathbf{H}\mathbf{H}'$ for some $s(d) \times k$ matrix $\mathbf{H}$, and thus,

$$g(\mathbf{x}) = \mathbf{v}_d(\mathbf{x})'\mathbf{H}\mathbf{H}'\mathbf{v}_d(\mathbf{x}) = \sum_{i=1}^{k} (\mathbf{H}'\mathbf{v}_d(\mathbf{x}))_i^2, \qquad \forall \mathbf{x} \in \mathbb{R}^n.$$

Since $\mathbf{x} \mapsto (\mathbf{H}'\mathbf{v}_d(\mathbf{x}))_i$ is a polynomial, then g is expressed as a sum of squares of the polynomials $\mathbf{x} \mapsto (\mathbf{H}'\mathbf{v}_d(\mathbf{x}))_i$, $i = 1, \ldots, k$.

Conversely, suppose that g (of degree $2d$) has a s.o.s. decomposition $g(\mathbf{x}) = \sum_{i=1}^{k} [h_i(\mathbf{x})]^2$ for some family $\{h_i : i = 1, \ldots, k\} \subset \mathbb{R}[\mathbf{x}]$. Then necessarily, the degree of each h_i is bounded by d. Let $\mathbf{h}_i$ be the vector of coefficients of the polynomial h_i, i.e., $h_i(\mathbf{x}) = \mathbf{h}_i'\mathbf{v}_d(\mathbf{x})$, $i = 1, \ldots, k$. Thus,

$$g(\mathbf{x}) = \sum_{i=1}^{k} \mathbf{v}_d(\mathbf{x})'\mathbf{h}_i\mathbf{h}_i'\mathbf{v}_d(\mathbf{x}) = \mathbf{v}_d(\mathbf{x})'\mathbf{Q}\mathbf{v}_d(\mathbf{x}), \qquad \forall \mathbf{x} \in \mathbb{R}^n,$$

with $\mathbf{Q} \in \mathbb{R}^{s(d)\times s(d)}$, $\mathbf{Q} := \sum_{i=1}^{k} \mathbf{h}_i\mathbf{h}_i' \succeq \mathbf{0}$, and the proposition follows. □

Given a s.o.s. polynomial $g \in \mathbb{R}[\mathbf{x}]_{2d}$, the identity $g(\mathbf{x}) = \mathbf{v}_d(\mathbf{x})'\mathbf{Q}\mathbf{v}_d(\mathbf{x})$ for all $\mathbf{x}$, provides linear equations that the coefficients of the matrix $\mathbf{Q}$ must satisfy. Hence, writing

$$\mathbf{v}_d(\mathbf{x})\mathbf{v}_d(\mathbf{x})' = \sum_{\alpha \in \mathbb{N}^n} \mathbf{B}_\alpha \mathbf{x}^\alpha,$$

for appropriate $s(d) \times s(d)$ real symmetric matrices $(\mathbf{B}_\alpha)$, checking whether the polynomial $\mathbf{x} \mapsto g(\mathbf{x}) = \sum_\alpha g_\alpha \mathbf{x}^\alpha$ is s.o.s. reduces to solving the **semidefinite optimization[1] (feasibility) problem**:

$$\begin{array}{l} \text{Find } \mathbf{Q} \in \mathbb{R}^{s(d)\times s(d)} \text{ such that:} \\ \mathbf{Q} = \mathbf{Q}', \quad \mathbf{Q} \succeq 0, \quad \langle \mathbf{Q}, \mathbf{B}_\alpha \rangle = g_\alpha, \qquad \forall \alpha \in \mathbb{N}^n, \end{array} \tag{2.1}$$

[1] For a brief background on Semidefinite Programming, the reader is referred to Section C.2.

a tractable convex optimization problem for which efficient software packages are available. Indeed, up to arbitrary precision $\epsilon > 0$ fixed, a semidefinite program can be solved in a computational time that is polynomial in the input size of the problem. Observe that the size $s(d) = \binom{n+d}{n}$ of the semidefinite program (2.1) is bounded by n^d.

On the other hand, *nonnegativity* of a polynomial $g \in \mathbb{R}[\mathbf{x}]$ can be checked also by solving a single semidefinite program. Indeed, if g is nonnegative then it can be written as a sum of squares of rational functions, and so, clearing denominators,

$$hg = f, \tag{2.2}$$

for some s.o.s. $h,\ f \in \Sigma[\mathbf{x}]$, and there exist bounds on the degree of h and f. Conversely, if (2.2) has a nontrivial s.o.s. solution $h,\ f \in \Sigma[\mathbf{x}]$, then g is obviously nonnegative. Therefore, using (2.1) for h and f in (2.2), finding a certificate $h,\ f$ for nonnegativity of g reduces to solving a single semidefinite program. Unfortunately, the available bounds for the size of this semidefinite program are by far too large for practical implementation. This is what makes the sum of squares property very attractive computationally, in contrast with the weaker nonnegativity property which is much harder to check (if not impossible).

Example 2.1. Consider the polynomial in $\mathbb{R}[\mathbf{x}] = \mathbb{R}[x_1, x_2]$

$$f(\mathbf{x}) = 2x_1^4 + 2x_1^3x_2 - x_1^2x_2^2 + 5x_2^4.$$

Suppose we want to check whether f is a sum of squares. As f is homogeneous, we attempt to write f in the form

$$\begin{aligned} f(x_1, x_2) &= 2x_1^4 + 2x_1^3x_2 - x_1^2x_2^2 + 5x_2^4 \\ &= \begin{pmatrix} x_1^2 \\ x_2^2 \\ x_1x_2 \end{pmatrix}' \begin{bmatrix} q_{11} & q_{12} & q_{13} \\ q_{12} & q_{22} & q_{23} \\ q_{13} & q_{23} & q_{33} \end{bmatrix} \begin{pmatrix} x_1^2 \\ x_2^2 \\ x_1x_2 \end{pmatrix} \\ &= q_{11}\, x_1^4 + q_{22}\, x_2^4 + (q_{33} + 2q_{12})\, x_1^2x_2^2 + 2q_{13}\, x_1^3x_2 + 2q_{23}\, x_1x_2^3, \end{aligned}$$

for some $\mathbf{Q} \succeq 0$. In order to have an identity, we obtain

$$\mathbf{Q} \succeq 0; \quad q_{11} = 2, \; q_{22} = 5, \; q_{33} + 2q_{12} = -1, \; 2q_{13} = 2, \; q_{23} = 0.$$

In this case we easily find the particular solution

$$0 \preceq \mathbf{Q} = \begin{bmatrix} 2 & -3 & 1 \\ -3 & 5 & 0 \\ 1 & 0 & 5 \end{bmatrix} = \mathbf{H}\mathbf{H}', \qquad \mathbf{H} = \frac{1}{\sqrt{2}} \begin{bmatrix} 2 & 0 \\ -3 & 1 \\ 1 & 3 \end{bmatrix},$$

and so

$$f(x_1, x_2) = \frac{1}{2}(2x_1^2 - 3x_2^2 + x_1x_2)^2 + \frac{1}{2}(x_2^2 + 3x_1x_2)^2,$$

which is indeed a sum of squares.

Sufficient condition for being s.o.s.

Observe that checking whether a polynomial $g \in \mathbb{R}[\mathbf{x}]_{2d}$ is s.o.s. via (2.1) requires introducing the auxiliary (symmetric matrix) variable $\mathbf{Q} \in \mathbb{R}^{s(d)\times s(d)}$, i.e., we do not have "if and only if" conditions expressed directly in terms of the coefficients $(g_{\boldsymbol{\alpha}})$ of g. In fact such conditions on g exist. They define the set

$$G := \{\, g = (g_{\boldsymbol{\alpha}}) \in \mathbb{R}^{s(2d)} \; : \; \exists\, \mathbf{Q} \in \mathbb{R}^{s(d)\times s(d)} \text{ such that (2.1) holds}\},$$

which is the orthogonal projection on $\mathbb{R}^{s(2d)}$ of the set of elements $(g, \mathbf{Q})$ that satisfy (2.1), a basic semi-algebraic set of $\mathbb{R}^{s(2d)+s(d)(s(d)+1)/2}$. Indeed the semidefinite constraint $\mathbf{Q} \succeq 0$ can be stated as polynomial inequality constraints on the entries of $\mathbf{Q}$ (use determinants of its principal minors). Hence, by the Projection Theorem of real algebraic geometry, the set G is a semi-algebraic set (but not a basic semi-algebraic set in general); see Theorem A.7.

Hence, in general, G is not a basic semi-algebraic set but a finite union $\cup_{i\in I} G_i$ of basic semi-algebraic sets $G_i \subset \mathbb{R}^{s(2d)}$ and it is very hard to obtain the polynomials in the variables $g_{\boldsymbol{\alpha}}$ that define each G_i.

The next result states a sufficient condition for $g \in \mathbb{R}[\mathbf{x}]$ to be s.o.s., directly in terms of the coefficients $(g_{\boldsymbol{\alpha}})$. Let $\mathbb{N}^n_d := \{\boldsymbol{\alpha} \in \mathbb{N}^n : \sum_{i=1}^n \alpha_i \leq d\}$ and let $\Gamma_d := \{2\boldsymbol{\beta} : \boldsymbol{\beta} \in \mathbb{N}^n_d\}$.

Theorem 2.2. *Let* $\mathbf{x} \mapsto g(\mathbf{x}) = \sum_{\alpha} g_{\alpha} \mathbf{x}^{\alpha}$ *be a polynomial of degree* $2d$ *with* $d \geq 1$, *and write*

$$g = \sum_{i=1}^{n} g_{id}\, x_i^{2d} + h + g_0,$$

where $h \in \mathbb{R}[\mathbf{x}]$ *contains none of the monomials* $(x_i^{2d})_{i=1}^n$. *Then* g *is s.o.s. if*

$$g_0 \geq \sum_{\alpha \notin \Gamma_d} |g_{\alpha}| - \sum_{\alpha \in \Gamma_d} \min[0, g_{\alpha}] \tag{2.3}$$

$$g_{id} \geq \sum_{\alpha \notin \Gamma_d} |g_{\alpha}| \frac{|\alpha|}{2d} - \sum_{\alpha \in \Gamma_d} \min[0, g_{\alpha}] \frac{|\alpha|}{2d}, \quad \forall i = 1, \ldots, n. \tag{2.4}$$

On the one hand, the conditions (2.3)-(2.4) are only sufficient but on the other hand, they define a convex polyhedron in the space $\mathbb{R}^{s(2d)}$ of coefficients (g_{α}) of $g \in \mathbb{R}[\mathbf{x}]$. This latter property may be interesting if one has to optimize in the space of s.o.s. polynomials of degree at most d.

2.2 Nonnegative Versus s.o.s. Polynomials

It is important to compare nonnegative and s.o.s. polynomials because we have just seen that one knows how to check efficiently whether a polynomial is s.o.s. but not whether it is only nonnegative. There are two kinds of results for that comparison, depending on whether or not one keeps the degree fixed.

The first result is rather negative as it shows that when the degree is fixed and the number of variables grows, then the gap between nonnegative and s.o.s. polynomials increases and is unbounded. Namely, let $\mathscr{P}[\mathbf{x}]_d$ (resp. $\mathcal{H}[\mathbf{x}]_d$) denote the cone of homogeneous and nonnegative (resp. homogeneous and s.o.s.) polynomials of degree $2d$. (Recall that a polynomial $p \in \mathbb{R}[\mathbf{x}]$ of degree d is homogeneous if $p(\lambda \mathbf{x}) = \lambda^d p(\mathbf{x})$ for all $\lambda \in \mathbb{R}$ and all $\mathbf{x} \in \mathbb{R}^n$.)

To compare both sets we need subsets of finite volume. So let $\mathscr{H}$ be the hyperplane $\{f \in \mathbb{R}[\mathbf{x}] \,:\, \int_{\mathbf{S}_{n-1}} f\, d\mu = 1\}$, where μ is the rotation invariant measure on the unit sphere $\mathbf{S}_{n-1} \subset \mathbb{R}^n$. Finally, let $\widehat{\mathscr{P}}[\mathbf{x}]_d := \mathscr{H} \cap \mathscr{P}[\mathbf{x}]_d$ and $\widehat{\mathcal{H}}[\mathbf{x}]_d := \mathscr{H} \cap \mathcal{H}[\mathbf{x}]_d$.

Theorem 2.3. *There exist constants $C_1, C_2 > 0$ depending only on d and such that*

$$C_1 n^{(d/2-1)/2} \leq \left(\frac{\mathrm{vol}(\widehat{\mathscr{P}}[\mathbf{x}]_d)}{\mathrm{vol}(\widehat{\mathcal{H}}[\mathbf{x}]_d)} \right) \leq C_2 n^{(d/2-1)/2}. \tag{2.5}$$

Therefore, if d is fixed and $n \to \infty$, the gap between $\widehat{\mathscr{P}}[\mathbf{x}]_d$ and $\widehat{\mathcal{H}}[\mathbf{x}]_d$ can become as large as desired.

On the other hand, while a nonnegative polynomial $f \in \mathbb{R}[\mathbf{x}]$ may not be a s.o.s. (Exercise 2.4), we next show that we can perturb f to make it a sum of squares. The price to pay is consistent with Theorem 2.3 in that the approximation of f we need consider does *not* have same degree as f.

Given $r \in \mathbb{N}$ arbitrary, let $\Theta_r, \theta_r \in \mathbb{R}[\mathbf{x}]$ be the polynomials

$$\Theta_r(\mathbf{x}) := 1 + \sum_{i=1}^{n} x_i^{2r}; \quad \theta_r(\mathbf{x}) := \sum_{i=1}^{n} \sum_{k=0}^{r} \frac{x_i^{2k}}{k!}. \tag{2.6}$$

Given $f \in \mathbb{R}[\mathbf{x}]$ let $\|f\|_1 := \sum_{\alpha \in \mathbb{N}^n} |f_\alpha|$ if $\mathbf{f} = (f_\alpha)$ is the vector of coefficients of f. Next, with $\epsilon > 0$, we define

$$f^1_{\epsilon r} := f + \epsilon\, \Theta_r; \quad f^2_{\epsilon r} := f + \epsilon\, \theta_r.$$

Theorem 2.4.
(a) *If $f \in \mathbb{R}[\mathbf{x}]$ is nonnegative on $[-1,1]^n$, then for every $\epsilon > 0$ there exists r^1_ϵ such that $f^1_{\epsilon r} \in \Sigma[\mathbf{x}]$ for all $r \geq r^1_\epsilon$ and $\|f - f^1_{\epsilon r}\|_1 \to 0$ as $\epsilon \downarrow 0$ (and $r \geq r^1_\epsilon$).*
(b) *If $f \in \mathbb{R}[\mathbf{x}]$ is nonnegative, then for every $\epsilon > 0$ there exists r^2_ϵ such that $f^2_{\epsilon r} \in \Sigma[\mathbf{x}]$ for all $r \geq r^2_\epsilon$ and $\|f - f^2_{\epsilon r}\|_1 \to 0$ as $\epsilon \downarrow 0$ (and $r \geq r^2_\epsilon$).*

Theorem 2.4 is a denseness result with respect to the l_1-norm of coefficients. Indeed, it states that a polynomial f which is nonnegative on $[-1,1]^n$ (resp. on $\mathbb{R}^n$) can be perturbed to a s.o.s. polynomial $f^1_{\epsilon r}$ (resp. $f^2_{\epsilon r}$) such that $\|f - f^1_{\epsilon r}\|_1 \to 0$ (resp. $\|f - f^2_{\epsilon r}\|_1 \to 0$). It also provides a certificate of nonnegativity of f on $[-1,1]^n$ (resp. on $\mathbb{R}^n$). Indeed fix

$\mathbf{x} \in [-1,1]^n$ (resp. $\mathbf{x} \in \mathbb{R}^n$) and let $\epsilon \to 0$ to obtain that $0 \leq f^1_{\epsilon r}(\mathbf{x}) \to f(\mathbf{x})$ (resp. $0 \leq f^2_{\epsilon r}(\mathbf{x}) \to f(\mathbf{x})$).

Concerning Theorem 2.4(b), observe that in addition to the l_1-norm convergence $\|f - f^2_{\epsilon r}\|_1 \to 0$, the convergence is also *uniform* on compact sets. Notice that a polynomial f nonnegative on the whole $\mathbb{R}^n$ (hence on the box $[-1,1]^n$) can also be approximated by the s.o.s. polynomial $f^1_{\epsilon r}$ of Theorem 2.4(a) which is simpler than the s.o.s. approximation $f^2_{\epsilon r}$. However, in contrast to the latter, the s.o.s. approximation $f \approx f^1_{\epsilon r}$ is *not* uniform on compact sets, and is really more appropriate for polynomials nonnegative on $[-1,1]^n$ only (and indeed the approximation $f \approx f^1_{\epsilon r}$ is uniform on $[-1,1]^n$). In addition, in Theorem 2.4(a) the integer r^1_ϵ does *not* depend on the *explicit choice* of the polynomial f but only on:

(a) ϵ and the dimension n,

(b) the degree and the size of the coefficients of f.

Therefore, if one fixes these four parameters, we find an r such that the statement of Theorem 2.4(a) holds for any f nonnegative on $[-1,1]^n$, whose degree and size of coefficients do not exceed the fixed parameters.

2.3 Representation Theorems: Univariate Case

In this section, we review the major representation theorems for nonnegative univariate polynomials, for which the results are quite complete. Let $\mathbb{R}[x]$ be the ring of real polynomials of the single variable x, and let $\Sigma[x] \subset \mathbb{R}[x]$ be its subset of polynomials that are sums of squares of elements of $\mathbb{R}[x]$. We first prove that if $p \in \mathbb{R}[x]$ is nonnegative, then $p \in \Sigma[x]$.

Theorem 2.5. *A polynomial $p \in \mathbb{R}[x]$ of even degree is nonnegative if and only if it can be written as a sum of squares of other polynomials, i.e., $p(x) = \sum_{i=1}^k [h_i(x)]^2$, with $h_i \in \mathbb{R}[x]$, $i = 1, \ldots, k$.*

Proof. Clearly, if $p(x) = \sum_{j=1}^k [h_j(x)]^2$, then $p(x) \geq 0$ for all $x \in \mathbb{R}$. Conversely, suppose that a polynomial $p \in \mathbb{R}[x]$ of degree $2d$ (and with highest degree term $p_{2d}x^{2d}$) is nonnegative on the real line $\mathbb{R}$. Then, the real roots of p should have even multiplicity, otherwise p would alter its sign in a neighborhood of a root. Let λ_j, $j = 1, \ldots, r$, be its real roots with corresponding multiplicity $2m_j$. Its complex roots can be arranged in conjugate pairs, $a_l + ib_l$, $a_l - ib_l$, $l = 1 \ldots, h$ (with $i^2 = -1$). Then p can be written in the form:

$$x \mapsto \quad p(x) = p_{2d} \prod_{j=1}^{r} (x - \lambda_j)^{2m_j} \prod_{l=1}^{h} \left((x - a_l)^2 + b_l^2 \right).$$

Note that the leading coefficient p_{2d} needs to be positive. By expanding the terms in the products, we see that p can be written as a sum of squares of $k = 2^h$ polynomials. (In fact, one may also show that p is a sum of only two squares.) □

We next concentrate on polynomials $p \in \mathbb{R}[x]$ that are nonnegative on an interval $I \subset \mathbb{R}$. Moreover, the three cases $I = (-\infty, b]$, $I = [a, \infty)$ and $I = [a, b]$ (with $a, b \in \mathbb{R}$) all reduce to the basic cases $I = [0, \infty)$ and $I = [-1, +1]$ using the change of variable

$$f(x) := p(b - x), \;\; f(x) := p(x - a), \;\; \text{and} \; f(x) := p\left(\frac{2x - (a + b)}{b - a}\right),$$

respectively.

The representation results that one obtains depend on the particular choice of the polynomials used in the description of the interval. The main result in the one-dimensional case can be summarized in the next theorem.

Theorem 2.6. *Let $p \in \mathbb{R}[x]$ be of degree n.*
(a) *Let $g \in \mathbb{R}[x]$ be the polynomial $x \mapsto g(x) := 1 - x^2$. Then $p \geq 0$ on $[-1, 1]$ if and only if*

$$p = f + g\,h, \quad f, h \in \Sigma[x],$$

and with both summands of degree less than $2n$.
(b) *Let $x \mapsto g_1(x) := 1 - x$, $x \mapsto g_2(x) := 1 + x$, $x \mapsto g_3(x) := g_1(x)g_2(x)$. Then, $p \geq 0$ on $I = [-1, 1]$ if and only if*

$$p = f_0 + g_1\, f_1 + g_2\, f_2 + g_3\, f_3, \quad f_0, f_1, f_2, f_3 \in \Sigma[x].$$

In addition, all the summands have degree at most n, and $f_1, f_2 = 0$, if n is even, whereas $f_0, f_3 = 0$, if n is odd.

The case $[0, \infty)$ reduces to the case $[-1, 1]$ (see Exercise 2.1) and we have:

Theorem 2.7. *Let $p \in \mathbb{R}[x]$ be nonnegative on $[0, +\infty)$. Then*

$$p = f_0 + x\, f_1,$$

for two s.o.s. polynomials $f_0, f_1 \in \Sigma[x]$ and the degree of both summands is bounded by $\deg p$.

It is important to emphasize that Theorem 2.6 (resp. Theorem 2.7) explicitly use the specific representation of the interval $[-1, 1]$ (resp. $[0, +\infty)$) as the basic semi-algebraic set: $\{x : 1-x \geq 0;\ 1+x \geq 0\}$ or $\{x : 1-x^2 \geq 0\}$ (resp. $\{x : x \geq 0\}$).

In Exercise 2.2 we consider the case $[-1, 1] = \{x \in \mathbb{R} : h(x) \geq 0\}$, with h not equal to the polynomial $x \mapsto (1 + x)(1 - x)$, and a weaker result is obtained.

The next result also considers the interval $[-1, 1]$ but provides another representation that does not use s.o.s.

Theorem 2.8. *Let $p \in \mathbb{R}[x]$. Then $p > 0$ on $[-1, 1]$ if and only if*

$$p = \sum_{i+j \leq d} c_{ij} (1 - x)^i (1 + x)^j, \quad c_{ij} \geq 0, \tag{2.7}$$

for some sufficiently large d.

Notice that Theorem 2.8 leads to a linear optimization feasibility problem to determine the coefficients c_{ij} in the representation (2.7) (Exercise 2.3).

2.4 Representation Theorems: Mutivariate Case

In this section we consider the multivariate case. As already mentioned, a nonnegative polynomial $p \in \mathbb{R}[\mathbf{x}]$ does not necessarily have a sum of squares representation. In Exercise 2.4 we show that the polynomials

$$p(x_1, x_2) = x_1^2 x_2^2 (x_1^2 + x_2^2 - 1) + 1,$$
$$p(x_1, x_2, x_3) = x_1^2 x_2^2 (x_1^2 + x_2^2 - 3x_3^2) + 6x_3^6$$

are nonnegative but they do not have a s.o.s. representation. On the other hand, nonnegative quadratic polynomials, and nonnegative fourth degree homogeneous polynomials of three variables have a s.o.s. representation (Exercise 2.5).

The next celebrated theorem due to Polyá provides a certificate of positivity for homogeneous polynomials that are positive on the simplex.

Theorem 2.9 (Polyá). *If $p \in \mathbb{R}[\mathbf{x}]$ is homogeneous and $p > 0$ on $\mathbb{R}^n_+ \setminus \{\mathbf{0}\}$, then for sufficienly large $k \in \mathbb{N}$, all non zero coefficients of the polynomial $\mathbf{x} \mapsto (x_1 + \ldots + x_n)^k p(\mathbf{x})$ are positive.*

As a consequence of Theorem 2.9 we next obtain the following representation result for nonhomogeneous polynomials that are strictly positive on $\mathbb{R}^n_+$. If $p \in \mathbb{R}[\mathbf{x}]$, we denote by $\tilde{p} \in \mathbb{R}[\mathbf{x}, x_0]$ the homogeneous polynomial associated with p, that is, $\tilde{p}(\mathbf{x}, x_0) := x_0^d\, p(\mathbf{x}/x_0)$, with d being the degree of p, and denote by $p_d(\mathbf{x}) = \tilde{p}(\mathbf{x}, 0)$ for all $\mathbf{x} \in \mathbb{R}^n$, the homogeneous part of p, of degree d.

Theorem 2.10. *Let $\mathbf{x} \mapsto g(\mathbf{x}) := 1 + \sum_{j=1}^n x_j$ and $p \in \mathbb{R}[\mathbf{x}]$ with degree d. If $p > 0$ on $\mathbb{R}^n_+$ and $p_d > 0$ on $\mathbb{R}^n_+ \setminus \{\mathbf{0}\}$, then for sufficienly large $k \in \mathbb{N}$, the polynomial $g^k p$ has positive coefficients.*

Proof. We first prove that $\tilde{p} > 0$ on $\mathbb{R}^{n+1}_+ \setminus \{\mathbf{0}\}$. Let $(\mathbf{x}, x_0) \in \mathbb{R}^{n+1}_+ \setminus \{\mathbf{0}\}$. If $x_0 = 0$, then $\mathbf{x} \in \mathbb{R}^n_+ \setminus \{\mathbf{0}\}$ and $\tilde{p}(\mathbf{x}, 0) = p_d(\mathbf{x}) > 0$. If $x_0 > 0$, then $\mathbf{y} := \mathbf{x}/x_0 \in \mathbb{R}^n_+$, and thus $p(\mathbf{y}) > 0$ implies that $\tilde{p}(\mathbf{x}, x_0) = x_0^d p(\mathbf{y}) > 0$. Therefore, $\tilde{p} > 0$ on $\mathbb{R}^{n+1}_+ \setminus \{\mathbf{0}\}$. Hence by Theorem 2.9, for all sufficiently large $k \in \mathbb{N}$, the polynomial

$$(\mathbf{x}, x_0) \;\mapsto\; q(\mathbf{x}, x_0) := \left(\sum_{j=1}^n x_j + x_0 \right)^k \tilde{p}(\mathbf{x}, x_0),$$

has positive coefficients. But then, as $\tilde{p}(\mathbf{x}, 1) = p(\mathbf{x})$, it follows that the polynomial

$$\mathbf{x} \;\mapsto\; q(\mathbf{x}, 1) \;=\; \left(\sum_{j=1}^n x_j + 1 \right)^k \tilde{p}(\mathbf{x}, 1) = g^k \cdot p,$$

has positive coefficients, for all sufficiently large $k \in \mathbb{N}$. □

We next characterize when a semi-algebraic set described by polynomial inequalities, equalities and non-equalities is empty. In order to achieve this, we need the following definition.

Definition 2.1. For $F := \{f_1, \ldots, f_m\} \subset \mathbb{R}[\mathbf{x}]$, and a set $J \subseteq \{1, \ldots, m\}$ we denote by $f_J \in \mathbb{R}[\mathbf{x}]$ the polynomial $\mathbf{x} \mapsto f_J(\mathbf{x}) := \prod_{j \in J} f_j(\mathbf{x})$, with the convention that $f_\emptyset = 1$. The set

$$P(f_1, \ldots, f_m) := \left\{ \sum_{J \subseteq \{1,\ldots,m\}} q_J f_J \quad : \quad q_J \in \Sigma[\mathbf{x}] \right\} \tag{2.8}$$

is called (by algebraic geometers) a **preordering**. It is also a convex cone of $\mathbb{R}[\mathbf{x}]$; see Appendix A.

We first state a key result from Stengle:

Theorem 2.11 (Stengle). *Let k be a real closed field, and let*

$$F := (f_i)_{i \in I_1}, \; G := (g_i)_{i \in I_2}, \; H := (h_i)_{i \in I_3} \subset k[\mathbf{x}]$$

be finite families of polynomials. Let
(a) *$P(F)$ be the preordering generated by the family F,*
(b) *$M(G)$ be the set of all finite products of the g_i's, $i \in I_2$ (the empty product being the constant polynomial 1), and*
(c) *$I(H)$ be the ideal generated by H.*
Consider semi-algebraic set

$$\mathbb{K} = \{\mathbf{x} \in k^n : f_i(\mathbf{x}) \geq 0, \quad \forall i \in I_1; \quad g_i(\mathbf{x}) \neq 0, \quad \forall i \in I_2; \\ h_i(\mathbf{x}) = 0, \quad \forall i \in I_3\}.$$

The set $\mathbb{K}$ is empty if and only if there exist $f \in P(F)$, $g \in M(G)$ and $h \in I(H)$ such that

$$f + g^2 + h = 0. \tag{2.9}$$

The polynomials $f, g, h \in k[\mathbf{x}]$ in (2.9) provide a Stengle *certificate* of $\mathbb{K} = \emptyset$. In (2.9), there is also a (very large) bound on the degree of $g \in M(G)$, the degree of the s.o.s. weights $q_J \in \Sigma[\mathbf{x}]$ in the representation

(2.8) of $f \in P(F)$, as well as the degree of the weights $p_j \in \mathbb{R}[\mathbf{x}]$ in the representation of $h = \sum_j p_j h_j$.

Therefore in view of Proposition 2.1, in principle checking existence of a certificate (f, g, h) requires solving a *single* semidefinite program. But unfortunately, the available degree bound being huge, the size of such a semidefinite program is by far too large for practical implementation.

Stengle and Farkas certificates

The Stengle certificate $f, g, h \in k[\mathbf{x}]$ in (2.9) for $\mathbb{K} = \emptyset$, is a nonlinear generalization of the celebrated Farkas Lemma (or Theorem of the alternative) in linear algebra which provides a certificate of emptyness for a polyhedral set $\mathbb{K} = \{\mathbf{x} : \mathbf{A}\mathbf{x} \leq \mathbf{b}\}$ (for some matrix $\mathbf{A} \in \mathbb{R}^{m \times n}$ and some vector $\mathbf{b} \in \mathbb{R}^m$).

In fact, a Farkas certificate is a particularly simple Stengle certificate for convex polyhedra. Indeed, a Farkas certificate of $\emptyset = \{\mathbf{x} : \mathbf{A}\mathbf{x} \leq \mathbf{b}\}$ is a nonnegative vector $\mathbf{u} \in \mathbb{R}^m_+$ such that $\mathbf{u}'\mathbf{A} = 0$ and $\mathbf{u}'\mathbf{b} < 0$. So let $\mathbf{x} \mapsto f_j(\mathbf{x}) = (\mathbf{b} - \mathbf{A}\mathbf{x})_j$, $j = 1, \ldots, m$, $\mathbf{x} \mapsto g_1(\mathbf{x}) = 1$ (so that $M(G)$ is identical to the constant polynomial 1) and $\mathbf{x} \mapsto h_1(\mathbf{x}) = 0$ (so that $I(H)$ is the 0 polynomial). The polynomial

$$\mathbf{x} \mapsto f(\mathbf{x}) := \frac{-1}{\mathbf{u}'\mathbf{b}} \sum_{j=1}^{m} u_j (\mathbf{b} - \mathbf{A}\mathbf{x})_j$$

is an element of $P(F)$ because $u_j \geq 0$ for every $j = 1, \ldots, m$ and $\mathbf{u}'\mathbf{b} < 0$. But then

$$1 + f(\mathbf{x}) = 1 - \frac{1}{\mathbf{u}'\mathbf{b}} \sum_{j=1}^{m} u_j (\mathbf{b} - \mathbf{A}\mathbf{x})_j = 1 - \frac{\mathbf{u}'\mathbf{b} - \mathbf{u}'\mathbf{A}\mathbf{x}}{\mathbf{u}'\mathbf{b}} = 0,$$

that is, (2.9) holds.

We next consider basic semi-algebraic sets, that is, semi-algebraic sets defined by inequalities only. As a direct consequence of Theorem 2.11, one obtains:

Theorem 2.12. (Stengle's Positivstellensatz and Nullstellensatz). *Let k be a real closed field, $f \in k[\mathbf{x}]$, and let*

$$\mathbb{K} = \{\, \mathbf{x} \in k^n : \quad f_j(\mathbf{x}) \geq 0, \; j = 1, \ldots, m \,\}.$$

(a) Nichtnegativstellensatz. *$f \geq 0$ on $\mathbb{K}$ if and only if there exists $\ell \in \mathbb{N}$, and $g, h \in P(f_1, \ldots, f_m)$ such that $fg = f^{2\ell} + h$.*
(b) Positivstellensatz. *$f > 0$ on $\mathbb{K}$ if and only if there exist $g, h \in P(f_1, \ldots, f_m)$ such that $fg = 1 + h$.*
(c) Nullstellensatz. *$f = 0$ on $\mathbb{K}$ if and only if there exists $\ell \in \mathbb{N}$, and $g \in P(f_1, \ldots, f_m)$ such that $f^{2\ell} + g = 0$.*

Again, as for Theorem 2.11, there is also a bound on ℓ and the degree of the s.o.s. weights $q_J \in \Sigma[\mathbf{x}]$ in the representation (2.8) of $g, h \in P(f_1, \ldots, f_m)$. This bound depends only on the dimension n and on the degree of the polynomials $(f, f_1, \ldots, f_m)$. Therefore in principle, checking existence of a certificate (l, g, h) in Theorem 2.12(a)-(c) requires solving a single semidefinite program (but of huge size). In practice, one fixes an *a priori* (much smaller) degree bound and solves the corresponding semidefinite program. If the latter has a feasible solution then one obtains a certificate $g, h \in P(f_1, \ldots, f_m)$. However, such a certificate is *numerical* and so can be obtained only up to some machine precision, because of numerical inaccuracies inherent to semidefinite programming solvers.

2.5 Polynomials Positive on a Compact Basic Semi-algebraic Set

In this section, we restrict our attention to compact basic semi-algebraic sets $\mathbb{K} \subset \mathbb{R}^n$ and obtain certificates of positivity on $\mathbb{K}$ that have certain algorithmic advantages. In fact, Putinar's Positivstellensatz below is the key result that we will later use extensively to solve the GMP.

2.5.1 *Representations via sums of squares*

The first representation result, known as Schmüdgen's Positivstellensatz, was an important breakthrough as it was the first to provide a simple characterization of polynomials positive on a compact basic semi-algebraic set $\mathbb{K}$, and with *no* additional assumption on $\mathbb{K}$ (or on its description).

Theorem 2.13 (Schmüdgen's Positivstellensatz). *Let* $(g_j)_{j=1}^m \subset \mathbb{R}[\mathbf{x}]$ *be such that the basic semi-algebraic set*

$$\mathbb{K} := \{\mathbf{x} \in \mathbb{R}^n : \quad g_j(\mathbf{x}) \geq 0, \ j = 1, \ldots, m\} \tag{2.10}$$

is compact. If $f \in \mathbb{R}[\mathbf{x}]$ *is strictly positive on* $\mathbb{K}$ *then* $f \in P(g_1, \ldots, g_m)$*, that is,*

$$f = \sum_{J \subseteq \{1,\ldots,m\}} f_J \, g_J, \quad \text{for some s.o.s. } f_J \in \Sigma[\mathbf{x}], \tag{2.11}$$

and with $g_J = \prod_{j \in J} g_j$.

Theorem 2.13 is a very powerful result, but note that the number of terms in (2.11) is exponential in the number of polynomials that define the set $\mathbb{K}$. However, a major improvement is possible under a relatively weak assumption on the polynomials that define the compact set $\mathbb{K}$. Associated with the finite family $(g_j) \subset \mathbb{R}[\mathbf{x}]$, the subset of $P(g_1, \ldots, g_m)$ defined by

$$Q(g) = Q(g_1, \ldots, g_m) := \left\{ q_0 + \sum_{j=1}^m q_j g_j \ : \quad (q_j)_{j=0}^m \subset \Sigma[\mathbf{x}] \right\} \tag{2.12}$$

is a convex cone called the *quadratic module* generated by the family (g_j).

Assumption 2.1. With $(g_j)_{j=1}^m \subset \mathbb{R}[\mathbf{x}]$, there exists $u \in Q(g)$ such that the level set $\{\mathbf{x} \in \mathbb{R}^n : u(\mathbf{x}) \geq 0\}$ is compact.

Theorem 2.14 (Putinar's Positivstellensatz). *Let* $\mathbb{K} \subset \mathbb{R}^n$ *be as in (2.10) and let Assumption 2.1 hold. If* $f \in \mathbb{R}[\mathbf{x}]$ *is strictly positive on* $\mathbb{K}$ *then* $f \in Q(g)$*, that is,*

$$f = f_0 + \sum_{j=1}^m f_j \, g_j, \tag{2.13}$$

for some s.o.s. polynomials $f_j \in \Sigma[\mathbf{x}]$, $j = 0, 1, \ldots, m$.

In contrast to Theorem 2.13, the number of terms in the representation (2.13) is linear in the number of polynomials that define $\mathbb{K}$, a crucial improvement from a computational point of view. The condition on $\mathbb{K}$

(Assumption 2.1) is not very restrictive. For instance, it is satisfied in the following cases:

(a) All the g_i's are affine and $\mathbb{K}$ is a polytope.
(b) The set $\{\mathbf{x} \in \mathbb{R}^n : g_j(\mathbf{x}) \geq 0\}$ is compact for some $j \in \{1, \ldots, m\}$.

Also, suppose that we know some $N > 0$ such that $\mathbb{K} \subset \{\mathbf{x} \in \mathbb{R}^n : \|\mathbf{x}\|^2 \leq N\}$. Let $\mathbf{x} \mapsto g_{m+1}(\mathbf{x}) := N - \|\mathbf{x}\|^2$. Adding the quadratic constraint $g_{m+1}(\mathbf{x}) \geq 0$ in the definition (2.10) of $\mathbb{K}$ does not change $\mathbb{K}$ as this last constraint is redundant. But with this new representation, $\mathbb{K}$ satisfies the required condition in Theorem 2.14.

The following theorem provides further understanding on the condition in Theorem 2.14.

Theorem 2.15. *Let $(g_j)_{i=1}^m \subset \mathbb{R}[\mathbf{x}]$, assume that $\mathbb{K} \subset \mathbb{R}^n$ defined in (2.10) is compact, and let $Q(g)$ be as in (2.12). The following conditions are equivalent:*
(a) *There exist finitely many $p_1, \ldots, p_s \in Q(g)$ such that the level set $\{\mathbf{x} \in \mathbb{R}^n : p_j(\mathbf{x}) \geq 0,\ j = 1, \ldots, s\}$ which contains $\mathbb{K}$ is compact and $\prod_{i \in J} p_j \in Q(g)$ for all $J \subseteq \{1, \ldots, s\}$.*
(b) *Assumption 2.1 holds.*
(c) *There exists $N \in \mathbb{N}$ such that the polynomial $\mathbf{x} \mapsto N - \|\mathbf{x}\|^2 \in Q(g)$.*
(d) *For all $p \in \mathbb{R}[\mathbf{x}]$, there is some $N \in \mathbb{N}$ such that both polynomials $\mathbf{x} \mapsto N + p(\mathbf{x})$ and $\mathbf{x} \mapsto N - p(\mathbf{x})$ are in $Q(g)$.*

Both Theorems 2.13 and 2.14 have significant computational advantages. Indeed, from Proposition 2.1, given a polynomial $f > 0$ on $\mathbb{K}$, checking whether f has the representation (2.11) or (2.13), and assuming an *a priori* bound on the degree of the unknown s.o.s. polynomials, reduces to solving a semidefinite optimization problem, as we saw in Section 2.3.

Example 2.2. Let $\mathbf{x} \mapsto f(\mathbf{x}) = x_1^3 - x_1^2 + 2x_1x_2 - x_2^2 + x_2^3$ and $\mathbb{K} = \{\mathbf{x} : g_1(\mathbf{x}) = x_1 \geq 0,\ g_2(\mathbf{x}) = x_2 \geq 0,\ g_3(\mathbf{x}) = x_1 + x_2 - 1 \geq 0\}$. To check whether $f \geq 0$ on $\mathbb{K}$ we attempt to write $f = f_0 + \sum_{i=1}^3 f_i g_i$, where $f_i \in \Sigma[\mathbf{x}]$ and each has degree 2, that is,

$$f_i = (1, x_1, x_2)\mathbf{Q}_i(1, x_1, x_2)', \quad \mathbf{Q}_i \succeq \mathbf{0},\ i = 0, 1, 2, 3.$$

Solving the semidefinite feasiblity problem, we find that

$$\mathbf{Q}_0 = \mathbf{0},\ \mathbf{Q}_1 = \begin{bmatrix} 0\,0\,0 \\ 0\,0\,0 \\ 0\,0\,1 \end{bmatrix},\ \mathbf{Q}_2 = \begin{bmatrix} 0\,0\,0 \\ 0\,1\,0 \\ 0\,0\,0 \end{bmatrix},\ \mathbf{Q}_3 = \begin{bmatrix} 0 & 0 & 0 \\ 0 & 1 & -1 \\ 0 & -1 & 1 \end{bmatrix},$$

and so,

$$f(\mathbf{x}) = x_2^2 x_1 + x_1^2 x_2 + (x_1 - x_2)^2 (x_1 + x_2 - 1)$$

which proves that $f \geq 0$ on $\mathbb{K}$.

Degree bound

The following result provides a bound on the degree of the weights $(f_j) \subset \Sigma[\mathbf{x}]$ in the representation (2.13). For $f \in \mathbb{R}[\mathbf{x}]$ written $\mathbf{x} \mapsto f(\mathbf{x}) = \sum_{\alpha} f_{\alpha} \mathbf{x}^{\alpha}$ let

$$\|f\|_0 := \max_{\alpha} \frac{|f_{\alpha}|}{\binom{|\alpha|}{\alpha}} \quad \text{with} \quad \binom{|\alpha|}{\alpha} := \frac{|\alpha|!}{\alpha_1! \cdots \alpha_n!}.$$

Theorem 2.16 (Degree bound). *Let $\mathbb{K} \subset \mathbb{R}^n$ in (2.10) satisfy Assumption 2.1 and assume that $\emptyset \neq \mathbb{K} \subset (-1,1)^n$. Then there is some $c > 0$ such that for all $f \in \mathbb{R}[\mathbf{x}]$ of degree d, and positive on $\mathbb{K}$ (i.e. such that $f^* := \min\{f(\mathbf{x}) : \mathbf{x} \in \mathbb{K}\} > 0$), the representation (2.13) holds with*

$$\deg f_j g_j \leq c \exp\left(\left(d^2 n^d \frac{\|f\|_0}{f^*}\right)^c\right), \qquad \forall j = 1, \ldots, m.$$

In view of the importance of Theorem 2.14 in the book, we next provide a proof due to M. Schweighofer.

Proof of Theorem 2.14

We first need some preliminary results. A quadratic module $Q \subset \mathbb{R}[\mathbf{x}]$ satisfies $1 \in Q$, $Q + Q \subset Q$ and $\Sigma[\mathbf{x}]Q \subset Q$. In addition Q is proper if $-1 \notin Q$, that is, if $Q \neq \mathbb{R}[\mathbf{x}]$. Finally Q is said to be Archimedean if for every $f \in Q$ there is some $N \in \mathbb{N}$ such that $N \pm f \in Q$.

Lemma 2.17. *If $Q \subset \mathbb{R}[\mathbf{x}]$ is a quadratic module then $I := Q \cap -Q$ is an ideal.*

Proof. Let $f \in \mathbb{R}[\mathbf{x}]$ and $g \in \mathscr{I}$. Then $f\,g = g\,(f+1)^2/2 + (-g)(f-1)^2/2 \in I$. □

Lemma 2.18. *If $Q \subset \mathbb{R}[\mathbf{x}]$ is a maximal proper quadratic module then $Q \cup -Q = \mathbb{R}[\mathbf{x}]$.*

Proof. Let $f \in \mathbb{R}[\mathbf{x}]$ and assume that $f \notin Q \cup -Q$. As Q is maximal, the quadratic modules $Q + f\Sigma[\mathbf{x}]$ and $Q - f\Sigma[\mathbf{x}]$ are not proper; that is there exists $g_1, g_2 \in Q$ and $s_1, s_2 \in \Sigma[\mathbf{x}]$ such that $-1 = g_1 + s_1 f$ and $-1 = g_2 - s_2 f$. Multiplying the first by s_2 and the second by s_1 yield $s_1 + s_2 + s_1 g_1 + s_2 g_2 = 0$. Therefore $s_1, s_2 \in I := Q \cap -Q$. But then $s_1 f \in I$ because I is an ideal and so we obtain the contradiction $-1 = g_1 + s_1 f \in Q$. □

Lemma 2.19. *Let $Q \subset \mathbb{R}[\mathbf{x}]$ be an Archimedean maximal proper quadratic module, let $I := Q \cap -Q$, and let $f \in \mathbb{R}[\mathbf{x}]$. Then there is a unique $a \in \mathbb{R}$ such that $f - a \in I$.*

Proof. As Q is Archimedean both sets $A := \{a \in \mathbb{R} : f - a \in Q\}$ and $B := \{b \in \mathbb{R} : b - f \in Q\}$ are not empty. It suffices to show that $A \cap B$ is a singleton. Being proper, Q does not contain any negative number and therefore $a \leq b$ for all $a \in A$, $b \in B$. Set $a_0 := \sup\{a : a \in A\}$ and $b_0 := \sup\{b : b \in B\}$ so that $a_0 \leq b_0$. In fact $a_0 = b_0$ because if $a_0 < c < b_0$ then $f - c \notin Q \cup -Q$, which contradicts Lemma 2.18. Then it suffices to show that $a_0 \in A$ and $b_0 \in B$ in which case $A \cap B = a_0$.

Assume that $a_0 \notin A$, i.e., $f - a_0 \notin Q$. Then by maximality of Q, the quadratic module $Q' := Q + (f - a_0)\Sigma[\mathbf{x}]$ cannot be proper and so $-1 = g + (f - a_0)s$ for some $g \in Q$ and $s \in \Sigma[\mathbf{x}]$. As Q is Archimedean, choose $N \in \mathbb{N}$ such that $N - s \in Q$ and $\epsilon \in \mathbb{R}$ such that $0 < \epsilon < N^{-1}$. As $a_0 - \epsilon \in A$ one has $f - (a_0 - \epsilon) \in Q$ and so $-1 + \epsilon s = g + (f - a_0 + \epsilon)s \in Q$ and $\epsilon(N - s) \in Q$. Adding up yields the contradiction $\epsilon N - 1 \in Q$ and $\epsilon N - 1 < 0$. Therefore $a_0 \in A$ and similar arguments also yield $b_0 \in B$, the desired result. □

So now let $f \in \mathbb{R}[\mathbf{x}]$ be (strictly) positive on $\mathbb{K}$. We will prove that $f \in Q(g)$.

Proposition 2.20. *There exists $s \in \Sigma[\mathbf{x}]$ such that $sf \in 1 + Q(g)$.*

Proof. We shall prove that the quadratic module $Q_0 := Q(g) - f\Sigma[\mathbf{x}]$ is not proper. Assume that Q_0 is proper. By Zorn's lemma one may extend Q_0 to a maximal proper quadratic module $Q \supset Q_0$ and Q is Archimedean because $Q \supset Q(g)$. By Lemma 2.19 there exists $a \in \mathbb{R}^n$ such that $x_i - a_i \in I := Q \cap -Q$ for every $i = 1, \ldots, n$. Since I is an ideal, $f - f(a) \in I$ for any $f \in \mathbb{R}[\mathbf{x}]$. Choosing $f := g_j$ yields that $g_j(a) = g_j - (g_j - g_j(a)) \in Q$ because $g_j \in Q(g) \subset Q$ and $-(g_j - g_j(a)) \in Q$. This implies $g_j(a) \geq 0$ and as j was arbitrary, $a \in \mathbb{K}$. Finally, $-f(a) = (f - f(a)) - f \in Q$ because $f - f(a) \in I \subset Q$ and $-f \in Q_0 \subset Q$; therefore $-f(a) \geq 0$, the desired result. $\square$

Proposition 2.21. *There exist $h \in Q(g)$ and $N \in \mathbb{N}$ such that $N - h \in \Sigma[\mathbf{x}]$ and $hf \in 1 + Q(g)$.*

Proof. Choose $s \in \Sigma[\mathbf{x}]$ as in Proposition 2.20. Using Theorem 2.15(d) there exists $k \in \mathbb{N}$ such that $2k - s, 2k - s^2 f - 1 \in Q(g)$. Let $h := s(2k - s)$ and $N := k^2$; then $h \in Q(g)$ and $N - h = k^2 - 2sk + s^2 = (k - s)^2 \in \Sigma[\mathbf{x}]$. In addition, $hf - 1 = s(2k - s)f - 1 = 2k(sf - 1) + (2k - s^2 f - 1) \in Q(g)$ because $sf - 1, 2k - s^2 f - 1 \in Q(g)$. $\square$

To conclude, choosing h, N as in Proposition 2.21 with $N > 0$ and $k \in \mathbb{N}$ such that $k + f \in Q(g)$, one obtains

$$(k - \frac{1}{N}) + f = \frac{1}{N}((N - h)(k + f) + (hf - 1) + kh) \in Q(g).$$

Iterate this process with $(k - N^{-1})$ in lieu of k to obtain that $(k - 2N^{-1}) + f \in Q(g)$. After kN iterations of this process one obtains the desired result $f \in Q(g)$.

2.5.2 *A matrix version of Putinar's Positivstellensatz*

For $f \in \mathbb{R}[\mathbf{x}]$ written $\mathbf{x} \mapsto f(\mathbf{x}) = \sum_{\boldsymbol{\alpha}} f_{\boldsymbol{\alpha}} \mathbf{x}^{\boldsymbol{\alpha}}$, recall that

$$\|f\|_0 = \max_{\boldsymbol{\alpha}} |f_{\boldsymbol{\alpha}}| \frac{\alpha_1! \cdots \alpha_n!}{|\boldsymbol{\alpha}|!}. \tag{2.14}$$

Definition 2.2. (a) The norm of a real symmetric matrix-polynomial $\mathbf{F} \in \mathbb{R}[\mathbf{x}]^{p \times p}$ is defined by $\|\mathbf{F}\| := \max_{\|\xi\|=1} \|\xi' \mathbf{F}(\mathbf{x}) \xi\|_0$.

(b) A real symmetric matrix-polynomial $\mathbf{F} \in \mathbb{R}[\mathbf{x}]^{p \times p}$ is said to be a sum of squares (in short s.o.s.) if $\mathbf{F} = \mathbf{L}\mathbf{L}'$ for some $q \in \mathbb{N}$ and some real matrix-polynomial $\mathbf{L} \in \mathbb{R}[\mathbf{x}]^{p \times q}$.

Let $\mathbf{I}$ denote the $p \times p$ identity matrix.

Theorem 2.22. *Let $\mathbb{K} \subset \mathbb{R}^n$ be the basic semi-algebraic set in (2.10) and let Assumption 2.1 hold. Let $\mathbf{F} \in \mathbb{R}[\mathbf{x}]^{p\times p}$ be a real symmetric matrix-polynomial of degree d. If for some $\delta > 0$, $\mathbf{F}(\mathbf{x}) \succeq \delta\mathbf{I}$ for all $\mathbf{x} \in \mathbb{K}$, then*

$$\mathbf{F}(\mathbf{x}) = \mathbf{F}_0(\mathbf{x}) + \sum_{j=1}^{m} \mathbf{F}_i(\mathbf{x})\, g_j(\mathbf{x}) \tag{2.15}$$

for some s.o.s. matrix-polynomials $(\mathbf{F}_j)_{j=0}^m$, and

$$\deg \mathbf{F}_0,\ \deg \mathbf{F}_1 g_1, \ldots, \deg \mathbf{F}_m g_m \leq c\left(d^2 n^d \frac{\|\mathbf{F}\|}{\delta}\right)^c.$$

Obviously, Theorem 2.22 is a matrix-polynomial analogue of Theorem 2.14. In fact, one might have characterized the property $\mathbf{F}(\mathbf{x}) \succeq \delta\mathbf{I}$ on $\mathbb{K}$ by using Theorem 2.14 as follows. Let $\mathbf{S} := \{\xi \in \mathbb{R}^p : \xi'\xi = 1\}$ and notice that the compact basic semi-algebraic set $\mathbb{K}\times\mathbf{S} \subset \mathbb{R}^n \times \mathbb{R}^p$ satisfies Assumption 2.1 whenever $\mathbb{K}$ does. If $f \in \mathbb{R}[\mathbf{x},\xi]$ denotes the polynomial $(\mathbf{x},\xi) \mapsto \xi'\mathbf{F}(\mathbf{x})\xi$, then

$$\mathbf{F} \succeq \delta\mathbf{I} \quad \text{on } \mathbb{K} \quad \Leftrightarrow \quad f \geq \delta \quad \text{on } \mathbb{K}\times\mathbf{S},$$

and so by Theorem 2.14, $\mathbf{F} \succeq \delta\mathbf{I}$ on $\mathbb{K}$ implies

$$\xi'\mathbf{F}(\mathbf{x})\xi = \sigma_0(\mathbf{x},\xi) + \sum_{j=1}^{m} \sigma_j(\mathbf{x},\xi)\, g_j(\mathbf{x}) + \sigma_{m+1}(\mathbf{x},\xi)\,(\xi'\xi - 1), \tag{2.16}$$

for some s.o.s. polynomials $(\sigma_0)_0^m \subset \Sigma[\mathbf{x},\xi]$ and some polynomial $\sigma_{m+1} \in \mathbb{R}[\mathbf{x},\xi]$.

However, in general, the degree of weights in the representation (2.15) and (2.16) is not the same. It is not clear whether one should be preferred.

2.5.3 *An alternative representation*

We next present an alternative representation not based on s.o.s. polynomials. We make the following assumption:

Assumption 2.2. The set $\mathbb{K}$ in (2.10) is compact and the polynomials $(1, g_j)_{j=1}^m$ generate the algebra $\mathbb{R}[\mathbf{x}]$.

Note that if $(1, g_j)_{j=1}^m$ do not generate $\mathbb{R}[\mathbf{x}]$, we add redundant inequalities as follows. Let $\underline{x}_k \leq \min\{x_k \,:\, \mathbf{x} \in \mathbb{K}\}$ for all $k = 1, \ldots, n$. Then, with $g_{m+k}(\mathbf{x}) := x_k - \underline{x}_k$, we introduce the additional (redundant) constraints $g_{m+k}(\mathbf{x}) \geq 0$, $k = 1, \ldots, n$, in the definition (2.10) of $\mathbb{K}$, and reset $m := m + n$. With this new equivalent definition, Assumption 2.2 holds.

For every $j = 1, \ldots, m$, let $\overline{g}_j := \max_{\mathbf{x} \in \mathbb{K}} g_j(\mathbf{x})$ (well-defined because $\mathbb{K}$ is compact), and let $(\widehat{g}_j)_{j=1}^m$ be the polynomials g_j's, normalized with respect to $\mathbb{K}$, that is, for $j = 1, \ldots, m$

$$\widehat{g}_j := \begin{cases} g_j/\overline{g}_j, & \text{if } \overline{g}_j > 0, \\ g_j, & \text{if } \overline{g}_j = 0. \end{cases} \tag{2.17}$$

We next let $G := (0, 1, \widehat{g}_1, \ldots, \widehat{g}_m, 1 - \widehat{g}_1, \ldots, 1 - \widehat{g}_m) \subset \mathbb{R}[\mathbf{x}]$, and let $\Delta_G \subset \mathbb{R}[\mathbf{x}]$ be the set of all products of the form $q_1 \cdots q_k$, for polynomials $(q_j)_{j=1}^k \subset G$, and integer $k \geq 1$. Denote by C_G the *cone* generated by Δ_G, i.e., $f \in C_G$ if

$$f = \sum_{\boldsymbol{\alpha}, \boldsymbol{\beta} \in \mathbb{N}^m} c_{\boldsymbol{\alpha}\boldsymbol{\beta}}\, \widehat{g}_1^{\alpha_1} \cdots \widehat{g}_m^{\alpha_m} (1 - \widehat{g}_1)^{\beta_1} \cdots (1 - \widehat{g}_m)^{\beta_m},$$

for finitely many nonnegative coefficients $(c_{\boldsymbol{\alpha}\boldsymbol{\beta}}) \subset \mathbb{R}_+$ (and with $\widehat{g}_k^0 = 1$), or using the vector notation

$$\widehat{\mathbf{g}} = \begin{pmatrix} \widehat{g}_1 \\ \vdots \\ \widehat{g}_m \end{pmatrix}, \quad \mathbf{1} - \widehat{\mathbf{g}} = \begin{pmatrix} 1 - \widehat{g}_1 \\ \vdots \\ 1 - \widehat{g}_m \end{pmatrix},$$

$f \in C_G$ if f has the compact form

$$f = \sum_{\boldsymbol{\alpha}, \boldsymbol{\beta} \in \mathbb{N}^m} c_{\boldsymbol{\alpha}\boldsymbol{\beta}}\, \widehat{\mathbf{g}}^{\boldsymbol{\alpha}}\, (\mathbf{1} - \widehat{\mathbf{g}})^{\boldsymbol{\beta}}. \tag{2.18}$$

Equivalently, $f \in C_G$ if f is a polynomial of $\mathbb{R}[\widehat{g}_1, \ldots, \widehat{g}_m, 1 - \widehat{g}_1, \ldots, 1 - \widehat{g}_m]$, with nonnegative coefficients.

Theorem 2.23. *Let $(g_i)_{i=1}^m \subset \mathbb{R}[\mathbf{x}]$, $\mathbb{K} \subset \mathbb{R}^n$ be as in (2.10). Under Assumption 2.2, if $f \in \mathbb{R}[\mathbf{x}]$ is strictly positive on $\mathbb{K}$, then $f \in C_G$. Equivalently, (2.18) holds for finitely many nonnegative coefficients $(c_{\boldsymbol{\alpha}\boldsymbol{\beta}})_{\boldsymbol{\alpha}, \boldsymbol{\beta} \in \mathbb{N}^m} \in \mathbb{R}_+$.*

In contrast to Theorems 2.13 and 2.14, the representation (2.18) involves some nonnegative scalar coefficients ($c_{\alpha\beta}$) rather than s.o.s. polynomials in $\Sigma[\mathbf{x}]$. Determining if $f > 0$ on $\mathbb{K}$ using Theorem 2.23 leads to a linear optimization feasibility problem for which extremely efficient software packages are available. On the other hand, it involves products of arbitrary powers of the g_i's and $(1-g_j)$'s, a highly undesirable feature. In particular, the presence of large binomial coefficients is source of ill-conditioning and numerical instability.

The case of polytopes

If $\mathbb{K}$ is a convex polytope then Theorem 2.23 simplifies and we obtain a generalization to a polytope in $\mathbb{R}^n$ of Theorem 2.8 for $[-1,1] \subset \mathbb{R}$.

Theorem 2.24. *Let $g_j \in \mathbb{R}[\mathbf{x}]$ be affine for every $j = 1, \ldots, m$ and assume that $\mathbb{K}$ in (2.10) is compact with a nonempty interior. If $f \in \mathbb{R}[\mathbf{x}]$ is strictly positive on $\mathbb{K}$ then*

$$f = \sum_{\alpha \in \mathbb{N}^m} c_\alpha \, g_1^{\alpha_1} \cdots g_m^{\alpha_m}, \tag{2.19}$$

for finitely many nonnegative scalars (c_α).

Notice that Theorem 2.24 is of the same flavor as Theorem 2.23, except it does not require to introduce the polynomials $1 - g_j/\overline{g}_j$, $j = 1, \ldots, m$.

Remarks

There are three features that distinguish the case $n > 1$ from the case $n = 1$ treated in the previous section.

(a) Theorems 2.13, 2.14, and 2.23, all deal with **compact** sets $\mathbb{K}$, whereas Theorem 2.6 can handle the (non compact) interval $[0, \infty)$.

(b) In Theorems 2.13, 2.14, and 2.23, f is restricted to be **strictly positive**, instead of **nonnegative** in Theorem 2.6.

(c) In Theorems 2.13, 2.14, and 2.23, nothing is said on the **degree** of the polynomials involved in (2.11), (2.13), or in (2.18), whereas in Theorem 2.6, the degree is bounded and known in advance. In fact, bounds exist for the representations (2.11) and (2.13); see e.g. Theorem 2.16. However they are not practical from a computational viewpoint. This is the reason why

Theorems 2.13, 2.14, and 2.23 do not lead to a polynomial time algorithm to check whether a polynomial f is positive on $\mathbb{K}$.

2.6 Polynomials Nonnegative on Real Varieties

In this section, we introduce representations of polynomials on a real variety. The first result considers an arbitrary real variety, whereas the second considers a finite variety associated with a zero-dimensional ideal I of $\mathbb{R}[\mathbf{x}]$ which is radical. A brief background on basic definitions and results of algebraic geometry can be found in Appendix A.

Let $V \subset \mathbb{R}^n$ be the real variety defined by:

$$V := \{\mathbf{x} \in \mathbb{R}^n : \quad g_j(\mathbf{x}) = 0, \quad j = 1, \ldots, m\}, \tag{2.20}$$

for some family of real polynomials $(g_j) \subset \mathbb{R}[\mathbf{x}]$, and given $f \in \mathbb{R}[\mathbf{x}]$ and $\epsilon > 0$, let $f_{\epsilon r} \in \mathbb{R}[\mathbf{x}]$ be the polynomial:

$$f_{\epsilon r} := f + \epsilon \sum_{k=0}^{r} \sum_{i=1}^{n} \frac{x_i^{2k}}{k!}, \qquad \epsilon \geq 0, \quad r \in \mathbb{N}. \tag{2.21}$$

Theorem 2.25. *Let $V \subset \mathbb{R}^n$ be as in (2.20), and let $f \in \mathbb{R}[\mathbf{x}]$ be nonnegative on V. Then, for every $\epsilon > 0$, there exists $r_\epsilon \in \mathbb{N}$ and nonnegative scalars $(\lambda_j)_{j=1}^m$, such that, for all $r \geq r_\epsilon$,*

$$f_{\epsilon r} + \sum_{j=1}^{m} \lambda_j \, g_j^2 \quad \textit{is s.o.s.} \tag{2.22}$$

In addition, $\|f - f_{\epsilon r}\|_1 \to 0$, as $\epsilon \downarrow 0$ (and $r \geq r_\epsilon$).

Theorem 2.25 is a denseness result, the analogue for varieties of Theorem 2.4(b) for $V = \mathbb{R}^n$, and provides a certificate of nonnegativity of f on V. Notice that in contrast with Theorems 2.13 and 2.14 (letting an equality constraint being two reverse inequality constraints) Theorem 2.25 makes *no* assumption on the variety V; in addition one has *scalar* multipliers $(\lambda_j\}$ in (2.22) instead of s.o.s. multipliers in (2.11) or (2.13). On the other hand, the former theorems state that if V is compact and f is nonnegative on V, then $f + \epsilon = f_{\epsilon/n,0}$ has the sum of squares representation (2.11) (or (2.13)), and $\|f - f_{\epsilon/n0}\|_\infty \to 0$ as $\epsilon \downarrow 0$, instead of the weaker $\|f - f_{\epsilon r}\|_1 \downarrow 0$ in Theorem 2.25.

Polynomials nonnegative on a finite variety

We recall that a zero-dimensional ideal $I \subset \mathbb{R}[\mathbf{x}]$ is an ideal[2] such that the associated variety

$$V_{\mathbb{C}}(I) := \{\, \mathbf{x} \in \mathbb{C}^n \,:\, g(\mathbf{x}) = 0 \quad \forall g \in I \,\}$$

is *finite*. In such a case, the quotient ring $\mathbb{R}[\mathbf{x}]/I$ is a finite-dimensional $\mathbb{R}$-vector space whose dimension is larger than $|V_{\mathbb{C}}(I)|$ and equal to $|V_{\mathbb{C}}(I)|$ if and only if I is radical.

This is an important special case when one deals with discrete sets as in discrete optimization; for instance when the set $\mathbb{K} \subset \mathbb{R}^n$ consists of the grid points (x_{ij}) that are solutions of the polynomial equations

$$\mathbb{K} = \left\{ \mathbf{x} \in \mathbb{R}^n \,:\, \prod_{j=1}^{2r_i} (x_i - x_{ij}) = 0; \quad i = 1, \ldots, n \right\}. \tag{2.23}$$

Binary optimization deals with the case when $r_i = 1$ for all $i = 1, \ldots, n$ and $x_{ij} \in \{0, 1\}$.

Theorem 2.26. *Let $I = \langle g_1, \ldots, g_m \rangle$ be a zero-dimensional ideal of $\mathbb{R}[\mathbf{x}]$ with associated (finite) variety $V_{\mathbb{C}}(I) \subset \mathbb{C}^n$. Assume that I is radical ($I = \sqrt{I}$), and let $\mathbf{S} := V_{\mathbb{C}}(I) \cap \mathbb{R}^n$. Let $(h_i)_{i=1}^r \subset \mathbb{R}[\mathbf{x}]$ and*

$$\mathbb{K} := \{\mathbf{x} \in \mathbf{S} : \quad h_i(\mathbf{x}) \geq 0, \quad i = 1, \ldots, r\}. \tag{2.24}$$

Then $f \in \mathbb{R}[\mathbf{x}]$ is nonnegative on $\mathbb{K}$ if and only if

$$f = f_0 + \sum_{i=1}^{r} f_i h_i + \sum_{j=1}^{m} v_i g_i \tag{2.25}$$

where $(f_i)_{i=0}^r \subset \Sigma[\mathbf{x}]$ and $(v_i)_{i=1}^m \subset \mathbb{R}[\mathbf{x}]$.

In the case where $\mathbb{K}$ is the set (2.23), the degree of the polynomials (f_i) is bounded by $(\sum_{i=1}^n r_i) - n$.

For finite varieties, observe that Theorem 2.26 provides a representation result stronger than that of Theorem 2.14. In particular, f is not required to be strictly positive and sometimes, a degree bound is also available.

---

[2]For definitions and properties of ideal, radical ideal, etc. the reader is referred to Section A.2.

2.7 Representations with Sparsity Properties

In this section, we introduce sparse representations for polynomials f nonnegative on a basic semi-algebraic set $\mathbb{K}$, when there is *weak coupling* between some subsets of variables in the polynomials g_j that define the set $\mathbb{K}$, and f. By weak coupling (to be detailed later) we mean that (a) each polynomial in the definition of $\mathbb{K}$ contains a few variables only, and (b), the polynomial f is a sum of polynomials, each containing also a few variables only. This sparse representation is computationally important as it translates to smaller semidefinite programs for computing the s.o.s. polynomials that define the representation. In fact, given the current state of semidefinite optimization, it is absolutely critical to exploit sparsity in order to solve problems involving a large number of variables.

We first consider the simple case of three sets of variables. Denote by $\mathbb{R}[\mathbf{x},\mathbf{y},\mathbf{z}]$ the ring of real polynomial in the variables $(x_1,\ldots,x_n)$, $(y_1,\ldots,y_m)$ and $(z_1,\ldots,z_p)$. Let $\mathbb{K}_{\mathbf{xy}} \subset \mathbb{R}^{n+m}$, $\mathbb{K}_{\mathbf{yz}} \subset \mathbb{R}^{m+p}$, and $\mathbb{K} \subset \mathbb{R}^{n+m+p}$ be the compact basic semi-algebraic sets defined by:

$$\mathbb{K}_{\mathbf{xy}} = \{\, (\mathbf{x},\mathbf{y}) \in \mathbb{R}^{n+m} : \quad g_j(\mathbf{x},\mathbf{y}) \geq 0, \quad j \in \mathbf{I}_{\mathbf{xy}}\} \tag{2.26}$$

$$\mathbb{K}_{\mathbf{yz}} = \{\, (\mathbf{y},\mathbf{z}) \in \mathbb{R}^{m+p} : \quad h_k(\mathbf{y},\mathbf{z}) \geq 0, \quad k \in \mathbf{I}_{\mathbf{yz}}\} \tag{2.27}$$

$$\mathbb{K} = \{\, (\mathbf{x},\mathbf{y},\mathbf{z}) \in \mathbb{R}^{n+m+p} : (\mathbf{x},\mathbf{y}) \in \mathbb{K}_{\mathbf{xy}}; \quad (\mathbf{y},\mathbf{z}) \in \mathbb{K}_{\mathbf{yz}}\,\} \tag{2.28}$$

for some polynomials $(g_j) \subset \mathbb{R}[\mathbf{x},\mathbf{y}]$, $(h_k) \subset \mathbb{R}[\mathbf{y},\mathbf{z}]$, and some finite index sets $\mathbf{I}_{\mathbf{xy}}, \mathbf{I}_{\mathbf{yz}} \subset \mathbb{N}$. Denote by $\Sigma[\mathbf{x},\mathbf{y}]$ (resp. $\Sigma[\mathbf{y},\mathbf{z}]$) the set of sums of squares in $\mathbb{R}[\mathbf{x},\mathbf{y}]$ (resp. $\mathbb{R}[\mathbf{y},\mathbf{z}]$).

Let $P(g) \subset \mathbb{R}[\mathbf{x},\mathbf{y}]$ and $P(h) \subset \mathbb{R}[\mathbf{y},\mathbf{z}]$ be the preorderings generated by $(g_j)_{j\in\mathbf{I}_{\mathbf{xy}}}$ and $(h_k)_{k\in\mathbf{I}_{\mathbf{yz}}}$, respectively, that is

$$P(g) = \left\{ \sum_{J\subseteq\mathbf{I}_{\mathbf{xy}}} \sigma_J \left(\prod_{j\in J} g_j \right) \;:\; \sigma_J \in \Sigma[\mathbf{x},\mathbf{y}] \right\}$$

$$P(h) = \left\{ \sum_{J\subseteq\mathbf{I}_{\mathbf{yz}}} \sigma_J \left(\prod_{k\in J} h_k \right) \;:\; \sigma_J \in \Sigma[\mathbf{y},\mathbf{z}] \right\}.$$

Similarly, let $Q(g) \subset \mathbb{R}[\mathbf{x},\mathbf{y}]$ and $Q(h) \subset \mathbb{R}[\mathbf{y},\mathbf{z}]$ be the quadratic modules

$$Q(g) = \left\{ \sigma_0 + \sum_{j\in\mathbf{I}_{\mathbf{xy}}} \sigma_j\, g_j \;:\quad \sigma_0, \sigma_j \in \Sigma[\mathbf{x},\mathbf{y}] \right\}$$

$$Q(h) = \left\{ \sigma_0 + \sum_{k \in \mathbf{I}_{\mathbf{yz}}} \sigma_k \, h_k \ : \quad \sigma_0, \sigma_k \in \Sigma[\mathbf{y}, \mathbf{z}] \right\}.$$

Theorem 2.27. *Let $\mathbb{K}_{\mathbf{xy}} \subset \mathbb{R}^n$, $\mathbb{K}_{\mathbf{yz}} \subset \mathbb{R}^m$, and $\mathbb{K} \subset \mathbb{R}^{n+m+p}$ be the compact basic semi-algebraic sets defined in (2.26)-(2.10), and assume that $\mathbb{K}$ has nonempty interior. Let $f \in \mathbb{R}[\mathbf{x}, \mathbf{y}] + \mathbb{R}[\mathbf{y}, \mathbf{z}]$.*
(a) *If f is positive on $\mathbb{K}$, then $f \in P(g) + P(h)$.*
(b) *If $(\mathbf{x}, \mathbf{y}) \mapsto N - \|(\mathbf{x}, \mathbf{y})\|^2 \in Q(g)$ and $(\mathbf{y}, \mathbf{z}) \mapsto N - \|(\mathbf{y}, \mathbf{z})\|^2 \in Q(h)$ for some scalar N, and if f is positive on $\mathbb{K}$, then $f \in Q(g) + Q(h)$.*

Let us contrast Theorem 2.14 with Theorem 2.27(b). Theorem 2.14 gives the representation:

$$f = p_0 + \sum_j p_j g_j + \sum_k q_k h_k$$

with $p_0, p_j, q_k \in \Sigma(\mathbf{x}, \mathbf{y}, \mathbf{z})$, while Theorem 2.27(b) gives the representation

$$f = p_0 + \sum_j p_j g_j + q_0 + \sum_k q_k h_k,$$

with $p_0, p_j \in \Sigma(\mathbf{x}, \mathbf{y})$, and $q_0, q_k \in \Sigma(\mathbf{y}, \mathbf{z})$. This sparser representation leads to semidefinite optimization problems of significantlty lower dimension, which in turn lead to significant computational advantages. In other words, Theorem 2.27 implies that the absence of coupling between the two sets of variables $\mathbf{x}$ and $\mathbf{z}$ in the original data f, g_j, h_k, is also reflected in the specialized *sparse* representations of Theorem 2.27(a)-(b).

Example 2.3. Let $\mathbb{K} = \{(x, y, z) | \ 1 - x^2 - y^2 \geq 0, 1 - y^2 - z^2 \geq 0\}$ and

$$f(x, y, z) = 1 + x^2y^2 - x^2y^4 + y^2 + z^2 - y^2z^4.$$

Theorem 2.27(b) allows the sparse representation

$$f = 1 + x^4y^2 + y^4z^2 + x^2y^2(1 - x^2 - y^2) + y^2z^2(1 - y^2 - z^2),$$

with $p_0(x, y) = 1 + x^4y^2$, $p_1(x, y) = x^2y^2$, $q_0(y, z) = y^4z^2$ and $q_1(y, z) = y^2z^2$. Note that $p_0, p_1 \in \Sigma(x, y)$ and $q_0, q_1 \in \Sigma(y, z)$.

We next extend Theorem 2.27 to more general sparsity patterns, when f and g_j satisfy certain sparsity conditions themselves. With $\mathbb{R}[\mathbf{x}] = \mathbb{R}[x_1, \ldots, x_n]$, let $I_0 := \{1, \ldots, n\}$ be the union $\cup_{k=1}^p I_k$ of p subsets I_k, $k = 1, \ldots, p$ (with possible overlaps). Let $n_k = |I_k|$. Let $\mathbb{R}[\mathbf{x}(I_k)]$ denote the ring of polynomials in the n_k variables $\mathbf{x}(I_k) = \{x_i : i \in I_k\}$, and so $\mathbb{R}[\mathbf{x}(I_0)] = \mathbb{R}[\mathbf{x}]$.

Assumption 2.3. Let $\mathbb{K} \subset \mathbb{R}^n$ be as in (2.10). A scalar $M > 0$ is known and such that $\|\mathbf{x}\|_\infty < M$ for all $\mathbf{x} \in \mathbb{K}$.

Note that under Assumption 2.3, we have $\sum_{i \in I_k} x_i^2 \leq n_k M^2$, $k = 1, \ldots, p$, and therefore, in the Definition (2.10) of $\mathbb{K}$, we add the p redundant quadratic constraints

$$0 \leq g_{m+k}(\mathbf{x}) \quad (:= n_k M^2 - \sum_{i \in I_k} x_i^2), \quad k = 1, \ldots, p, \tag{2.29}$$

and set $m' = m + p$, so that $\mathbb{K}$ is now defined by:

$$\mathbb{K} := \{\mathbf{x} \in \mathbb{R}^n : \quad g_j(\mathbf{x}) \geq 0, \quad j = 1, \ldots, m'\}. \tag{2.30}$$

Note that $g_{m+k} \in \mathbb{R}[\mathbf{x}(I_k)]$, for all $k = 1, \ldots, p$.

Assumption 2.4. Let $\mathbb{K} \subset \mathbb{R}^n$ be as in (2.30). The index set $J = \{1, \ldots, m'\}$ is partitioned into p disjoint sets J_k, $k = 1, \ldots, p$, and the collections $\{I_k\}$ and $\{J_k\}$ satisfy:
(a) For every $j \in J_k$, $g_j \in \mathbb{R}[\mathbf{x}(I_k)]$, that is, for every $j \in J_k$, the constraint $g_j(\mathbf{x}) \geq 0$ only involves the variables $\mathbf{x}(I_k) = \{x_i : i \in I_k\}$.
(b) The objective function $f \in \mathbb{R}[\mathbf{x}]$ can be written as

$$f = \sum_{k=1}^{p} f_k, \quad \text{with } f_k \in \mathbb{R}[\mathbf{x}(I_k)], \quad k = 1, \ldots, p. \tag{2.31}$$

The main result about sparsity is as follows.

Theorem 2.28. *Let $\mathbb{K} \subset \mathbb{R}^n$ be as in (2.30) (i.e. $\mathbb{K}$ as in (2.10) with the additional redundant quadratic constraints (2.29)). Let Assumptions 2.3 and 2.4 hold and in addition, assume that for every $k = 1, \ldots, p-1$,*

$$\left(I_{k+1} \cap (\cup_{j=1}^{k} I_j)\right) \subseteq I_s \quad \text{for some } s \leq k. \tag{2.32}$$

If $f \in \mathbb{R}[\mathbf{x}]$ is strictly positive on $\mathbb{K}$, then

$$f = \sum_{k=1}^{p} \left(q_k + \sum_{j \in J_k} q_{jk}\, g_j \right), \tag{2.33}$$

for some sums of squares polynomials $(q_k, q_{jk}) \subset \mathbb{R}[\mathbf{x}(I_k)]$, $k = 1, \ldots, p$.

The key property (2.32) that allows the sparse representation (2.33) is called the *running intersection property.* Under this property the absence of coupling of variables in the original data is preserved in the representation (2.33). So Theorem 2.28 provides a representation that is more specific than that of Theorem 2.14. Let us illustrate Theorem 2.28 with an example.

Example 2.4. Let $\mathbf{x} = (x_1, \ldots, x_5)$ and

$$\begin{aligned}
\mathbf{x} &\mapsto g_1(\mathbf{x}) = 1 - x_1^2 - x_2^2 \\
\mathbf{x} &\mapsto g_2(\mathbf{x}) = 1 - x_2^4 x_3^4 - x_3^2 \\
\mathbf{x} &\mapsto g_3(\mathbf{x}) = 1 - x_3^2 - x_4^2 - x_5^2 \\
\mathbf{x} &\mapsto f(\mathbf{x}) = 1 + x_1^2 x_2^2 + x_2^2 x_3^2 + x_3^2 - x_3^2 x_4^2 - x_3^2 x_5^2 \\
\mathbb{K} &= \{\mathbf{x} : \quad g_1(x_1, x_2), g_2(x_2, x_3), g_3(x_3, x_4, x_5) \geq 0\}.
\end{aligned}$$

Then $I_1 = \{1, 2\}$ meaning that the polynomial g_1 only involves variables x_1, x_2, $I_2 = \{2, 3\}$ and $I_3 = \{3, 4, 5\}$. Moreover, $f = f_1 + f_2 + f_3$, with $f_1 = 1 + x_1^2 x_2^2$, $f_2 = x_2^2 x_3^2$ and $f_3 = x_3^2 - x_3^2 x_4^2 - x_3^2 x_5^2$. Let us check property (2.32). For $k = 1$, $I_2 \cap I_1 = \{2\} \subset I_1$. For $k = 2$, $I_3 \cap (I_1 \cup I_2) = \{3\} \subset I_2$ and thus property (2.32) holds. Thus, Theorem 2.28 allows the sparse representation

$$\begin{aligned}
f(\mathbf{x}) = 1 &+ x_1^4 x_2^2 + x_1^2 x_2^4 + x_1^2 x_2^2 \, g_1(\mathbf{x}) \\
&+ x_2^6 x_3^6 + x_2^2 x_3^4 + x_2^2 x_3^2 \, g_2(\mathbf{x}) \\
&+ x_3^4 + x_3^2 \, g_3(\mathbf{x}).
\end{aligned}$$

2.8 Representation of Convex Polynomials

If $C \subseteq \mathbb{R}^n$ is a nonempty convex set, a function $f : C \to \mathbb{R}$ is convex on C if

$$f(\lambda x + (1 - \lambda) y) \leq \lambda f(x) + (1 - \lambda) f(y), \qquad \forall \lambda \in (0, 1), \; x, y \in C.$$

Similarly, f is strictly convex on C if the above inequality is strict for every $x, y \in C$, $x \neq y$, and all $\lambda \in (0, 1)$.

If $C \subseteq \mathbb{R}^n$ is an open convex set and f is twice differentiable on C, then f is convex on C if and only if its Hessian[3] $\nabla^2 f$ is positive semidefinite on C (denoted $\nabla^2 f \succeq 0$ on C). Finally, if $\nabla^2 f$ is positive definite on C (denoted $\nabla^2 f \succ 0$ on C) then f is strictly convex on C.

[3] Recall that the Hessian $\nabla^2 f(\mathbf{x})$ is the $n \times n$ symmetric matrix whose entry (i, j) is $\partial^2 f / \partial x_i \partial x_j$ evaluated at $\mathbf{x}$.

Definition 2.3. A polynomial $f \in \mathbb{R}[\mathbf{x}]_{2d}$ is said to be s.o.s.-convex if its Hessian $\nabla^2 f$ is a s.o.s. matrix polynomial, that is, $\nabla^2 f = \mathbf{L}\mathbf{L}'$ for some real matrix polynomial $\mathbf{L} \in \mathbb{R}[\mathbf{x}]^{n\times s}$ (for some $s \in \mathbb{N}$).

Example 2.5. A polynomial $f \in \mathbb{R}[\mathbf{x}]$ is said to be *separable* if $f = \sum_{i=1}^n f_i$ for some univariate polynomials $f_i \in \mathbb{R}[x_i]$. And so, every convex separable polynomial f is s.o.s.-convex because its Hessian $\nabla^2 f(\mathbf{x})$ is a positive semidefinite diagonal matrix, with all diagonal entries $(f_i^{''}(x_i))_i$, nonnegative for all $\mathbf{x} \in \mathbb{R}^n$. Hence, being univariate and nonnegative, $f_i^{''}$ is s.o.s. for every $i = 1, \ldots, n$, from which one easily concludes that f is s.o.s.-convex.

An important feature of s.o.s.-convexity is that it can be can be checked numerically by solving a semidefinite program. Of course every s.o.s.-convex polynomial is convex. Also, the s.o.s.-convex polynomials have the following interesting property.

Lemma 2.29. *If a symmetric matrix polynomial* $\mathbf{P} \in \mathbb{R}[\mathbf{x}]^{r\times r}$ *is s.o.s. then the double integral*

$$(\mathbf{x}, \mathbf{u}) \mapsto \quad \mathbf{F}(\mathbf{x}, \mathbf{u}) := \int_0^1 \int_0^t \mathbf{P}(\mathbf{u} + s(\mathbf{x} - \mathbf{u}))\, ds\, dt \tag{2.34}$$

is also a symmetric s.o.s. matrix polynomial $\mathbf{F} \in \mathbb{R}[\mathbf{x}, \mathbf{u}]^{r\times r}$.

Proof. Writing $\mathbf{P} = (p_{ij})_{1\le i,j\le n}$ with $p_{ij} \in \mathbb{R}[\mathbf{x}]$ for every $1 \le i, j \le n$, let $d := \max_{i,j} \deg p_{ij}$. With $\mathbf{x}, \mathbf{u}$ fixed, introduce the univariate matrix polynomial $s \mapsto \mathbf{Q}(s) := \mathbf{P}(\mathbf{u} + s(\mathbf{x} - \mathbf{u}))$ so that $\mathbf{Q} = (q_{ij})$ where $q_{ij} \in \mathbb{R}[s]$ has degree at most d for every $1 \le i, j \le n$. Observe that

$$\int_0^1 \int_0^t \mathbf{P}(\mathbf{u} + s(\mathbf{x} - \mathbf{u}))\, dsdt \;=\; \int_0^1 \int_0^t \mathbf{Q}(s)\, dsdt \;=\; \int_\Delta \mathbf{Q}(s)\, d\mu,$$

where μ is uniformly distibuted on the set $\Delta := \{(s,t) \,:\, 0 \le t \le 1;\, 0 \le s \le t\}$. By Tchakaloff's theorem (see Tchakaloff (1957) and Theorem B.12), there exists a measure φ finitely supported on, say m, points $\{(s_k, t_k)\}_{k=1}^m \subset \Delta$ whose moments up to degree d match exactly those of μ. That is, there exist positive weights h_k, $k = 1, \ldots, m$, such that

$$\int_\Delta f\, d\mu \;=\; \sum_{k=1}^m h_k\, f(s_k, t_k), \qquad \forall\, f \in \mathbb{R}[s,t],\ \deg f \le d.$$

So let $\mathbf{x}, \mathbf{u} \in \mathbb{R}^n$ be fixed. If $\mathbf{P} = \mathbf{L}\mathbf{L}'$ for some $\mathbf{L} \in \mathbb{R}[\mathbf{x}]^{r \times q}$, one obtains:

$$\begin{aligned} \int_0^1 \int_0^t \mathbf{P}(\mathbf{u} + s(\mathbf{x} - \mathbf{u}))\, dsdt &= \sum_{k=1}^m h_k\, \mathbf{Q}(s_k) \\ &= \sum_{k=1}^m h_k \mathbf{L}(\mathbf{u} + s_k(\mathbf{x} - \mathbf{u}))\, \mathbf{L}'(\mathbf{u} + s_k(\mathbf{x} - \mathbf{u})) \\ &= \mathbf{A}\mathbf{A}' \end{aligned}$$

for some $\mathbf{A} \in \mathbb{R}[\mathbf{x}, \mathbf{u}]^{r \times mq}$. □

And as a consequence:

Lemma 2.30. *For a polynomial $f \in \mathbb{R}[\mathbf{x}]$ and every $\mathbf{x}, \mathbf{u} \in \mathbb{R}^n$:*

$$\begin{aligned} f(\mathbf{x}) &= f(\mathbf{u}) + \nabla f(\mathbf{u})'(\mathbf{x} - \mathbf{u}) \\ &\quad + (\mathbf{x} - \mathbf{u})' \underbrace{\int_0^1 \int_0^t \nabla^2 f(\mathbf{u} + s(\mathbf{x} - \mathbf{u}))dsdt}_{\mathbf{F}(\mathbf{x},\mathbf{u})} (\mathbf{x} - \mathbf{u}). \end{aligned}$$

And so if f is s.o.s.-convex and $f(\mathbf{u}) = 0, \nabla f(\mathbf{u}) = 0$, then f is a s.o.s. polynomial.

Let $\mathbb{K}$ be the basic semi-algebraic set in (2.10), $Q(g)$ be as in (2.12) and define $Q_c(g) \subset Q(g)$ to be the set:

$$Q_c(g) := \left\{ \sigma_0 + \sum_{j=1}^m \lambda_j\, g_j \;:\quad \lambda \in \mathbb{R}^m_+\,;\ \sigma_0 \in \Sigma[\mathbf{x}],\ \sigma_0 \text{ convex} \right\}. \quad (2.35)$$

The set $Q_c(g)$ is a specialization of the quadratic module $Q(g)$ to the convex case, in that the weights associated with the g_j's are nonnegative scalars, i.e., s.o.s. polynomials of degree 0, and the s.o.s. polynomial σ_0 is convex. In particular, every $f \in Q_c(g)$ is nonnegative on $\mathbb{K}$. Let $\mathbf{F}_{\mathbb{K}} \subset \mathbb{R}[\mathbf{x}]$ be the convex cone of convex polynomials nonnegative on $\mathbb{K}$. Recall that $\|f\|_1$ denote the l_1-norm of the vector of coefficients, i.e., $\|f\|_1 = \sum_\alpha |f_\alpha|$.

Theorem 2.31. *Let $\mathbb{K}$ be as in (2.10), Slater's condition*[4] *hold and g_j be concave for every $j = 1, \ldots, m$.*

Then with $Q_c(g)$ as in (2.35), the set $Q_c(g) \cap \mathbf{F}_{\mathbb{K}}$ is dense in $\mathbf{F}_{\mathbb{K}}$ for the l_1-norm $\|\cdot\|_1$. In particular, if $\mathbb{K} = \mathbb{R}^n$ (so that $\mathbf{F}_{\mathbb{R}^n} =: \mathbf{F}$ is now the set of nonnegative convex polynomials), then $\Sigma[\mathbf{x}] \cap \mathbf{F}$ is dense in $\mathbf{F}$.

Theorem 2.31 states that if f is convex and nonnegative on $\mathbb{K}$ (including the case $\mathbb{K} \equiv \mathbb{R}^n$) then one may approximate f by a sequence $(f_{\epsilon r}) \subset Q_c(g) \cap \mathbf{F}_{\mathbb{K}}$ with $\|f - f_{\epsilon r}\|_1 \to 0$ as $\epsilon \to 0$ (and $r \to \infty$). For instance, with $r_0 := \lfloor (\deg f)/2 \rfloor + 1$, the polynomial $f_{\epsilon r}$ may be defined as $\mathbf{x} \mapsto f_{\epsilon r}(\mathbf{x}) := f + \epsilon(\theta_{r_0}(\mathbf{x}) + \theta_r(\mathbf{x}))$ with θ_r as in (2.6). Observe that Theorem 2.31 provides f with a *certificate* of nonnegativity on $\mathbb{K}$. Indeed, let $\mathbf{x} \in \mathbb{K}$ be fixed arbitrary. Then as $f_{\epsilon r} \in Q_c(g)$ one has $f_{\epsilon r}(\mathbf{x}) \geq 0$. Letting $\epsilon \downarrow 0$ yields $0 \leq \lim_{\epsilon \to 0} f_{\epsilon r}(\mathbf{x}) = f(\mathbf{x})$. And as $\mathbf{x} \in \mathbb{K}$ was arbitray, $f \geq 0$ on $\mathbb{K}$.

For the class of s.o.s.-convex polynomials, we have the more precise result:

Theorem 2.32. *Let $\mathbb{K}$ be as in (2.10) and Slater's condition hold. Let $f \in \mathbb{R}[\mathbf{x}]$ be such that $f^* := \inf_{\mathbf{x}}\{f(\mathbf{x}) : \mathbf{x} \in \mathbb{K}\} = f(\mathbf{x}^*)$ for some $\mathbf{x}^* \in \mathbb{K}$. If f and $(-g_j)_{j=1}^m$ are s.o.s.-convex then $f - f^* \in Q_c(g)$.*

Proof. As Slater's condition holds, there exists a vector of Lagrange-KKT multipliers $\boldsymbol{\lambda} \in \mathbb{R}^m_+$ such that the Lagrangian polynomial

$$L_f := f - f^* - \sum_{j=1}^{m} \lambda_j \, g_j \tag{2.36}$$

is nonnegative and satisfies $L_f(\mathbf{x}^*) = 0$ as well as $\nabla L_f(\mathbf{x}^*) = 0$; see Section C.1. Moreover, as f and $(-g_j)$ are s.o.s.-convex then so is L_f because $\nabla^2 L_f = \nabla^2 f - \sum_j \lambda_j \nabla^2 g_j$ with $\boldsymbol{\lambda} \geq 0$. Therefore, by Lemma 2.30, $L_f \in \Sigma[\mathbf{x}]$, i.e.,

$$f - f^* - \sum_{j=1}^{m} \lambda_j \, g_j \; = \; f_0,$$

for some $f_0 \in \Sigma[\mathbf{x}]$. As L_f is convex then so is f_0 and so $f - f^* \in Q_c(g)$. □

---

[4]Slater's considion holds if there exists $\mathbf{x}_0 \in \mathbb{K}$ such that $g_j(\mathbf{x}_0) > 0$ for every $j = 1, \ldots, m$.

Hence the class of s.o.s.-convex polynomials is very interesting because a nice representation result is available. Notice that the above representation holds with $f \geq 0$ on $\mathbb{K}$ and $\mathbb{K}$ is not required to be compact. Another interesting case is when the Hessian of f is positive definite on $\mathbb{K}$.

Theorem 2.33. *Let $\mathbb{K}$ be as in (2.10) and let Assumption 2.1 and Slater's condition both hold. Let g_j be concave, $j = 1, \ldots, m$, and let $f \in \mathbb{R}[\mathbf{x}]$ be convex and such that $\nabla^2 f \succ 0$ on $\mathbb{K}$. If $f \geq 0$ on $\mathbb{K}$ then $f \in Q(g)$, i.e.,*

$$f = f_0 + \sum_{j=}^{m} f_j \, g_j, \tag{2.37}$$

for some s.o.s. polynomials $(f_j)_{j=0}^m \subset \Sigma[\mathbf{x}]$.

Proof. Under Assumption 2.1 the set $\mathbb{K}$ is compact. Hence, let $f^* := \min_{\mathbf{x} \in \mathbb{K}} f(\mathbf{x})$ and let $\mathbf{x}^* \in \mathbb{K}$ be a minimizer. As $f, -g_j$ are convex, $j = 1, \ldots, m$, and Slater's condition holds, there exists a vector of nonnegative multipliers $\boldsymbol{\lambda} \in \mathbb{R}^m_+$ such that the Lagrangian polynomial L_f in (2.36) is nonnegative on $\mathbb{R}^n$ with $L_f(\mathbf{x}^*) = 0$ and $\nabla L_f(\mathbf{x}^*) = 0$. Next, as $\nabla^2 f \succ 0$ and $-\nabla^2 g_j \succeq 0$ on $\mathbb{K}$, $\nabla^2 L_f \succ 0$ on $\mathbb{K}$; this is because $\mathbb{K}$ is compact, $\nabla^2 L_f \succeq \nabla^2 f$, and the smallest eigenvalue of $\nabla^2 L_f$ is a continuous function of $\mathbf{x}$. Hence there is some $\delta > 0$ such that $\nabla^2 L_f(\mathbf{x}) \succeq \delta \mathbf{I}_n$ for all $\mathbf{x} \in \mathbb{K}$ (where $\mathbf{I}_n$ denotes the $n \times n$ identity matrix). Next, by Lemma 2.30,

$$L_f(\mathbf{x}) = \langle (\mathbf{x} - \mathbf{x}^*), \mathbf{F}(\mathbf{x}, \mathbf{x}^*)(\mathbf{x} - \mathbf{x}^*) \rangle$$

where $\mathbf{F}(\mathbf{x}, \mathbf{x}^*)$ is the matrix polynomial defined by:

$$\mathbf{x} \mapsto \quad \mathbf{F}(\mathbf{x}, \mathbf{x}^*) := \left(\int_0^1 \int_0^t \nabla^2 L_f(\mathbf{x}^* + s(\mathbf{x} - \mathbf{x}^*)) \, ds \, dt \right).$$

As $\mathbb{K}$ is convex, $\mathbf{x}^* + s(\mathbf{x} - \mathbf{x}^*) \in \mathbb{K}$ for all $s \in [0, 1]$, and so for every $\xi \in \mathbb{R}^n$,

$$\xi' \mathbf{F}(\mathbf{x}, \mathbf{x}^*) \xi \geq \delta \int_0^1 \int_0^t \xi' \xi ds dt = \frac{\delta}{2} \xi' \xi.$$

Therefore, $\mathbf{F}(\mathbf{x},\mathbf{x}^*) \succeq \frac{\delta}{2}\,\mathbf{I}_n$ for every $\mathbf{x} \in \mathbb{K}$, and by Theorem 2.22,

$$\mathbf{F}(\mathbf{x},\mathbf{x}^*) \;=\; \mathbf{F}_0(\mathbf{x}) + \sum_{j=1}^{n} \mathbf{F}_j(\mathbf{x})\,\mathbf{g}_j(\mathbf{x}),$$

for some real symmetric s.o.s. matrix polynomials $(\mathbf{F}_j) \subset \mathbb{R}[\mathbf{x}]^{n\times n}$. Therefore,

$$\begin{aligned} L_f(\mathbf{x}) &= \langle \mathbf{x}-\mathbf{x}^*, \mathbf{F}(\mathbf{x},\mathbf{x}^*)(\mathbf{x}-\mathbf{x}^*)\rangle\,, \\ &= \langle \mathbf{x}-\mathbf{x}^*, \mathbf{F}_0(\mathbf{x})(\mathbf{x}-\mathbf{x}^*)\rangle + \sum_{j=1}^{n} g_j(\mathbf{x})\,\langle \mathbf{x}-\mathbf{x}^*, \mathbf{F}_j(\mathbf{x})(\mathbf{x}-\mathbf{x}^*)\rangle \\ &= \sigma_0(\mathbf{x}) + \sum_{j=1}^{n} \sigma_j(\mathbf{x})\,g_j(\mathbf{x}) \end{aligned}$$

for some s.o.s. polynomials $(\sigma_j)_{j=0}^{m} \subset \Sigma[\mathbf{x}]$. Hence, recalling the definition of L_f,

$$\begin{aligned} f(\mathbf{x}) &= f^* + L_f(\mathbf{x}) + \sum_{j=1}^{m} \lambda_j\, g_j(\mathbf{x}) \\ &= (f^* + \sigma_0(\mathbf{x})) + \sum_{j=1}^{m} (\lambda_j + \sigma_j(\mathbf{x}))\,g_j(\mathbf{x}), \qquad \mathbf{x} \in \mathbb{R}^n, \end{aligned}$$

that is, $f \in Q(g)$ because $f^* \geq 0$ and $\boldsymbol{\lambda} \geq 0$. □

So for the class of compact basic semi-algebraic sets $\mathbb{K}$ defined by concave polynomials, one obtains a refinement of Putinar's Positivstellensatz Theorem 2.14, for the class of convex functions whose Hessian is positive definite on $\mathbb{K}$, and that are only nonnegative on $\mathbb{K}$.

2.9 Summary

In this chapter, we have presented representation results for positive polynomials, and polynomials positive on semi-algebraic sets or varieties. Once we assume a degree for the polynomials involved, these representations can be computed via semidefinite or linear optimization methods, and thus their computation can be done efficiently. Especially important is Theorem 2.28 that allows the use of sparse semidefinite relaxations of Chapter 5 for polynomial optimization problems with a large number of variables and a sparsity pattern that satisfies Assumption 2.4. Also a nice and simple representation result holds for the very interesting class of s.o.s.-convex polynomials.

2.10 Exercises

Exercise 2.1. **(Goursat's Lemma)** With $f \in \mathbb{R}[x]$ and $\deg f = m$, the polynomial $\hat{f}$

$$x \mapsto \hat{f}(x) := (1+x)^m f\left(\frac{1-x}{1+x}\right),$$

is called the Goursat transform of f.
(a) What is the Goursat transform of $\hat{f}$?
(b) Prove that $f > 0$ on $[-1,1]$ if and only if $\hat{f} > 0$ on $[0,\infty)$ and degree $\hat{f} = m$.
(c) $f \geq 0$ on $[-1,1]$ if and only if $\hat{f} \geq 0$ on $[0,\infty)$ and degree $\hat{f} \leq m$.

Exercise 2.2. Consider a polynomial $h \in \mathbb{R}[x]$ with $[-1,1] = \{x \in \mathbb{R} : h(x) \geq 0\}$, that is, h is not necessarily the polynomial $x \mapsto g(x) := (1+x)(1-x)$. Prove that if $p > 0$ on $[-1,1]$, then $p = s_0 + hs_1$ for some $s_0, s_1 \in \Sigma[x]$.

Exercise 2.3. Given a polynomial $p \in \mathbb{R}[x]$, use Theorem 2.8 to show that the problem of checking whether $p > 0$ on $[-1,1]$ reduces to a linear optimization feasibility problem.

Exercise 2.4. Show that the following polynomials are nonnegative but not s.o.s.
(a) $p(x_1,x_2) = x_1^2x_2^2(x_1^2+x_2^2-1)+1$.
(b) $p(x_1,x_2,x_3) = x_1^2x_2^2(x_1^2+x_2^2-3x_3^2)+6x_3^6$.

Exercise 2.5. Show that the following classes of polynomials have the property that if they are nonnegative, then they are s.o.s.
(a) Quadratic polynomials
(b) Fourth degree homogeneous polynomials on 3 variables.

Exercise 2.6. Let $\mathbf{A} \in \mathbb{R}^{m\times n}, \mathbf{b} \in \mathbb{R}^m$. Show that a Farkas certificate of $\emptyset = \{\mathbf{x} : \mathbf{A}\mathbf{x} = \mathbf{b};\ \mathbf{x} \geq 0\}$ is a particularly simple Stengle certificate (2.9).

Exercise 2.7. Prove Theorem 2.12 by using Theorem 2.11

Exercise 2.8. Let $f \in \mathbb{R}[\mathbf{x}]$ be separable, i.e., $\mathbf{x} \mapsto f(\mathbf{x}) = \sum_{i=1}^{n} f_i(x_i)$ for some univariate polynomials $f_i \in \mathbb{R}[x_i]$, $i = 1, \ldots, n$. Show that if f_i is convex for every $i = 1, \ldots, n$, then f is s.o.s.-convex.

2.11 Notes and Sources

2.2. For a nice exposition on degree bounds for Hilbert's 17th problem see Schmid (1998). Theorem 2.3 is from Blekherman (2006) whereas Theorem 2.4(a) is from Lasserre and Netzer (2007) and provides an explicit construction of an approximating sequence of sums of squares for the denseness result of Berg (1987). Theorem 2.4(b) is from Lasserre (2006e) and Theorem 2.2 is from Lasserre (2007b).

2.3. Most of the material for the one-dimensional case is taken from Powers and Reznick (2000). Theorem 2.6(a) is due to Fekete (1935) whereas Theorem 2.6(b) is attributed F. Luckàcs. Theorem 2.7 is from Pólya and Szegö (1976). Theorem 2.8 is due to Hausdorff (1915) and Bernstein (1921).

2.4. The important Theorem 2.9 is due to Pólya (1974). The Positivstellensatz (Theorem 2.12) is credited to Stengle (1974) but was proved earlier by Krivine (1964a). For a complete and recent exposition of representation of positive polynomials, the reader is referred to Prestel and Delzell (2001); see also Kuhlmann *et al.* (2005) and the more recent Helton and Putinar (2007), Scheiderer (2008) and Marshall (2008). For a nice discussion on historical aspects on Hilbert's 17th problem the reader is referred to Reznick (2000).

2.5. Theorem 2.13 is due to Schmüdgen (1991). The machinery uses the spectral theory of self-adjoint operators in Hilbert spaces. The important Theorem 2.14 is due to Putinar (1993) and Jacobi and Prestel (2001), whereas Theorem 2.15 is due to Schmüdgen (1991). Concerning degree bounds for s.o.s. terms that appear in those representation results, Theorem 2.16 is from Nie and Schweighofer (2007); see also Marshall (2009) and Schweighofer (2005a). For the non compact case, negative results are provided in Scheiderer (2008). However, some nice representation results have been obtained in some specific cases; see e.g. Marshall (2010). Theorem 2.24 for the case of polytopes is due to Cassier (1984) and Handelman (1988) and Theorem 2.23 for compact semi-algebraic sets follows from a result due to Krivine (1964a,b), and re-stated later in Becker and Schwartz (1983), and Vasilescu (2003). The proof of Theorem 2.14 due to Schweighofer (2005b) is taken from Laurent (2008) and Proposition 2.21 is from (Marshall, 2008, 5.4.4). Theorem 2.22, the matrix-polynomial version of Putinar's Positivstellensatz, was first proved in Scherer and Hol (2004) and Kojima and Maramatsu (2007), independently (with no degree bound). The version with degree bound is from Helton and Nie (2010).

2.6. Theorem 2.25 is from Lasserre (2005) whereas Theorem 2.26 is from Parrilo (2002) who extended previous results in Lasserre (2002c) for the grid case.

2.7. Theorem 2.27 is from Lasserre (2006c). Theorem 2.28 was first proved in Lasserre (2006a) under the assumption that the feasible set $\mathbb{K}$ has a nonempty interior. This assumption was later removed in Kojima and Maramatsu (2009). For extensions of Theorem 2.28 to some non compact cases, see the recent Kuhlmann and Putinar (2007, 2009).

2.8. Lemma 2.29 and 2.30 are from Helton and Nie (2010) who introduced the very interesting class of s.o.s.-convex polynomials. Theorems 2.31 and 2.32 are from Lasserre (2008a). Some properties of convex polynomials with applications to polynomial programming have been investigated in Andronov *et al.* (1982) and Belousov and Klatte (2002).

2.9. Exercise 2.1 is from Goursat (1894). Exercise 2.2 is from Powers and Reznick (2000), who also provide a priori bounds on the degrees of s_0, s_1 based on the degree of p, h and the smallest absolute value of the roots of p.

Chapter 3

Moments

Most results of this chapter are the *dual* analogues of those described in Chapter 2. Indeed the problem of representing polynomials that are positive on a set $\mathbb{K}$ has a dual facet which is the problem of characterizing sequences of reals that are moment sequences of some finite Borel measure supported on $\mathbb{K}$. Moreover, as we shall see, this beautiful duality is nicely captured by standard duality in convex analysis, applied to some appropriate convex cones of $\mathbb{R}[\mathbf{x}]$. We review basic results in the moment problem and also particularize to some specific important cases like in Chapter 2.

Let $\overline{\mathbf{z}} \in \mathbb{C}^n$ denote the complex conjugate vector of $\mathbf{z} \in \mathbb{C}^n$. Let $\{\mathbf{z}^{\boldsymbol{\alpha}}\overline{\mathbf{z}}^{\boldsymbol{\beta}}\}_{\mathbf{z}\in\mathbb{C}^n,\boldsymbol{\alpha},\boldsymbol{\beta}\in\mathbb{N}^n}$ be the basis of monomials for the ring of polynomials $\mathbb{C}[\mathbf{z},\overline{\mathbf{z}}] = \mathbb{C}[z_1,\ldots,z_n,\overline{z}_1,\ldots,\overline{z}_n]$ of the $2n$ variables $\{z_j,\overline{z}_j\}$, with coefficients in $\mathbb{C}$. Recall that for every $\mathbf{z} \in \mathbb{C}^n$, $\boldsymbol{\alpha} \in \mathbb{N}^n$, the notation $\mathbf{z}^{\boldsymbol{\alpha}}$ stands for the monomial $z_1^{\alpha_1}\cdots z_n^{\alpha_n}$ of $\mathbb{C}[\mathbf{z}]$.

The *support* of a Borel measure μ on $\mathbb{R}^n$ is a closed set, the complement in $\mathbb{R}^n$ of the largest open set $O \subset \mathbb{R}^n$ such that $\mu(O) = 0$ (and recall that $\mathbb{C}^n$ may be identified with $\mathbb{R}^{2n}$).

Definition 3.1. **(a) The full moment problem:** Let $(g_i)_{i=1}^m \subset \mathbb{C}[\mathbf{z},\overline{\mathbf{z}}]$ be such such that each $g_i(\mathbf{z},\overline{\mathbf{z}})$ is real, and let $\mathbb{K} \subset \mathbb{C}^n$ be the set defined by:

$$\mathbb{K} := \{\mathbf{z} \in \mathbb{C}^n \;:\; g_i(\mathbf{z},\overline{\mathbf{z}}) \geq 0,\; i = 1,\ldots,m\}.$$

Given an infinite sequence $\mathbf{y} = (y_{\boldsymbol{\alpha}\boldsymbol{\beta}}) \subset \mathbb{C}$ of complex numbers, with $\boldsymbol{\alpha},\boldsymbol{\beta} \in \mathbb{N}^n$, is $\mathbf{y}$ a $\mathbb{K}$-moment sequence, i.e., does there exist a measure μ supported on $\mathbb{K}$, such that

$$y_{\boldsymbol{\alpha}\boldsymbol{\beta}} = \int_{\mathbb{K}} \mathbf{z}^{\boldsymbol{\alpha}}\overline{\mathbf{z}}^{\boldsymbol{\beta}}\, d\mu, \qquad \forall\, \boldsymbol{\alpha},\boldsymbol{\beta} \in \mathbb{N}^n? \tag{3.1}$$

(b) The truncated moment problem: Given (g_i), $\mathbb{K} \subset \mathbb{C}^n$ as above, a finite subset $\Delta \subset \mathbb{N}^n \times \mathbb{N}^n$, and a finite sequence $\mathbf{y} = (y_{\alpha\beta})_{(\alpha,\beta)\in\Delta} \subset \mathbb{C}$ of complex numbers, is $\mathbf{y}$ a $\mathbb{K}$-moment sequence, i.e., does there exist a measure μ supported on $\mathbb{K}$, such that

$$y_{\alpha\beta} = \int_{\mathbb{K}} \mathbf{z}^{\alpha}\overline{\mathbf{z}}^{\beta}\, d\mu, \qquad \forall\, \alpha, \beta \in \Delta? \tag{3.2}$$

Definition 3.2. In both full and truncated cases, a measure μ as in (3.1) or (3.2), is said to be a **representing** measure of the sequence $\mathbf{y}$.
If the representing measure μ is unique then μ is said to be **determinate** (i.e., determined by its moments), and **indeterminate** otherwise.

Example 3.1. For instance, the probability measure μ on the real line $\mathbb{R}$, with density with respect to the Lebesgue measure given by:

$$x \mapsto f(x) := \begin{cases} (x\sqrt{2\pi})^{-1} \exp\left(-\ln(x)^2/2\right) & \text{if } x > 0, \\ 0 \text{ otherwise,} \end{cases}$$

and called the log-normal distribution, is *not* determinate. Indeed, for each a with $-1 \leq a \leq 1$, the probability measure with density

$$x \mapsto \quad f_a(x) := f(x)\left[\,1 + a\sin(2\pi\ln x)\,\right]$$

has exactly the same moments as μ.

The above moment problem encompasses all the classical one-dimensional moment problems of the 20th century.

(a) The Hamburger problem refers to $\mathbb{K} = \mathbb{R}$ and $(y_\alpha)_{\alpha\in\mathbb{N}} \subset \mathbb{R}$.
(b) The Stieltjes problem refers to $\mathbb{K} = \mathbb{R}_+$, and $(y_\alpha)_{\alpha\in\mathbb{N}} \subset \mathbb{R}_+$.
(c) The Hausdorff problem refers to $\mathbb{K} = [a, b]$, and $(y_\alpha)_{\alpha\in\mathbb{N}} \subset \mathbb{R}$.
(d) The Toeplitz problem refers to $\mathbb{K}$ being the unit circle in $\mathbb{C}$, and $(y_\alpha)_{\alpha\in\mathbb{Z}} \subset \mathbb{C}$.

In all chapters of this book, we only consider *real* moment problems, that is, moment problems characterized with:

- A set $\mathbb{K} \subset \mathbb{R}^n$, and
- A sequence $\mathbf{y} = (y_{\boldsymbol{\alpha}})$, $\boldsymbol{\alpha} \in \mathbb{N}^n$.

The multi-dimensional moment problem is significantly more difficult than the one-dimensional case for which the results are fairly complete. This is because, in view of Theorem 3.1 below, obtaining conditions for a sequence to be moments of a representing measure with support on a given subset $\Omega \subseteq \mathbb{R}^n$, is related to characterizing polynomials that are nonnegative on Ω. When the latter characterization is available, it will translate into conditions on the sequence. But as we have seen in Chapter 2, and in contrast to the univariate case, polynomials that are nonnegative on a given set $\Omega \subseteq \mathbb{R}^n$ have no simple characterization, except for compact basic semi-algebraic sets as detailed in Section 2.5. Thus, for instance, the full multi-dimensional $\mathbb{K}$-moment problem is still unsolved for general sets $\mathbb{K} \subset \mathbb{C}^n$, including $\mathbb{K} = \mathbb{C}$.

Before we proceed further, we first state the important Riesz-Haviland theorem. Let $\mathbf{y} = (y_{\boldsymbol{\alpha}}) \subset \mathbb{R}$ be an infinite sequence, and let $L_{\mathbf{y}} : \mathbb{R}[\mathbf{x}] \to \mathbb{R}$, be the linear functional

$$f(\mathbf{x}) = \sum_{\boldsymbol{\alpha} \in \mathbb{N}^n} f_{\boldsymbol{\alpha}} \mathbf{x}^{\boldsymbol{\alpha}} \quad \mapsto L_{\mathbf{y}}(f) = \sum_{\boldsymbol{\alpha} \in \mathbb{N}^n} f_{\boldsymbol{\alpha}} y_{\boldsymbol{\alpha}}. \tag{3.3}$$

Theorem 3.1. (Riesz-Haviland) *Let $\mathbf{y} = (y_{\boldsymbol{\alpha}})_{\boldsymbol{\alpha} \in \mathbb{N}^n} \subset \mathbb{R}$ and let $\mathbb{K} \subset \mathbb{R}^n$ be closed. There exists a finite Borel measure μ on $\mathbb{K}$ such that*

$$\int_{\mathbb{K}} \mathbf{x}^{\boldsymbol{\alpha}}\, d\mu = y_{\boldsymbol{\alpha}}, \qquad \forall \boldsymbol{\alpha} \in \mathbb{N}^n, \tag{3.4}$$

if and only if $L_{\mathbf{y}}(f) \geq 0$ for all polynomials $f \in \mathbb{R}[\mathbf{x}]$ nonnegative on $\mathbb{K}$.

Note that Theorem 3.1 is not very practical as we do not have an explicit characterization of polynomials that are nonnegative on a general closed set $\mathbb{K} \subset \mathbb{R}^n$. However, we have seen in Chapter 2 some nice representations for the subclass of compact basic semi-algebraic sets $\mathbb{K} \subset \mathbb{R}^n$. Theorem 3.1 will serve as our primary proof tool in the next sections.

3.1 The One-dimensional Moment Problem

Given an infinite sequence $\mathbf{y} = (y_j) \subset \mathbb{R}$, introduce the Hankel matrices $\mathbf{H}_n(\mathbf{y})$, $\mathbf{B}_n(\mathbf{y})$ and $\mathbf{C}_n(\mathbf{y}) \in \mathbb{R}^{(n+1)\times(n+1)}$, defined by

$$\mathbf{H}_n(\mathbf{y})(i,j) := y_{i+j-2}; \quad \mathbf{B}_n(\mathbf{y})(i,j) := y_{i+j-1}; \quad \mathbf{C}_n(\mathbf{y})(i,j) := y_{i+j},$$

for all $i, j \in \mathbb{N}$, with $1 \leq i, j \leq n+1$. The Hankel matrix $\mathbf{H}_n(\mathbf{y})$ is the one-dimensional (or univariate) version of what we later call a **moment** matrix in Section 3.2.1.

3.1.1 *The full moment problem*

For any two real-valued square symmetric matrices $\mathbf{A}, \mathbf{B}$, recall that the notation $\mathbf{A} \succeq \mathbf{B}$ (resp. $\mathbf{A} \succ \mathbf{B}$) stands for $\mathbf{A} - \mathbf{B}$ being positive semidefinite (resp. $\mathbf{A} - \mathbf{B}$ being positive definite).

For the full Hamburger, Stieltjes, and Hausdorff moment problems, we have:

Theorem 3.2. *Let $\mathbf{y} = (y_j)_{j \in \mathbb{N}} \subset \mathbb{R}$. Then*
(a) $\mathbf{y}$ *has a representing Borel measure μ on $\mathbb{R}$ if and only if the quadratic form*

$$\mathbf{x} \mapsto s_n(\mathbf{x}) := \sum_{i,j=0}^{n} y_{i+j} x_i x_j \tag{3.5}$$

is positive semidefinite for all $n \in \mathbb{N}$. Equivalently, $\mathbf{H}_n(\mathbf{y}) \succeq \mathbf{0}$ for all $n \in \mathbb{N}$.
(b) $\mathbf{y}$ *has a representing Borel measure μ on $\mathbb{R}_+$ if and only if the quadratic forms (3.5) and*

$$\mathbf{x} \mapsto u_n(\mathbf{x}) := \sum_{i,j=0}^{n} y_{i+j+1} x_i x_j \tag{3.6}$$

are positive semidefinite for all $n \in \mathbb{N}$. Equivalently, $\mathbf{H}_n(\mathbf{y}) \succeq \mathbf{0}$, and $\mathbf{B}_n(\mathbf{y}) \succeq \mathbf{0}$ for all $n \in \mathbb{N}$.
(c) $\mathbf{y}$ *has a representing Borel measure μ on $[a, b]$ if and only if the quadratic forms (3.5) and*

$$\mathbf{x} \mapsto v_n(\mathbf{x}) := \sum_{i,j=0}^{n} (-y_{i+j+2} + (b+a)\, y_{i+j+1} - ab\, y_{i+j})\, x_i x_j \quad n \in \mathbb{N}, \tag{3.7}$$

are positive semidefinite for all $n \in \mathbb{N}$. Equivalently, $\mathbf{H}_n(\mathbf{y}) \succeq \mathbf{0}$ and $-\mathbf{C}_n(\mathbf{y}) + (a+b)\,\mathbf{B}_n(\mathbf{y}) - ab\,\mathbf{H}_n(\mathbf{y}) \succeq \mathbf{0}$, for all $n \in \mathbb{N}$.

Proof. **(a)** If $y_n = \int_{\mathbb{R}} z^n d\mu(z)$, then

$$s_n(\mathbf{x}) = \sum_{i,j=0}^{n} x_i x_j \int_{\mathbb{R}} z^{i+j} d\mu(z) = \int_{\mathbb{R}} \left(\sum_{i=0}^{n} x_i z^i \right)^2 d\mu(z) \geq 0.$$

Conversely, we assume that (3.5) holds, or, equivalently, $\mathbf{H}_n(\mathbf{y}) \succeq \mathbf{0}$, for all $n \in \mathbb{N}$. Therefore, for every $\mathbf{q} \in \mathbb{R}^{n+1}$ we have $\mathbf{q}'\mathbf{H}_n(\mathbf{y})\mathbf{q} \geq 0$. Let $p \in \mathbb{R}[x]$ be nonnegative on $\mathbb{R}$ so that it is s.o.s. and can be written $p = \sum_{j=1}^{r} q_j^2$ for some $r \in \mathbb{N}$ and some polynomials $(q_j)_{j=1}^{r} \subset \mathbb{R}[x]$. But then

$$\sum_{k=0}^{2n} p_k y_k = L_{\mathbf{y}}(p) = L_{\mathbf{y}}(\sum_{j=1}^{r} q_j^2) = \sum_{j=1}^{r} \mathbf{q}_j'\mathbf{H}_n(\mathbf{y})\mathbf{q}_j \geq 0,$$

where $\mathbf{q}_j$ is the vector of coefficients of the polynomial $q_j \in \mathbb{R}[x]$. As $p \geq 0$ was arbitrary, by Theorem 3.1, (3.4) holds with $\mathbb{K} := \mathbb{R}$.

(b) This is similar to part (a).

(c) One direction is immediate. For the converse, we assume that both (3.5) and (3.7) hold. Then, for every $\mathbf{q} \in \mathbb{R}^{n+1}$ we have

$$\mathbf{q}' \left[-\mathbf{C}_n(\mathbf{y}) + (a+b)\,\mathbf{B}_n(\mathbf{y}) - ab\,\mathbf{H}_n(\mathbf{y}) \right] \mathbf{q} \geq 0.$$

Let $p \in \mathbb{R}[x]$ be nonnegative on $[a, b]$ of even degree $2n$ and thus, by Theorem 2.6(b), it can be written $x \mapsto p(x) = u(x) + (b-x)(x-a)q(x)$ with both polynomials u, q being s.o.s. with $\deg q \leq 2n-2$, $\deg u \leq 2n$. If $\deg p = 2n-1$ then again by Theorem 2.6(b), p can be written $x \mapsto p(x) = v(x)(x-a) + w(x)(b-x)$ for some s.o.s. polynomials v, w of degree less than $2n-2$. But then $p(x) = ((x-a)p(x)+(b-x)p(x))/(b-a) = u(x) + (b-x)(x-a)q(x)$ for some s.o.s. polynomials u, q with degree less than $2n$.

Thus, in both even and odd cases, writing $u = \sum_j u_j^2$ and $q = \sum_k q_k^2$ for some polynomials (u_j, q_k) of degree less than n, and with associated vectors of coefficients $(\mathbf{u}_j, \mathbf{q}_k) \subset \mathbb{R}^{n+1}$, one obtains

$$L_{\mathbf{y}}(p) = \sum_j \mathbf{u}_j' \mathbf{H}_n(\mathbf{y}) \mathbf{u}_j + \sum_k \left(\mathbf{q}_k' \left[-\mathbf{C}_n(\mathbf{y}) + (a+b)\,\mathbf{B}_n(\mathbf{y}) - ab\,\mathbf{H}_n(\mathbf{y}) \right] \mathbf{q}_k \right) \geq 0.$$

Therefore $L_{\mathbf{y}}(p) \geq 0$ for every polynomial p nonnegative on $[a, b]$. By Theorem 3.1, Eq. (3.4) holds with $\mathbb{K} := [a, b]$. □

Observe that Theorem 3.2 provides a criterion directly in terms of the sequence (y_n).

3.1.2 *The truncated moment problem*

We now state the analogue of Theorem 3.2 for the truncated moment problem for a sequence $\mathbf{y} = (y_k)_{k=0}^{2n}$ (even case), and $\mathbf{y} = (y_k)_{k=0}^{2n+1}$ (odd case). We first introduce some notation.

For an infinite sequence $\mathbf{y} = (y_j)_{j \in \mathbb{N}}$, write the Hankel moment matrix $\mathbf{H}_n(\mathbf{y})$ in the form

$$\mathbf{H}_n(\mathbf{y}) = [\mathbf{v}_0, \mathbf{v}_1, \ldots, \mathbf{v}_n] \tag{3.8}$$

where $(\mathbf{v}_j) \subset \mathbb{R}^{n+1}$ denote the column vectors of $\mathbf{H}_n(\mathbf{y})$. The **Hankel** rank of $\mathbf{y} = (y_j)_{j=0}^{2n}$, denoted by $\mathrm{rank}(\mathbf{y})$, is the smallest integer $1 \leq i \leq n$, such that $\mathbf{v}_i \in \mathrm{span}\{\mathbf{v}_0, \ldots, \mathbf{v}_{i-1}\}$. If $\mathbf{H}_n(\mathbf{y})$ is nonsingular, then its rank is $\mathrm{rank}(\mathbf{y}) = n + 1$. Given an $m \times n$ matrix $\mathbf{A}$, $\mathrm{Range}(\mathbf{A})$ denotes the image space of $\mathbf{A}$, i.e., $\mathrm{Range}(\mathbf{A}) = \{\mathbf{A}\mathbf{u},\ \mathbf{u} \in \mathbb{R}^n\}$.

Theorem 3.3 (The even case). *Let* $\mathbf{y} = (y_j)_{0 \leq j \leq 2n} \subset \mathbb{R}$.

(a) $\mathbf{y}$ *has a representing Borel measure* μ *on* $\mathbb{R}$ *if and only if* $\mathbf{H}_n(\mathbf{y}) \succeq \mathbf{0}$ *and* $\mathrm{rank}(\mathbf{H}_n(\mathbf{y})) = \mathrm{rank}(\mathbf{y})$.

(b) $\mathbf{y}$ *has a representing Borel measure* μ *on* $\mathbb{R}_+$ *if and only if* $\mathbf{H}_n(\mathbf{y}) \succeq 0$, $\mathbf{B}_{n-1}(\mathbf{y}) \succeq \mathbf{0}$, *and the vector* $(y_{n+1}, \ldots, y_{2n})$ *is in* $\mathrm{Range}(\mathbf{B}_{n-1}(\mathbf{y}))$.

(c) $\mathbf{y}$ *has a representing Borel measure* μ *on* $[a, b]$ *if and only if* $\mathbf{H}_n(\mathbf{y}) \succeq \mathbf{0}$ *and* $(a+b)\,\mathbf{B}_{n-1}(\mathbf{y}) \succeq ab\,\mathbf{H}_{n-1}(\mathbf{y}) + \mathbf{C}_{n-1}(\mathbf{y})$.

Theorem 3.4 (The odd case). *Let* $\mathbf{y} = (y_j)_{0 \leq j \leq 2n+1} \subset \mathbb{R}$.

(a) $\mathbf{y}$ *has a representing Borel measure* μ *on* $\mathbb{R}$ *if and only if* $\mathbf{H}_n(\mathbf{y}) \succeq \mathbf{0}$ *and* $\mathbf{v}_{n+1} \in \operatorname{Range} \mathbf{H}_n(\mathbf{y})$.

(b) $\mathbf{y}$ *has a representing Borel measure* μ *on* $\mathbb{R}_+$ *if and only if* $\mathbf{H}_n(\mathbf{y}) \succeq 0$, $\mathbf{B}_n(\mathbf{y}) \succeq \mathbf{0}$, *and the vector* $(y_{n+1}, \ldots, y_{2n+1})$ *is in* $\operatorname{Range} \mathbf{H}_n(\mathbf{y})$.

(c) $\mathbf{y}$ *has a representing Borel measure* μ *on* $[a, b]$ *if and only if* $b\,\mathbf{H}_n(\mathbf{y}) \succeq \mathbf{B}_n(\mathbf{y})$ *and* $\mathbf{B}_n(\mathbf{y}) \succeq a\,\mathbf{H}_n(\mathbf{y})$.

Example 3.2. In the univariate case $n = 1$, let $\mathbf{y} \in \mathbb{R}^5$ be the truncated sequence $\mathbf{y} = (1, 1, 1, 1, 2)$, hence with associated Hankel moment matrix

$$\mathbf{H}_2(\mathbf{y}) = \begin{bmatrix} 1 & 1 & 1 \\ 1 & 1 & 1 \\ 1 & 1 & 2 \end{bmatrix}.$$

One may easily check that $\mathbf{H}_2(\mathbf{y}) \succeq 0$, but it turns out that $\mathbf{y}$ has no representing Borel measure μ on the real line $\mathbb{R}$. However, observe that for sufficiently small $\epsilon > 0$, the perturbed sequence $\mathbf{y}_\epsilon := (1, 1, 1 + \epsilon, 1, 1, 2)$ satisfies $\mathbf{H}_2(\mathbf{y}_\epsilon) \succ 0$ and so, by Theorem 3.3(a), $\mathbf{y}_\epsilon$ has a finite Borel representing measure μ_ϵ. But then, necessarily, there is no compact interval $[a, b]$ such that μ_ϵ is supported on $[a, b]$ for every $\epsilon > 0$; see Exercise 3.1.

In truncated moment problems, i.e., given a finite sequence $\mathbf{y} = (y_k)_{k=0}^n$, the basic issue is to find conditions under which we may extend the sequence $\mathbf{y}$, so as to be able to build up positive semidefinite moment matrices of higher orders. These higher order moment matrices are called **positive extensions** or **flat extensions** when their rank does not increase with the size. The rank and range conditions in Theorems 3.3-3.4 are such conditions.

3.2 The Multi-dimensional Moment Problem

Most of the applications considered in later chapters of this book refer to real (and not complex) moment problems. Correspondingly, we introduce the basic concepts of moment and localizing matrices in the real case $\mathbb{R}^n$.

However, these concepts also have their natural counterparts in $\mathbb{C}^n$, with the usual scalar product $\langle \mathbf{u}, \mathbf{v}\rangle = \sum_j \overline{u}_j v_j$.

As already mentioned, the multi-dimensional case is significantly more difficult because of the lack of nice characterization of polynomials that are nonnegative on a given subset $\Omega \subseteq \mathbb{R}^n$. Fortunately, we have seen in Section 2.5 that such a characterization exists for the important case of compact basic semi-algebraic sets.

For an integer $r \in \mathbb{N}$, let $\mathbb{N}^n_r := \{\boldsymbol{\alpha} \in \mathbb{N}^n : |\boldsymbol{\alpha}| \leq r\}$ with $|\boldsymbol{\alpha}| := \sum_{i=1}^n \alpha_i \leq r$. Recall that

$$\mathbf{v}_r(\mathbf{x}) := (1, x_1, x_2, \ldots x_n, x_1^2, x_1 x_2, \ldots, x_1 x_n, \ldots, x_1^r, \ldots, x_n^r)' \qquad (3.9)$$

denotes the canonical basis of the real vector space $\mathbb{R}[\mathbf{x}]_r$ of real-valued polynomials of degree at most r (and let $s(r) := \binom{n+r}{n}$ denote its dimension). Then, a polynomial $p \in \mathbb{R}[\mathbf{x}]_r$ is written as

$$\mathbf{x} \mapsto p(\mathbf{x}) = \sum_{\boldsymbol{\alpha} \in \mathbb{N}^n} p_{\boldsymbol{\alpha}} \mathbf{x}^{\boldsymbol{\alpha}} = \langle \mathbf{p}, \mathbf{v}_r(\mathbf{x})\rangle,$$

where $\mathbf{p} = \{p_{\boldsymbol{\alpha}}\} \in \mathbb{R}^{s(r)}$ denotes its vector of coefficients in the basis (3.9). And so we may identify $p \in \mathbb{R}[\mathbf{x}]$ with its vector of coefficients $\mathbf{p} \in \mathbb{R}^{s(r)}$.

3.2.1 *Moment and localizing matrix*

We next define the important notions of moment matrix and localizing matrix.

Moment matrix

Given a $s(2r)$-sequence $\mathbf{y} = (y_{\boldsymbol{\alpha}})$, let $\mathbf{M}_r(\mathbf{y})$ be the **moment** matrix of dimension $s(r)$, with rows and columns labeled by (3.9), and constructed as follows:

$$\mathbf{M}_r(\mathbf{y})(\boldsymbol{\alpha}, \boldsymbol{\beta}) = L_{\mathbf{y}}(\mathbf{x}^{\boldsymbol{\alpha}} \mathbf{x}^{\boldsymbol{\beta}}) = y_{\boldsymbol{\alpha}+\boldsymbol{\beta}}, \qquad \forall \boldsymbol{\alpha}, \boldsymbol{\beta} \in \mathbb{N}^n_r, \qquad (3.10)$$

with $L_{\mathbf{y}}$ defined in (3.3). Equivalently, $\mathbf{M}_r(\mathbf{y}) = L_{\mathbf{y}}(\mathbf{v}_r(\mathbf{x})\mathbf{v}_r(\mathbf{x})')$ where the latter notation means that we apply $L_{\mathbf{y}}$ to each entry of the matrix $\mathbf{v}_r(\mathbf{x})\mathbf{v}_r(\mathbf{x})'$.

Let us consider an example with $n = r = 2$. In this case, $\mathbf{M}_2(\mathbf{y})$ becomes

$$\mathbf{M}_2(\mathbf{y}) = \left[\begin{array}{c|cc|ccc} y_{00} & y_{10} & y_{01} & y_{20} & y_{11} & y_{02} \\ \hline y_{10} & y_{20} & y_{11} & y_{30} & y_{21} & y_{12} \\ y_{01} & y_{11} & y_{02} & y_{21} & y_{12} & y_{03} \\ \hline y_{20} & y_{30} & y_{21} & y_{40} & y_{31} & y_{22} \\ y_{11} & y_{21} & y_{12} & y_{31} & y_{22} & y_{13} \\ y_{02} & y_{12} & y_{03} & y_{22} & y_{13} & y_{04} \end{array}\right].$$

In general, $\mathbf{M}_r(\mathbf{y})$ defines a bilinear form $\langle .,. \rangle_{\mathbf{y}}$ on $\mathbb{R}[\mathbf{x}]_r$ as follows:

$$\langle p, q \rangle_{\mathbf{y}} := L_{\mathbf{y}}(pq) = \langle \mathbf{p}, \mathbf{M}_r(\mathbf{y})\mathbf{q} \rangle = \mathbf{p}'\mathbf{M}_r(\mathbf{y})\mathbf{q}, \qquad \forall \mathbf{p}, \mathbf{q} \in \mathbb{R}^{s(r)},$$

where again, $p, q \in \mathbb{R}[\mathbf{x}]_r$, and $\mathbf{p}, \mathbf{q} \in \mathbb{R}^{s(r)}$ denote their vector of coefficients.

Recall from Definition 3.2 that if $\mathbf{y}$ is a sequence of moments for some measure μ, then μ is called a representing measure for $\mathbf{y}$, and if unique, then μ is said to be determinate, and indeterminate otherwise, whereas $\mathbf{y}$ is called a determinate (resp. indeterminate) moment sequence. In addition, for every $q \in \mathbb{R}[\mathbf{x}]$,

$$\langle \mathbf{q}, \mathbf{M}_r(\mathbf{y})\mathbf{q} \rangle = L_{\mathbf{y}}(q^2) = \int q^2 \, d\mu \geq 0, \tag{3.11}$$

so that $\mathbf{M}_r(\mathbf{y}) \succeq \mathbf{0}$. It is also immediate to check that if the polynomial q^2 is expanded as $q(\mathbf{x})^2 = \sum_{\boldsymbol{\alpha} \in \mathbb{N}^n} h_{\boldsymbol{\alpha}} \mathbf{x}^{\boldsymbol{\alpha}}$, then

$$\langle \mathbf{q}, \mathbf{M}_r(\mathbf{y})\mathbf{q} \rangle = L_{\mathbf{y}}(q^2) = \sum_{\boldsymbol{\alpha} \in \mathbb{N}^n} h_{\boldsymbol{\alpha}} y_{\boldsymbol{\alpha}}.$$

• Every measure with *compact* support (say $\mathbb{K} \subset \mathbb{R}^n$) is determinate, because by the Stone-Weierstrass theorem, the space of polynomials is dense (for the sup-norm) in the space of continuous functions on $\mathbb{K}$.

• Not every sequence $\mathbf{y}$ that satisfies $\mathbf{M}_i(\mathbf{y}) \succeq \mathbf{0}$ for every i, has a representing measure μ on $\mathbb{R}^n$. This is in contrast with the one-dimensional case where by Theorem 3.2(a), a sequence $\mathbf{y}$ such that $\mathbf{H}_n(\mathbf{y}) \succeq \mathbf{0}$ for all $n = 0, 1, \ldots$, has a representing measure. (Recall that in the one-dimensional case, the moment matrix is just the Hankel matrix $\mathbf{H}_n(\mathbf{y})$ in (3.8).) However, we have the following useful result:

Proposition 3.5. *Let* $\mathbf{y}$ *be a sequence indexed in the basis* $\mathbf{v}_\infty(\mathbf{x})$*, which satisfies* $\mathbf{M}_i(\mathbf{y}) \succeq \mathbf{0}$*, for all* $i = 0, 1, \ldots$.
(a) *If the sequence* $\mathbf{y}$ *satisfies*

$$\sum_{k=1}^{\infty} \left[L_{\mathbf{y}}(x_i^{2k}) \right]^{-1/2k} = +\infty, \qquad i = 1, \ldots, n, \tag{3.12}$$

then $\mathbf{y}$ *has a determinate representing measure on* $\mathbb{R}^n$.
(b) *If there exist* $c, a > 0$ *such that*

$$|\, y_{\boldsymbol{\alpha}} \,| \leq c\, a^{|\boldsymbol{\alpha}|}, \qquad \forall \boldsymbol{\alpha} \in \mathbb{N}^n, \tag{3.13}$$

then $\mathbf{y}$ *has a determinate representing measure with support contained in the box* $[-a, a]^n$.

In the one-dimensonial case (3.12) is called Carleman's condition. The moment matrix also has the following properties:

Proposition 3.6. *Let* $d \geq 1$*, and let* $\mathbf{y} = (y_{\boldsymbol{\alpha}}) \subset \mathbb{R}$ *be such that* $M_d(\mathbf{y}) \succeq 0$. *Then*

$$|\, y_{\boldsymbol{\alpha}} \,| \leq \max \left[y_0, \max_{i=1,\ldots,n} L_{\mathbf{y}}(x_i^{2d}) \right], \qquad \forall \boldsymbol{\alpha} \in \mathbb{N}^n_{2d}.$$

In addition, rescaling $\mathbf{y}$ *so that* $y_0 = 1$*, and letting* $\tau_d := \max_{i=1,\ldots,n} L_{\mathbf{y}}(x_i^{2d})$,

$$|\, y_{\boldsymbol{\alpha}} \,|^{\frac{1}{|\boldsymbol{\alpha}|}} \leq \tau_d^{\frac{1}{2d}}, \qquad \forall \boldsymbol{\alpha} \in \mathbb{N}^n_{2d}, \quad \boldsymbol{\alpha} \neq 0.$$

Localizing matrix

Given a polynomial $u \in \mathbb{R}[\mathbf{x}]$ with coefficient vector $\mathbf{u} = \{u_\gamma\}$, we define the **localizing** matrix with respect to $\mathbf{y}$ and u, to be the matrix $\mathbf{M}_r(u\, \mathbf{y})$ with rows and columns indexed by (3.9), and obtained from $\mathbf{M}_r(\mathbf{y})$ by :

$$\mathbf{M}_r(u\, \mathbf{y})(\boldsymbol{\alpha}, \boldsymbol{\beta}) = L_{\mathbf{y}}(u(\mathbf{x})\mathbf{x}^{\boldsymbol{\alpha}}\mathbf{x}^{\boldsymbol{\beta}}) = \sum_{\gamma \in \mathbb{N}^n} u_\gamma y_{\gamma+\boldsymbol{\alpha}+\boldsymbol{\beta}}, \quad \forall \boldsymbol{\alpha}, \boldsymbol{\beta} \in \mathbb{N}^n_r. \tag{3.14}$$

Equivalently, $\mathbf{M}_r(u\,\mathbf{y}) = L_{\mathbf{y}}(u\,\mathbf{v}_r(\mathbf{x})\mathbf{v}_r(\mathbf{x})')$, where the previous notation means that $L_{\mathbf{y}}$ is applied entrywise. For instance, when $n = 2$, with

$$\mathbf{M}_1(\mathbf{y}) = \begin{bmatrix} y_{00} & y_{10} & y_{01} \\ y_{10} & y_{20} & y_{11} \\ y_{01} & y_{11} & y_{02} \end{bmatrix} \quad \text{and } \mathbf{x} \mapsto u(\mathbf{x}) = a - x_1^2 - x_2^2,$$

we obtain

$$\mathbf{M}_1(u\,\mathbf{y}) = \begin{bmatrix} ay_{00} - y_{20} - y_{02} & ay_{10} - y_{30} - y_{12} & ay_{01} - y_{21} - y_{03} \\ ay_{10} - y_{30} - y_{12} & ay_{20} - y_{40} - y_{22} & ay_{11} - y_{31} - y_{13} \\ ay_{01} - y_{21} - y_{03} & ay_{11} - y_{31} - y_{13} & ay_{02} - y_{22} - y_{04} \end{bmatrix}.$$

Similar to (3.11) we have

$$\langle \mathbf{p}, \mathbf{M}_r(u\,\mathbf{y})\mathbf{q} \rangle = L_{\mathbf{y}}(u\,pq),$$

for all polynomials $p, q \in \mathbb{R}[\mathbf{x}]_r$ with coefficient vectors $\mathbf{p}, \mathbf{q} \in \mathbb{R}^{s(r)}$. In particular, if $\mathbf{y}$ has a representing measure μ, then

$$\langle \mathbf{q}, \mathbf{M}_r(u\,\mathbf{y})\mathbf{q} \rangle = L_{\mathbf{y}}(u\,q^2) = \int u\,q^2\,d\mu, \tag{3.15}$$

for every polynomial $q \in \mathbb{R}[\mathbf{x}]$ with coefficient vector $\mathbf{q} \in \mathbb{R}^{s(r)}$. Therefore, $\mathbf{M}_r(u\,\mathbf{y}) \succeq \mathbf{0}$ whenever μ has its support contained in the set $\{\mathbf{x} \in \mathbb{R}^n : u(\mathbf{x}) \geq 0\}$.

It is also immediate to check that if the polynomial uq^2 is expanded as $u(\mathbf{x})\,q(\mathbf{x})^2 = \sum_{\boldsymbol{\alpha} \in \mathbb{N}^n} h_{\boldsymbol{\alpha}} \mathbf{x}^{\boldsymbol{\alpha}}$ then

$$\langle \mathbf{q}, \mathbf{M}_r(u\,\mathbf{y})\mathbf{q} \rangle = \sum_{\boldsymbol{\alpha} \in \mathbb{N}^n} h_{\boldsymbol{\alpha}} y_{\boldsymbol{\alpha}} = L_{\mathbf{y}}(u\,q^2). \tag{3.16}$$

3.2.2 *Positive and flat extensions of moment matrices*

We next discuss the notion of positive extension for moment matrices.

Definition 3.3. Given a finite sequence $\mathbf{y} = (y_{\boldsymbol{\alpha}})_{|\boldsymbol{\alpha}| \leq 2r}$ with $\mathbf{M}_r(\mathbf{y}) \succeq \mathbf{0}$, the **moment extention problem** is defined as follows: extend the sequence $\mathbf{y}$ with new scalars $y_{\boldsymbol{\beta}}$, $2r < |\boldsymbol{\beta}| \leq 2(r+1)$, so as to obtain a new finite sequence $(y_{\boldsymbol{\alpha}})_{|\boldsymbol{\alpha}| \leq 2(r+1)}$ such that $\mathbf{M}_{r+1}(\mathbf{y}) \succeq \mathbf{0}$.

If such an extension $\mathbf{M}_{r+1}(\mathbf{y})$ is possible, it is called a **positive extension** of $\mathbf{M}_r(\mathbf{y})$. If in addition, $\operatorname{rank} \mathbf{M}_{r+1}(\mathbf{y}) = \operatorname{rank} \mathbf{M}_r(\mathbf{y})$, then $\mathbf{M}_{r+1}(\mathbf{y})$ is called a **flat extension** of $\mathbf{M}_r(\mathbf{y})$.

For truncated moment problems, flat extensions play an important role. We first introduce the notion of an atomic measure. An s-atomic measure is a measure with s **atoms**, that is, a linear positive combination of s Dirac measures.

Theorem 3.7 (Flat extension). *Let* $\mathbf{y} = (y_{\boldsymbol{\alpha}})_{|\boldsymbol{\alpha}|\leq 2r}$. *Then the sequence* $\mathbf{y}$ *admits a* rank $\mathbf{M}_r(\mathbf{y})$*-atomic representing measure* μ *on* $\mathbb{R}^n$ *if and only if* $\mathbf{M}_r(\mathbf{y}) \succeq \mathbf{0}$ *and* $\mathbf{M}_r(\mathbf{y})$ *admits a flat extension* $\mathbf{M}_{r+1}(\mathbf{y}) \succeq 0$.

Theorem 3.7 is useful as it provides a simple numerical means to check whether a finite sequence has a representing measure.

Example 3.3. Let μ be the measure on $\mathbb{R}$, defined to be $\mu = \delta_0 + \delta_1$, that is, μ is the sum of two Dirac measures at the points $\{0\}$ and $\{1\}$. Then

$$\mathbf{M}_1(\mathbf{y}) = \begin{bmatrix} 2 & 1 \\ 1 & 1 \end{bmatrix}, \quad \mathbf{M}_2(\mathbf{y}) = \begin{bmatrix} 2 & 1 & 1 \\ 1 & 1 & 1 \\ 1 & 1 & 1 \end{bmatrix},$$

and obvioulsy, rank $\mathbf{M}_2(\mathbf{y}) =$ rank $\mathbf{M}_1(\mathbf{y}) = 2$.

3.3 The $\mathbb{K}$-moment Problem

The (real) $\mathbb{K}$-moment problem identifies those sequences $\mathbf{y}$ that are moment-sequences of a measure with support contained in a set $\mathbb{K} \subset \mathbb{R}^n$.

Given m polynomials $g_i \in \mathbb{R}[\mathbf{x}]$, $i = 1, \ldots, m$, let $\mathbb{K} \subset \mathbb{R}^n$ be the basic semi-algebraic set

$$\mathbb{K} := \{\mathbf{x} \in \mathbb{R}^n \; : \; g_i(\mathbf{x}) \geq 0, \; i = 1, \ldots, m\}. \tag{3.17}$$

For notational convenience, we also define $g_0 \in \mathbb{R}[\mathbf{x}]$ to be the constant polynomial with value 1 (i.e., $g_0 = 1$).

Recall that given a family $(g_j)_{j=1}^m \subset \mathbb{R}[\mathbf{x}]$, we denote by g_J, $J \subseteq \{1, \ldots, m\}$, the polynomial $\mathbf{x} \mapsto g_J(\mathbf{x}) := \prod_{j \in J} g_j(\mathbf{x})$. In particular, when $J = \emptyset$, $g_\emptyset = 1$.

Let $\mathbf{y} = (y_{\boldsymbol{\alpha}})_{\boldsymbol{\alpha} \in \mathbb{N}^n}$ be a given infinite sequence. For every $r \in \mathbb{N}$ and every $J \subseteq \{1, \ldots, m\}$, let $\mathbf{M}_r(g_J\, \mathbf{y})$ be the localizing matrix of order r with respect to the polynomial $g_J := \prod_{j \in J} g_j$; in particular, with $J := \emptyset$, $\mathbf{M}_r(g_\emptyset\, \mathbf{y}) = \mathbf{M}_r(\mathbf{y})$ is the moment matrix (of order r) associated with $\mathbf{y}$.

As we have already seen, there is a duality between the theory of moments and the representation of positive polynomials. The following important theorem, which is the dual facet of Theorems 2.13 and 2.14, makes this statement more precise.

Theorem 3.8. *Let* $\mathbf{y} = (y_\alpha)_{\alpha \in \mathbb{N}^n}$ *be a given infinite sequence in* $\mathbb{R}$, $L_\mathbf{y} : \mathbb{R}[\mathbf{x}] \to \mathbb{R}$ *be the linear functional introduced in (3.3), and let* $\mathbb{K}$ *be as in (3.17), assumed to be compact.*
(a) *The sequence* $\mathbf{y}$ *has a finite Borel representing measure with support contained in* $\mathbb{K}$, *if and only if*

$$L_\mathbf{y}(f^2\, g_J) \geq 0, \qquad \forall\, J \subseteq \{1, \ldots, m\}, \quad \forall\, f \in \mathbb{R}[\mathbf{x}], \tag{3.18}$$

or, equivalently, if and only if

$$\mathbf{M}_r(g_J\, \mathbf{y}) \succeq \mathbf{0}, \qquad \forall\, J \subseteq \{1, \ldots, m\}, \quad \forall r \in \mathbb{N}. \tag{3.19}$$

(b) *Assume that there exists* $u \in \mathbb{R}[\mathbf{x}]$ *of the form*

$$u = u_0 + \sum_{j=1}^m u_i g_i, \quad u_i \in \Sigma[\mathbf{x}], \qquad i = 0, 1, \ldots, m,$$

and such that the level set $\{\mathbf{x} \in \mathbb{R}^n : u(\mathbf{x}) \geq 0\}$ *is compact. Then,* $\mathbf{y}$ *has a finite Borel representing measure with support contained in* $\mathbb{K}$, *if and only if*

$$L_\mathbf{y}(f^2\, g_j) \geq 0, \qquad \forall\, j = 0, 1, \ldots, m, \quad \forall\, f \in \mathbb{R}[\mathbf{x}], \tag{3.20}$$

or, equivalently, if and only if

$$\mathbf{M}_r(g_j\, \mathbf{y}) \succeq \mathbf{0}, \qquad \forall j = 0, 1, \ldots, m, \quad \forall r \in \mathbb{N}. \tag{3.21}$$

Proof. (a) For every $J \subseteq \{1, \ldots, m\}$ and $f \in \mathbb{R}[\mathbf{x}]_r$, the polynomial $f^2 g_J$ is nonnegative on $\mathbb{K}$. Therefore, if $\mathbf{y}$ is the sequence of moments of a measure μ supported on $\mathbb{K}$, then $\int f^2 g_J \, d\mu \geq 0$. Equivalently, $L_\mathbf{y}(f^2\, g_J) \geq 0$, or, in view of (3.16), $\mathbf{M}_r(g_J\, \mathbf{y}) \succeq \mathbf{0}$. Hence (3.18)-(3.19) hold.

Conversely, assume that (3.18), or equivalently, (3.19) holds. As $\mathbb{K}$ is compact, by Theorem 3.1, $\mathbf{y}$ is the moment sequence of a measure with support contained in $\mathbb{K}$ if and only if $\sum_{\alpha \in \mathbb{N}^n} f_\alpha y_\alpha \geq 0$ for all polynomials $f \geq 0$ on $\mathbb{K}$.

Let $f > 0$ on $\mathbb{K}$, so that by Theorem 2.13

$$f = \sum_{J\subseteq\{1,\ldots,m\}} p_J\, g_J, \tag{3.22}$$

for some polynomials $\{p_J\} \subset \mathbb{R}[\mathbf{x}]$, all sums of squares. Hence, since $p_J \in \Sigma[\mathbf{x}]$, from (3.18) and from the linearity of $L_{\mathbf{y}}$, we have $L_{\mathbf{y}}(f) \geq 0$. Hence, for all polynomials $f > 0$ on $\mathbb{K}$, we have $L_{\mathbf{y}}(f) = \sum_{\boldsymbol{\alpha}} f_{\boldsymbol{\alpha}} y_{\boldsymbol{\alpha}} \geq 0$. Next, let $f \in \mathbb{R}[\mathbf{x}]$ be nonnegative on $\mathbb{K}$. Then for arbitrary $\epsilon > 0$, $f + \epsilon > 0$ on $\mathbb{K}$, and thus, $L_{\mathbf{y}}(f+\epsilon) = L_{\mathbf{y}}(f) + \epsilon y_0 \geq 0$. As $\epsilon > 0$ was arbitrary, $L_{\mathbf{y}}(f) \geq 0$ follows. Therefore, $L_{\mathbf{y}}(f) \geq 0$ for all $f \in \mathbb{R}[\mathbf{x}]$, nonnegative on $\mathbb{K}$, which by Theorem 3.1, implies that $\mathbf{y}$ is the moment sequence of some measure with support contained in $\mathbb{K}$.
(b) The proof is similar to part (a) and is left as an exercise. □

Recall the definitions (2.8) (resp. (2.12)) of the preordering $P(g_1,\ldots, g_m)$ (resp. the quadratic module $Q(g_1,\ldots,g_m)$), which are both convex cones of $\mathbb{R}[\mathbf{x}]$, and let $\mathbb{R}[\mathbf{x}]^*$ be the algebraic dual of $\mathbb{R}[\mathbf{x}]$. Then in the language of convex analysis, (3.18) (resp. (3.20)) states that $L_{\mathbf{y}}$ (or $\mathbf{y}$) belongs to the *dual* cone $P(g_1,\ldots,g_m)^*$ (resp. $Q(g_1,\ldots,g_m)^*$) $\subset \mathbb{R}[\mathbf{x}]^*$.

Hence, the duality between polynomials positive on a basic semi-algebraic set $\mathbb{K} \subset \mathbb{R}^n$ and sequences $\mathbf{y} \in \mathbb{R}[\mathbf{x}]^*$ that have a finite Borel representing measure supported on $\mathbb{K}$, is nicely captured by standard duality of convex analysis, applied to the appropriate convex cones $P(g_1,\ldots,g_m)$ or $Q(g_1,\ldots,g_m)$ of $\mathbb{R}[\mathbf{x}]$.

Note that the conditions (3.19) and (3.21) of Theorem 3.8 are stated in terms of positive semidefiniteness of the localizing matrices associated with the polynomials g_J and g_j involved in the definition (3.17) of the compact set $\mathbb{K}$. Alternatively, we also have:

Theorem 3.9. *Let $\mathbf{y} = (y_{\boldsymbol{\alpha}})_{\boldsymbol{\alpha}\in\mathbb{N}^n} \subset \mathbb{R}$ be a given infinite sequence, $L_{\mathbf{y}} : \mathbb{R}[\mathbf{x}]\to\mathbb{R}$ be the linear functional introduced in (3.3), and let $\mathbb{K}$ be as in (3.17), assumed to be compact. Let $C_G \subset \mathbb{R}[\mathbf{x}]$ be the convex cone defined in (2.18), and let Assumption 2.2 hold. Then, $\mathbf{y}$ has a representing measure μ with support contained in $\mathbb{K}$ if and only if*

$$L_{\mathbf{y}}(f) \geq 0, \qquad \forall\, f \in C_G, \tag{3.23}$$

or, equivalently,

$$L_{\mathbf{y}}(\widehat{g}^{\boldsymbol{\alpha}}\,(1-\widehat{g})^{\boldsymbol{\beta}}) \geq 0, \qquad \forall\, \boldsymbol{\alpha},\boldsymbol{\beta} \in \mathbb{N}^m \tag{3.24}$$

with $\widehat{g}$ as in (2.17).

Proof. If $\mathbf{y}$ is the moment sequence of some measure with support contained in $\mathbb{K}$, then (3.24) follows directly from Theorem 3.1, because $\widehat{g}_j \geq 0$ and $1 - \widehat{g}_j \geq 0$ on $\mathbb{K}$, for all $j = 1, \ldots, m$.

Conversely, let (3.24) (and so (3.23)) hold, and let $f \in \mathbb{R}[\mathbf{x}]$, with $f > 0$ on $\mathbb{K}$. By Theorem 2.23, $f \in C_G$, and so f can be written as in (2.18). Therefore, $L_{\mathbf{y}}(f) \geq 0$ and (3.24) follows from the linearity of $L_{\mathbf{y}}$. Finally, let $f \geq 0$ on $\mathbb{K}$, so that $f + \epsilon > 0$ on $\mathbb{K}$ for every $\epsilon > 0$. Therefore, $0 \leq L_{\mathbf{y}}(f + \epsilon) = L_{\mathbf{y}}(f) + \epsilon y_0$ because $f + \epsilon \in C_G$. As $\epsilon > 0$ was arbitrary, we obtain $L_{\mathbf{y}}(f) \geq 0$. Therefore, $L_{\mathbf{y}}(f) \geq 0$ for all $f \in \mathbb{R}[\mathbf{x}]$, nonnegative on $\mathbb{K}$, which by Theorem 3.1, implies that $\mathbf{y}$ is the moment sequence of some measure with support contained in $\mathbb{K}$. □

Exactly as Theorem 3.8 was the dual facet of Theorems 2.13 and 2.14, Theorem 3.9 is the dual facet of Theorem 2.23. Again, in the language of convex analysis, (3.24) states that $L_{\mathbf{y}}$ (or $\mathbf{y}$) belongs to the dual cone $C_G^* \subset \mathbb{R}[\mathbf{x}]^*$ of $C_G \subset \mathbb{R}[\mathbf{x}]$.

Note that Eqs. (3.24) reduce to countably many **linear** conditions on the sequence $\mathbf{y}$. Indeed, for fixed $\boldsymbol{\alpha}, \boldsymbol{\beta} \in \mathbb{N}^m$, we write

$$\widehat{g}^{\boldsymbol{\alpha}}\,(1 - \widehat{g})^{\boldsymbol{\beta}} = \sum_{\gamma \in \mathbb{N}^n} q_\gamma(\boldsymbol{\alpha}, \boldsymbol{\beta})\, \mathbf{x}^\gamma,$$

for finitely many coefficients $(q_\gamma(\boldsymbol{\alpha}, \boldsymbol{\beta}))$. Then, (3.24) becomes

$$\sum_{\gamma \in \mathbb{N}^n} q_\gamma(\boldsymbol{\alpha}, \boldsymbol{\beta})\, y_\gamma \geq 0, \qquad \forall\, \boldsymbol{\alpha}, \boldsymbol{\beta} \in \mathbb{N}^m. \tag{3.25}$$

Eq. (3.25) is to be contrasted with the positive semidefiniteness conditions (3.20) of Theorem 3.8.

In the case where all the g_j's in (3.17) are affine (so that $\mathbb{K}$ is a convex polytope), we also have a specialized version of Theorem 3.9.

Theorem 3.10. *Let $\mathbf{y} = (y_{\boldsymbol{\alpha}})_{\boldsymbol{\alpha} \in \mathbb{N}^n} \subset \mathbb{R}^\infty$ be a given infinite sequence, $L_{\mathbf{y}} : \mathbb{R}[\mathbf{x}] \to \mathbb{R}$ be the linear functional introduced in (3.3). Assume that $\mathbb{K}$ is compact with nonempty interior, and all the g_j's in (3.17) are affine, so that $\mathbb{K}$ is a convex polytope. Then, $\mathbf{y}$ has a finite Borel representing measure with support contained in $\mathbb{K}$ if and only if*

$$L_{\mathbf{y}}(g^{\boldsymbol{\alpha}}) \geq 0, \qquad \forall\, \boldsymbol{\alpha} \in \mathbb{N}^m. \tag{3.26}$$

A sufficient condition for the truncated $\mathbb{K}$-moment problem

Finally, we present a very important sufficient condition for the truncated $\mathbb{K}$-moment problem. That is, we provide a condition on a finite sequence $\mathbf{y} = (y_{\boldsymbol{\alpha}})$ to admit a finite Borel representing measure supported on $\mathbb{K}$. Moreover, this condition can be checked numerically by standard techniques from linear algebra.

Theorem 3.11. *Let $\mathbb{K} \subset \mathbb{R}^n$ be the basic semi-algebraic set*

$$\mathbb{K} := \{\mathbf{x} \in \mathbb{R}^n \,:\, g_j(\mathbf{x}) \geq 0, \; j = 1, \ldots, m\},$$

for some polynomials $g_j \in \mathbb{R}[\mathbf{x}]$ of degree $2v_j$ or $2v_j - 1$, for all $j = 1, \ldots, m$. Let $\mathbf{y} = (y_{\boldsymbol{\alpha}})$ be a finite sequence with $|\alpha| \leq 2r$, and let $v := \max_j v_j$. Then $\mathbf{y}$ has a rank $\mathbf{M}_{r-v}(\mathbf{y})$*-atomic representing measure μ with support contained in $\mathbb{K}$ if and only if:*
(a) $\mathbf{M}_r(\mathbf{y}) \succeq \mathbf{0}$, $\quad \mathbf{M}_{r-v}(g_j\mathbf{y}) \succeq \mathbf{0}$, *$j = 1, \ldots, m$, and*
(b) $\operatorname{rank} \mathbf{M}_r(\mathbf{y}) = \operatorname{rank} \mathbf{M}_{r-v}(\mathbf{y})$.

In addition, μ has $\operatorname{rank} \mathbf{M}_r(\mathbf{y}) - \operatorname{rank} \mathbf{M}_{r-v}(g_i\mathbf{y})$ *atoms $\mathbf{x} \in \mathbb{R}^n$ that satisfy $g_i(\mathbf{x}) = 0$, for all $i = 1, \ldots, m$.*

Note that in Theorem 3.11, the set $\mathbb{K}$ is **not** required to be compact. The rank-condition can be checked by standard techniques from numerical linear algebra. However it is also important to remember that computing the rank is sensitive to numerical imprecisions.

3.4 Moment Conditions for Bounded Density

In this section, we consider the $\mathbb{K}$-moment problem with bounded density. That is, given a finite Borel (reference) measure μ on $\mathbb{K} \subseteq \mathbb{R}^n$ with moment sequence $\mathbf{y} = (y_{\boldsymbol{\alpha}})$, $\boldsymbol{\alpha} \in \mathbb{N}^n$, under what conditions on $\mathbf{y}$ do we have

$$y_{\boldsymbol{\alpha}} = \int_{\mathbb{K}} \mathbf{x}^{\boldsymbol{\alpha}} \, h \, d\mu, \qquad \forall \boldsymbol{\alpha} \in \mathbb{N}^n, \tag{3.27}$$

for some bounded density $0 \leq h \in L_\infty(\mathbb{K}, \mu)$?

The measure $d\nu := h d\mu$ is said to be *uniformly absolutely continuous* with respect to μ (denoted $\nu \ll \mu$) and h is called the Radom-Nikodym derivative of ν with respect to μ. This is a refinement of the general $\mathbb{K}$-moment problem where one only asks for existence of some finite Borel representing measure on $\mathbb{K}$ (not necessarily with a density with respect to some reference measure μ).

Recall that for two finite measures μ, ν on a σ-algebra $\mathscr{B}$, one has the natural partial order $\nu \leq \mu$ if and only if $\nu(B) \leq \mu(B)$ for every $B \in \mathscr{B}$; and observe that $\nu \leq \mu$ obviously implies $\nu \ll \mu$ but the converse does not hold.

3.4.1 *The compact case*

We first consider the case where the support of the reference measure μ is a compact basic semi-algebraic set $\mathbb{K} \subset \mathbb{R}^n$.

Theorem 3.12. *Let $\mathbb{K} \subset \mathbb{R}^n$ be compact and defined as in (3.17). Let $\mathbf{z} = (z_\alpha)$ be the moment sequence of a finite Borel measure μ on $\mathbb{K}$.*
(a) *A sequence $\mathbf{y} = (y_\alpha)$ has a finite Borel representing measure on $\mathbb{K}$, uniformly absolutely continuous with respect to μ, if and only if there is some scalar κ such that:*

$$0 \leq L_{\mathbf{y}}(f^2 g_J) \leq \kappa L_{\mathbf{z}}(f^2 g_J) \qquad \forall f \in \mathbb{R}[\mathbf{x}], \quad \forall J \subseteq \{1, \ldots, m\}. \tag{3.28}$$

(b) *In addition, if the polynomial $N - \|\mathbf{x}\|^2$ belongs to the quadratic module $Q(g)$ then one may replace (3.28) with the weaker condition*

$$0 \leq L_{\mathbf{y}}(f^2 g_j) \leq \kappa L_{\mathbf{z}}(f^2 g_j) \quad \forall f \in \mathbb{R}[\mathbf{x}]; \quad \forall j = 0, \ldots, m \tag{3.29}$$

(with the convention $g_0 = 1$).
(c) *Suppose that the g_j's are normalized so that*

$$0 \leq g_j \leq 1 \quad \text{on} \quad \mathbb{K}, \quad \forall j = 1, \ldots, m,$$

and that the family $(1, \{g_j\})$ generates the algebra $\mathbb{R}[\mathbf{x}]$. Then a sequence $\mathbf{y} = (y_\alpha)$ has a finite Borel representing measure on $\mathbb{K}$, uniformly absolutely continuous with respect to μ, if and only if there is some scalar κ such that:

$$0 \leq L_{\mathbf{y}}(g^{\boldsymbol{\alpha}} (1 - g)^{\boldsymbol{\beta}}) \leq \kappa L_{\mathbf{z}}(g^{\boldsymbol{\alpha}}(1 - g)^{\boldsymbol{\beta}}), \qquad \forall \boldsymbol{\alpha}, \boldsymbol{\beta} \in \mathbb{N}^m. \tag{3.30}$$

Proof. We only prove (a) as (b) and (c) can be proved with very similar arguments. The *only if* part: Let $d\nu = h d\mu$ for some $0 \leq h \in L_\infty(\mathbb{K}, \mu)$, and let $\kappa := \|h\|_\infty$. Observe that $g_J \geq 0$ on $\mathbb{K}$ for all $J \subseteq \{1, \ldots, m\}$.

Therefore, for every $J \subseteq \{1,\ldots,m\}$ and all $f \in \mathbb{R}[\mathbf{x}]$:

$$L_{\mathbf{y}}(f^2\, g_J) \,=\, \int_{\mathbb{K}} f^2 g_J\, d\nu = \int_{\mathbb{K}} f^2 g_J h\, d\mu \,\leq\, \kappa \int_{\mathbb{K}} f^2 g_J\, d\mu \,=\, \kappa\, L_{\mathbf{z}}(f^2 g_J),$$

and so (3.28) is satisfied.

The *if part*: Let $\mathbf{y}$ and $\mathbf{z}$ be such that (3.28) holds true. Then by Theorem 3.8, $\mathbf{y}$ has a finite Borel representing measure ν on $\mathbb{K}$. In addition, let $\boldsymbol{\gamma} := (\gamma_{\boldsymbol{\alpha}})$ with $\gamma_{\boldsymbol{\alpha}} := \kappa z_{\boldsymbol{\alpha}} - y_{\boldsymbol{\alpha}}$ for all $\boldsymbol{\alpha} \in \mathbb{N}^n$. From (3.28), one has

$$L_{\boldsymbol{\gamma}}(f^2 g_J) \geq 0, \qquad \forall\, f \in \mathbb{R}[\mathbf{x}]; \quad \forall J \subseteq \{1,\ldots,m\},$$

and so, by Theorem 3.8 again, $\boldsymbol{\gamma}$ has a finite Borel representing measure ψ on $\mathbb{K}$. Moreover, from the definition of the sequence $\boldsymbol{\gamma}$

$$\int_{\mathbb{K}} f\, d(\psi + \nu) \,=\, \int_{\mathbb{K}} f\, \kappa\, d\mu, \qquad \forall\, f \in \mathbb{R}[\mathbf{x}],$$

and therefore, as measures on compact sets are moment determinate, $\psi + \nu = \kappa\mu$. Hence $\kappa\mu \geq \nu$ which shows that $\nu \ll \mu$ and so one may write $d\nu = h\, d\mu$ for some $0 \leq h \in L_1(\mathbb{K}, \mu)$. From $\nu \leq \kappa\mu$ one obtains

$$\int_A (h - \kappa)\, d\mu \,\leq\, 0, \qquad \forall\, A \,\in\, \mathscr{B}(\mathbb{K})$$

(where $\mathscr{B}(\mathbb{K})$ is Borel σ-algebra associated with $\mathbb{K}$). And so $0 \leq h \leq \kappa$, μ-almost everywhere on $\mathbb{K}$. Equivalently, $\|h\|_\infty \leq \kappa$, the desired result. $\square$

Notice that using moment and localizing matrices,

$$\begin{aligned}
(3.28) &\Leftrightarrow 0 \preceq \mathbf{M}_i(g_J\, \mathbf{y}) \preceq \kappa\, \mathbf{M}_i(g_J\, \mathbf{z}), \quad i = 1,\ldots; \; J \subseteq \{1,\ldots,m\} \\
(3.29) &\Leftrightarrow 0 \preceq \mathbf{M}_i(g_j\, \mathbf{y}) \preceq \kappa\, \mathbf{M}_i(g_j\, \mathbf{z}), \quad i = 1,\ldots; \; j = 0,1,\ldots,m.
\end{aligned}$$

3.4.2 *The non compact case*

Let the reference measure μ be a finite Borel measure on $\mathbb{R}^n$, not supported on a compact set. As one wishes to find moment conditions, it is natural to consider the case where all the moments $\mathbf{z} = (z_{\boldsymbol{\alpha}})$ of μ are finite, and a simple sufficient condition is that μ satisfies the generalized Carleman condition (3.12) of Nussbaum.

Theorem 3.13. *Let $\mathbf{z} = (z_{\boldsymbol{\alpha}})$ be the moment sequence of a finite Borel measure μ on $\mathbb{R}^n$ which satisfies the generalized Carleman condition, i.e.:*

$$\sum_{k=1}^{\infty} L_{\mathbf{z}}(x_i^{2k})^{-1/2k} = \infty, \qquad \forall i = 1, \ldots, n. \tag{3.31}$$

A sequence $\mathbf{y} = (y_{\boldsymbol{\alpha}})$ has a finite Borel representing measure ν on $\mathbb{R}^n$, uniformly absolutely continuous with respect to μ, if there exists a scalar $0 < \kappa$ such that for all $i = 1, \ldots, n$:

$$0 \leq L_{\mathbf{y}}(f^2) \leq \kappa \, L_{\mathbf{z}}(f^2) \qquad \forall f \in \mathbb{R}[\mathbf{x}]. \tag{3.32}$$

Proof. For every $i = 1, \ldots, n$, (3.32) with $\mathbf{x} \mapsto f(\mathbf{x}) = x_i^k$, yields

$$L_{\mathbf{y}}(x_i^{2k})^{-1/2k} \geq \kappa^{-1/2k} \, L_{\mathbf{z}}(x_i^{2k})^{-1/2k}, \qquad \forall k = 1, 2, \ldots,$$

and so, using (3.31), one obtains

$$\sum_{k=1}^{\infty} L_{\mathbf{y}}(x_i^{2k})^{-1/2k} \geq \sum_{k=1}^{\infty} \kappa^{-1/2k} \, L_{\mathbf{z}}(x_i^{2k})^{-1/2k} = +\infty,$$

for every $i = 1, \ldots, n$, i.e., the generalized Carleman condition (3.12) holds for the sequence $\mathbf{y}$. Combining this with the first inequality in (3.32) yields that $\mathbf{y}$ has a unique finite Borel representing measure ν on $\mathbb{R}^n$. It remains to prove that $\nu \ll \mu$ and its density h is in $L_\infty(\mathbb{R}^n, \mu)$.

Let $\boldsymbol{\gamma} = (\gamma_{\boldsymbol{\alpha}})$ with $\gamma_{\boldsymbol{\alpha}} := \kappa z_{\boldsymbol{\alpha}} - y_{\boldsymbol{\alpha}}$ for all $\boldsymbol{\alpha} \in \mathbb{N}^n$. Then the second inequality in (3.32) yields

$$L_{\boldsymbol{\gamma}}(f^2) \geq 0 \qquad \forall f \in \mathbb{R}[\mathbf{x}]. \tag{3.33}$$

Next, observe that from (3.32), for every $i = 1, \ldots, n$, and every $k = 0, 1, \ldots$,

$$L_{\boldsymbol{\gamma}}(x_i^{2k}) \leq \kappa \, L_{\mathbf{z}}(x_i^{2k}),$$

which implies

$$L_{\boldsymbol{\gamma}}(x_i^{2k})^{-1/2k} \geq \kappa^{-1/2k} \, L_{\mathbf{z}}(x_i^{2k})^{-1/2k}, \tag{3.34}$$

and so, for every $i = 1, \ldots, n$,

$$\sum_{k=1}^{\infty} L_{\gamma}(x_i^{2k})^{-1/2k} \geq \sum_{k=1}^{\infty} \kappa^{-1/2k} L_{\mathbf{z}}(x_i^{2k})^{-1/2k} = +\infty,$$

i.e., γ satisfies the generalized Carleman condition. In view of (3.33), γ has a (unique) finite Borel representing measure ψ on $\mathbb{R}^n$. Next, from the definition of γ, one has

$$\int_{\mathbb{R}^n} f\, d(\psi + \nu) = \kappa \int_{\mathbb{R}^n} f\, d\mu, \qquad \forall f \in \mathbb{R}[\mathbf{x}].$$

But as μ (and so $\kappa\mu$) satisfies Carleman condition (3.31), $\kappa\mu$ is moment determinate and therefore, $\kappa\mu = \nu + \psi$.

Hence $\nu \ll \mu$ follows from $\nu \leq \kappa\mu$. Finally, writing $d\nu = h d\mu$ for some nonnegative $h \in L_1(\mathbb{R}^n, \mu)$, and using $\nu \leq \kappa\mu$, one obtains

$$\int_A (h - \kappa)\, d\mu \leq 0 \qquad \forall\, A \in \mathscr{B}(\mathbb{R}^n),$$

and so $0 \leq h \leq \kappa$, μ-almost everywhere on $\mathbb{R}^n$. Equivalently, $\|h\|_\infty \leq \kappa$, the desired result. □

Observe that (3.32) is extremely simple as it is equivalent to stating that

$$\kappa\, \mathbf{M}_r(\mathbf{z}) \succeq \mathbf{M}_r(\mathbf{y}) \succeq 0, \qquad \forall r = 0, 1, \ldots$$

Finally, a sufficient condition for (3.31) to hold is

$$\int \exp |x_i|\, d\mu < \infty, \qquad \forall i = 1, \ldots, n.$$

3.5 Summary

In this chapter, we have presented representation theorems for moment problems. The representations results involving moments are dual to the representation results of Chapter 2 involving positive polynomials, and illustrate the duality between moments and positive polynomials. From a practical perspective, Theorems 3.7 and 3.11 provide a computable sufficient condition to check whether a finite sequence of moments has a representing measure. As we will see in the next chapter, Algorithm 4.2 to extract measures given moment sequences, uses Theorem 3.11. Finally, we have also considered moment conditions for existence of a measure with bounded density with respect to some given reference measure.

3.6 Exercises

Exercise 3.1. Consider Example 3.2.
(a) Show that for sufficiently small $\epsilon > 0$, the sequence $\mathbf{y}_\epsilon = (1, 1, 1+\epsilon, 1, 1)$ has a finite Borel representing measure μ_ϵ on $\mathbb{R}$.
(b) Could a compact interval $[a, b] \subset \mathbb{R}$ be such that the support of μ_ϵ is contained in $[a, b]$ for every ϵ? Why? (Hint: One could extract a sequence μ_{ϵ_i} that converges weakly to some probability measure ν supported on $[a, b]$; see tightness of probability measures in Section B.1.)

Exercise 3.2. Show Theorem 3.8(b).

Exercise 3.3. Recall that for a convex cone $C \subset \mathbb{R}^n$, its dual cone C^* is defined by $C^* = \{\mathbf{z} \in \mathbb{R}^n : \langle \mathbf{z}, \mathbf{y} \rangle \geq 0, \forall \mathbf{y} \in C\}$. Let $\mathscr{P} \subset \mathbb{R}[\mathbf{x}]$ and $\Sigma[\mathbf{x}]$ be the convex cones of nonnegative and s.o.s. polynomials, respectively. Next, define the two convex cones:

$$\mathscr{M} = \{\mathbf{y} \in \mathbb{R}^{\mathbb{N}_n} : \mathbf{y} \text{ has a representing measure}\}\,;\ \mathscr{M}_{\succeq} = \{\mathbf{y} \in \mathbb{R}^{\mathbb{N}^n} : \mathbf{M}(\mathbf{y}) \succeq 0\}.$$

(a) Show that $\mathscr{M}^* = \mathscr{P}$ and $\mathscr{P}^* = \mathscr{M}$.
(b) Show that $\mathscr{M}_{\succeq}^* = \Sigma[\mathbf{x}]$ and $\mathscr{M}_{\succeq} = \Sigma[\mathbf{x}]^*$.

3.7 Notes and Sources

Moment problems have a long and rich history. For historical remarks and details on various approaches for the moment problem, the interested reader is referred to Landau (1987b). See also Akhiezer (1965), Curto and Fialkow (2000), and Simon (1998). Example 3.1 is from (Feller, 1966, p. 227) whereas Example 3.2 is from Laurent (2008). Theorem 3.1 was first proved by M. Riesz for closed sets $\mathbb{K} \subset \mathbb{R}$, and subsequently generalized to closed sets $\mathbb{K} \subset \mathbb{R}^n$ by Haviland (1935, 1936).

3.1. Most of this section is from Curto and Fialkow (1991). Theorem 3.3(c) and Theorem 3.4 were proved by Krein and Nudel'man (1977) who also gave the sufficient conditions $\mathbf{H}_n(\gamma), \mathbf{B}_{n-1}(\gamma) \succ \mathbf{0}$ for Theorem 3.3(b) and the sufficient condition $\mathbf{H}_n(\gamma), \mathbf{B}_n(\gamma) \succ 0$ for Theorem 3.4(b).

3.2. The localizing matrix was introduced in Curto and Fialkow (2000) and Berg (1987). The multivariate condition in Proposition 3.5 that generalizes an earlier result of Carleman (1926) in one dimension, is stated in Berg (1987), and was proved by Nussbaum (1966). Proposition 3.6 is taken from Lasserre (2007b). The infinite and truncated moment matrices (and in particular their kernel) have a lot of very interesting properties. For more details, the interested reader is referred to Laurent (2008).

3.3. Concerning the solution of the $\mathbb{K}$-moment problem, Theorem 3.8(a) was proved by Schmüdgen (1991) with a nice interplay between real algebraic geometry and functional analysis. Indeed, the proof uses Stengle's Positivstellensatz (Theorem 2.12) and the spectral theory of self-adjoint operators in Hilbert spaces. Its refinement (b) is due to Putinar (1993), and Jacobi and Prestel (2001). Incidently, in Schmüdgen (1991), the Positivstellensatz Theorem 2.13 appears as a Corollary of Theorem 3.8(a). Theorem 2.24 is due to Cassier (1984) and Handelman (1988), and appears prior to the more general Theorem 3.9 due to Vasilescu (2003). Theorems 3.7 and 3.11 are due to Curto and Fialkow (1991, 1996, 1998), where the results are stated for the complex plane $\mathbb{C}$, but generalize to $\mathbb{C}^n$ and $\mathbb{R}^n$. An alternative proof of some of these results can be found in Laurent (2005); for instance Theorem 3.11 follows from (Laurent, 2005, Theor. 5.23).

3.4. This section is from Lasserre (2006b). The moment problem with bounded density was initially studied by Markov on the interval $[0, 1]$ with μ the Lebesgue measure, a refinement of the Hausdorff moment problem where one only asks for existence of some finite Borel representing measure ν on $[0, 1]$. For an interesting discussion with historical details, the interested reader is referred to Diaconis and Freedman (2006) where, in particular, the authors have proposed a simplified proof as well as conditions for existence of density in $L_p([0, 1], \mu)$ with a similar flavor.

Chapter 4

Algorithms for Moment Problems

We describe a general methodology to solve the abstract generalized moment problem (GMP). That is, we provide a hierarchy of semidefinite relaxations whose associated sequence of optimal values converges to the optimal value of the basic GMP. Variants of this methodology are also provided to handle additional features such as countably many moment constraints, several measures, or GMP wih sparsity properties. This chapter should allow the reader to understand the basic idea underlying the methodology that permits to solve (or approximate) a problem formulated as a particular instance of the GMP.

In this chapter, we return to the generalized moment problem (1.1), and we utilize the results of the two previous chapters in order to provide computable bounds and extract solutions for Problem (1.1). The goal of this chapter is to make the transition from the characterizations of the previous two chapters to effective and efficient computation. The approach taken in this chapter is representative of the overall approach of the book.

4.1 The Overall Approach

Let Γ be a finite set of indices, $\{f, (h_j)_{j\in\Gamma}\} \subset \mathbb{R}[\mathbf{x}]$, with

$$f(\mathbf{x}) = \sum_{\boldsymbol{\alpha}\in\mathbb{N}^n} f_{\boldsymbol{\alpha}} \mathbf{x}^{\boldsymbol{\alpha}} \quad \text{and} \quad h_j(\mathbf{x}) = \sum_{\boldsymbol{\alpha}\in\mathbb{N}^n} h_{j\boldsymbol{\alpha}} \mathbf{x}^{\boldsymbol{\alpha}},$$

and let $\mathbb{K} \subset \mathbb{R}^n$ be the basic semi-algebraic set

$$\mathbb{K} = \{\, \mathbf{x} \in \mathbb{R}^n \,:\, g_j(\mathbf{x}) \geq 0,\ j = 1, \ldots, m \,\}, \tag{4.1}$$

with $g_j \in \mathbb{R}[\mathbf{x}]$, $j = 1, \ldots, m$. The generalized moment problem is given by

$$\begin{aligned}\rho_{\text{mom}} = \sup_{\mu \in \mathscr{M}(\mathbb{K})_+} & \int_{\mathbb{K}} f \, d\mu \\ \text{s.t.} & \int_{\mathbb{K}} h_j \, d\mu \lesseqqgtr \gamma_j, \quad j \in \Gamma.\end{aligned} \tag{4.2}$$

and let $\Gamma_+ \subseteq \Gamma$ be the set of indices corresponding to inequality constraints "$\leq$".

Let $\mathbf{y} = (y_{\boldsymbol{\alpha}})$ with $y_{\boldsymbol{\alpha}} = \int_{\mathbb{K}} x^{\boldsymbol{\alpha}} \, d\mu$, for some $\mu \in \mathscr{M}(\mathbb{K})_+$. With $L_{\mathbf{y}}$ as in (3.3), rewrite (4.2) in the equivalent form:

$$\begin{aligned}\rho_{\text{mom}} = \sup_{\mathbf{y}} \; & L_{\mathbf{y}}(f) \quad (= \sum_{\boldsymbol{\alpha} \in \mathbb{N}^n} f_{\boldsymbol{\alpha}} y_{\boldsymbol{\alpha}}) \\ \text{s.t. } & L_{\mathbf{y}}(h_j) = \sum_{\boldsymbol{\alpha} \in \mathbb{N}^n} h_{j\boldsymbol{\alpha}} y_{\boldsymbol{\alpha}} \lesseqqgtr \gamma_j, \quad j \in \Gamma \\ & y_{\boldsymbol{\alpha}} = \int_{\mathbb{K}} x^{\boldsymbol{\alpha}} \, d\mu, \quad \boldsymbol{\alpha} \in \mathbb{N}^n, \quad \text{for some } \mu \in \mathscr{M}(\mathbb{K})_+.\end{aligned} \tag{4.3}$$

Note that problem (4.3) is entirely described through the moments $y_{\boldsymbol{\alpha}}$ of μ only, and not μ itself. Therefore, we may replace the unknown measure μ by its sequence of moments $\mathbf{y}$, and conditions on $\mathbf{y}$ that state that $\mathbf{y}$ should be the moment sequence of some measure μ, with support contained in $\mathbb{K}$. We have seen in Chapter 3 that such conditions exist, and depending on the chosen type of conditions, are obtained with either **semidefinite** or **linear** constraints on $\mathbf{y}$.

We next develop the dual point of view via nonnegative polynomials. Recall that the dual of problem (4.2) is

$$\begin{aligned}\rho_{\text{pop}} = \inf_{\lambda} \; & \sum_{j \in \Gamma} \gamma_j \lambda_j \\ \text{s.t. } & \sum_{j \in \Gamma} \lambda_j h_j(\mathbf{x}) - f(\mathbf{x}) \geq 0, \quad \forall \, \mathbf{x} \in \mathbb{K} \\ & \lambda_j \geq 0, \quad \forall j \in \Gamma_+.\end{aligned} \tag{4.4}$$

To compute ρ_{pop}, we need an efficient characterization of the nonnegativity on $\mathbb{K}$ of the function $f_0 := \sum_j \lambda_j h_j - f$ in (4.4). In general, no such characterization exists. However, when $\mathbb{K}$ is the basic semi-algebraic set defined in (4.1), then the representations of polynomials positive on $\mathbb{K}$ described in Chapter 2, are available. Depending on the chosen representation, the nonnegativity constraint in (4.4) reduces to semidefinite or linear constraints on the coefficients of the polynomial f.

4.2 Semidefinite Relaxations

Recall from Chapter 3 that with a sequence $\mathbf{y} = (y_{\boldsymbol{\alpha}})_{\boldsymbol{\alpha}\in\mathbb{N}^n}$, and $g_j \in \mathbb{R}[\mathbf{x}]$, we can associate the localizing matrix $\mathbf{M}_r(g_j\mathbf{y})$, for all $j = 1, \ldots, m$.

Depending on parity, let $2v_j$ or $2v_j - 1$ be the degree of $g_j \in \mathbb{R}[\mathbf{x}]$, and $2v_0$ or $2v_0 - 1$ be that of $f \in \mathbb{R}[\mathbf{x}]$. Similarly, let $2w_j$ or $2w_j - 1$ be the degree of the polynomial $h_j \in \mathbb{R}[\mathbf{x}]$, for all $j \in \Gamma$.

The primal relaxation

For $i \geq i_0 := \max[\max_{j=0,\ldots,m} v_j, \max_{j\in\Gamma} w_j]$, consider the semidefinite optimization problem:

$$
\begin{aligned}
\rho_i = \sup_{\mathbf{y}}\ & L_{\mathbf{y}}(f) \\
\text{s.t. } & L_{\mathbf{y}}(h_j) \leqq \gamma_j, \quad j \in \Gamma \\
& \mathbf{M}_i(\mathbf{y}) \succeq \mathbf{0}, \\
& \mathbf{M}_{i-v_j}(g_j\mathbf{y}) \succeq \mathbf{0},\ j = 1, \ldots, m.
\end{aligned}
\tag{4.5}
$$

• The semidefinite constraints $\mathbf{M}_i(\mathbf{y}), \mathbf{M}_{i-v_j}(g_j\mathbf{y}) \succeq \mathbf{0}$ state necessary conditions on the variables $(\mathbf{y}_{\boldsymbol{\alpha}})$ with $|\boldsymbol{\alpha}| \leq 2r$, to be moments of some finite Borel measure supported on $\mathbb{K}$. Therefore, it immediately follows that $\rho_i \geq \rho_{\text{mom}}$ for all $i \geq i_0$ and so, Problem (4.5) is a **semidefinite relaxation** of Problem (4.3) (or (4.2)) for all $i \geq i_0$. In addition, $(\rho_i)_i$ forms a monotone nonincreasing sequence, because more constraints are added as i increases.

• We call (4.5) a *primal* relaxation of (4.3) as the unknown variable $\mathbf{y}$ is of same "flavor" as the variable μ in the moment problem (4.2).

• The semidefinite relaxation (4.5) is a particular convex optimization problem called a **semidefinite program**, that can be solved efficiently. Indeed if one uses certain interior point algorithms, then up to arbitrary precision $\epsilon > 0$ fixed, a semidefinite program can be solved in a computational time that is polynomial in the input size (i.e., the size needed to code its description); see Section C.2. This feature is essential to explain the power of the moment approach described in this book.

The dual relaxation

Write $\mathbf{M}_i(\mathbf{y}) = \sum_{\boldsymbol{\alpha}} \mathbf{B}_{\boldsymbol{\alpha}} y_{\boldsymbol{\alpha}}$ and $\mathbf{M}_{i-v_j}(g_j\mathbf{y}) = \sum_{\boldsymbol{\alpha}} \mathbf{C}_{j\boldsymbol{\alpha}} y_{\boldsymbol{\alpha}}$, for appropriate real symmetric matrices $(\mathbf{B}_{\boldsymbol{\alpha}}, \mathbf{C}_{j\boldsymbol{\alpha}})_{\boldsymbol{\alpha}}$. From standard duality in convex optimization (and more precisely, in semidefinite programming), the dual of the semidefinite program (4.5) reads:

$$
\begin{array}{rl}
\rho_i^* = \displaystyle\inf_{\boldsymbol{\lambda},\mathbf{X},\mathbf{Z}_j} & \displaystyle\sum_{j\in\Gamma} \lambda_j \gamma_j \\
& \text{s.t.} \ -\langle \mathbf{X}, \mathbf{B}_{\boldsymbol{\alpha}}\rangle - \displaystyle\sum_{j=1}^{m} \langle \mathbf{Z}_j, \mathbf{C}_{j\boldsymbol{\alpha}}\rangle + \sum_{j\in\Gamma} \lambda_j h_{j\boldsymbol{\alpha}} = f_{\boldsymbol{\alpha}}, \quad |\boldsymbol{\alpha}| \leq 2r, \\
& \mathbf{X}, \mathbf{Z}_j \succeq \mathbf{0}, \quad j = 1, \ldots, m; \quad \lambda_j \geq 0, \ \forall j \in \Gamma_+.
\end{array}
\tag{4.6}
$$

See Section C.2. The dual variable λ_j is associated with the constraint $L_{\mathbf{y}}(h_j) \leqq \gamma_j$ of the primal, whereas the dual matrix variable $\mathbf{X}$ (resp. $\mathbf{Z}_j$) is associated with the semidefinite constraint $\mathbf{M}_i(\mathbf{y}) \succeq 0$ (resp. $\mathbf{M}_{i-v_j}(g_j\,\mathbf{y}) \succeq 0$).

We next interpret the constraints of Problem (4.6). So let $(\boldsymbol{\lambda}, \mathbf{X}, \{\mathbf{Z}_j\})$ be a feasible solution of (4.6). Mutliplying each side of the constraint by $\mathbf{x}^{\boldsymbol{\alpha}}$ and summing up yields

$$
\sum_{j\in\Gamma} \lambda_j h_j(\mathbf{x}) - f(\mathbf{x}) = \left\langle \mathbf{X}, \sum_{\boldsymbol{\alpha}} \mathbf{B}_{\boldsymbol{\alpha}} \mathbf{x}^{\boldsymbol{\alpha}} \right\rangle + \sum_{j=1}^{m} \left\langle \mathbf{Z}_j, \sum_{\boldsymbol{\alpha}} \mathbf{C}_{j\boldsymbol{\alpha}} \mathbf{x}^{\boldsymbol{\alpha}} \right\rangle. \tag{4.7}
$$

Observe that with $\mathbf{v}_i(\mathbf{x})$ as in (3.9), and recalling the definition of moment and localizing matrices, we have

$$
\sum_{\boldsymbol{\alpha}} \mathbf{B}_{\boldsymbol{\alpha}} \mathbf{x}^{\boldsymbol{\alpha}} = \mathbf{v}_i(\mathbf{x})\mathbf{v}_i(\mathbf{x})'; \quad \text{and} \quad \sum_{\boldsymbol{\alpha}} \mathbf{C}_{j\boldsymbol{\alpha}} \mathbf{x}^{\boldsymbol{\alpha}} = g_j(\mathbf{x}) \mathbf{v}_{i-v_j}(\mathbf{x}) \mathbf{v}_{i-v_j}(\mathbf{x})',
$$

for all $j = 1, \ldots, m$. From the spectral decomposition of the positive semidefinite matrices $\mathbf{X}$ and $\{\mathbf{Z}_j\}$, we obtain

$$\mathbf{X} = \sum_k \mathbf{f}_{0k}\,\mathbf{f}_{0k}'; \quad \mathbf{Z}_j = \sum_k \mathbf{f}_{jk}\,\mathbf{f}_{jk}', \quad j = 1, \ldots, m,$$

for some vectors $\{\mathbf{f}_{jk}\}$ of appropriate dimensions. Substituting in (4.7), we obtain

$$\sum_{j\in\Gamma} \lambda_j h_j(\mathbf{x}) - f(\mathbf{x}) = \sum_k \langle \mathbf{f}_{0k}, \mathbf{v}_i(\mathbf{x})\rangle^2 + \sum_{j=1}^{m}\sum_k \langle \mathbf{f}_{jk}, \mathbf{v}_{i-v_j}(\mathbf{x})\rangle^2 g_j(\mathbf{x}),$$

that is,

$$\sum_{j\in\Gamma} \lambda_j h_j - f = f_0 + \sum_{j=1}^{m} f_j\, g_j, \tag{4.8}$$

with

$$f_0(\mathbf{x}) := \sum_k \langle \mathbf{f}_{0k}, \mathbf{v}_i(\mathbf{x})\rangle^2; \quad f_j(\mathbf{x}) := \sum_k \langle \mathbf{f}_{jk}, \mathbf{v}_{i-v_j}(\mathbf{x})\rangle^2,$$

for all $j = 1, \ldots, m$. Therefore, the f_j's are all sum of squares, $j = 0, 1, \ldots, m$, and in addition, $\deg f_j g_j \leq 2r$, for all j (with $g_0 = 1$). The identity (4.8) is a certificate of nonnegativity on $\mathbb{K}$ of the polynomial $\sum_{j\in\Gamma} \lambda_j h_j - f$.

In fact, we have just shown that the dual semidefinite relaxation (4.6) has the equivalent formulation

$$\begin{array}{rl} \rho_i^* = \displaystyle\inf_{\lambda, f_j} & \displaystyle\sum_{j\in\Gamma} \lambda_j \gamma_j \\ \text{s.t.} & \displaystyle\sum_{j\in\Gamma} \lambda_j\, h_j - f = f_0 + \sum_{j=1}^{m} f_j\, g_j, \\ & f_j \in \Sigma[\mathbf{x}] \text{ and } \deg f_j \leq 2(i - v_j), \quad j = 0, \ldots, m. \\ & \lambda_j \geq 0, \ \forall j \in \Gamma_+. \end{array} \tag{4.9}$$

Therefore the dual semidefinite program (4.9) is a **strengthening** of the dual (4.4) because we have replaced the nonnegativity on $\mathbb{K}$ by a specific (sufficient) certificate of positivity.

The above approach corresponds to a classical scheme in optimization. To a relaxation of constraints in the primal problem (i.e., replacing (4.3) with (4.5)) corresponds a **strengthening** of the initial dual (replacing (4.4) with (4.9)).

Let $v := \max_{j=1}^{m} v_j$. The moment approach algorithm for solving (or at least approximate) the moment problem (1.1) is a follows:

Algorithm 4.1. (The moment problem (1.1))

Input: A set of polynomials $\{f, (h_j)_{j\in\Gamma}\} \subset \mathbb{R}[\mathbf{x}]$ with degree $2v_0$ or $2v_0 - 1$, and $2w_j$ or $2w_j - 1$, for every $j \in \Gamma$.
A basic semi-algebraic set $\mathbb{K} = \{\mathbf{x} \in \mathbb{R}^n \ : \ g_j(\mathbf{x}) \geq 0, \ j = 1, \ldots, m\}$, where the polynomials g_j are of degree $2v_j$ or $2v_j - 1$, $j = 1, \ldots, m$; A number k, the index of the highest relaxation.

Output: The optimal value ρ_{mom} and a finite set of points of $\mathbb{K}$ (the support of an optimal solution μ of (1.1)), or an upper bound ρ_k on ρ_{mom}.

Algorithm:
1. Solve the semidefinite optimization problem (4.5) with optimal value ρ_i.
2. If no optimal solution $\mathbf{y}^*$ is found and $i < k$, then increase i by one and go to Step 1; if no optimal solution is found and $i = k$ stop and output ρ_k only provides an upper bound $\rho_k \geq \rho_{\text{mom}}$.
3. Let $\mathbf{y}^*$ be an optimal solution. If

$$\operatorname{rank} \mathbf{M}_{s-v}(\mathbf{y}^*) = \operatorname{rank} \mathbf{M}_s(\mathbf{y}^*) \quad \text{for some } i_0 \leq s \leq i \tag{4.10}$$

then $\rho_i = \rho_{\text{mom}}$ and $\operatorname{rank} \mathbf{M}_s(\mathbf{y}^*)$ points of $\mathbb{K}$ are extracted with Algorithm 4.2 (see Section 4.3). They form the (finite) support of a measure μ on $\mathbb{K}$, an optimal solution of the moment problem (1.1).
4. If (4.10) does not hold and $i < k$, then increase i by one and go to Step 1; otherwise, stop and output ρ_k only provides an upper bound $\rho_k \geq \rho_{\text{mom}}$.

We next address the convergence of the semidefinite relaxations (4.5)-(4.9).

Theorem 4.1. *Let $\{f, (h_j)_{j\in\Gamma}, (g_j)_{j=1}^m\} \subset \mathbb{R}[\mathbf{x}]$, with $h_0 > 0$ on $\mathbb{K}$, and let $\mathscr{M}(\mathbb{K})_+$ be the set of finite Borel measures on $\mathbb{K}$, with $\mathbb{K} \subset \mathbb{R}^n$ as in (4.1). Let Assumption 2.1 hold, and let ρ_{mom} be the optimal value of the generalized moment problem defined in (4.2), assumed to be finite. Consider the sequence of primal and dual semidefinite relaxations defined in (4.5) and (4.6), with respective sequences of optimal values $(\rho_i)_i$ and $(\rho_i^*)_i$. Then:*

(a) $\rho_i^* \downarrow \rho_{\text{mom}}$ *and* $\rho_i \downarrow \rho_{\text{mom}}$.

(b) *If (4.5) has an optimal solution* $\mathbf{y}$ *which satisfies*

$$\operatorname{rank} \mathbf{M}_s(\mathbf{y}) = \operatorname{rank} \mathbf{M}_{s-v}(\mathbf{y}) \quad \text{for some } i_0 \leq s \leq i \tag{4.11}$$

(with $v := \max_j v_j$) then $\rho_i = \rho_{\text{mom}}$ and the generalized moment problem (4.2) has an optimal solution $\mu \in \mathscr{M}(\mathbb{K})_+$ which is finitely supported on $\operatorname{rank} \mathbf{M}_s(\mathbf{y})$ *points of* $\mathbb{K}$.

Proof. (a) By Theorem 1.3, we know that $\rho_{\text{pop}} = \rho_{\text{mom}}$ and so

$$\rho_{\text{pop}} = \rho_{\text{mom}} \leq \rho_i \leq \rho_i^*, \qquad \forall i \geq i_0.$$

Let $\epsilon > 0$ and $\boldsymbol{\lambda}$ be a feasible solution of (4.4) with associated value $\rho_{\boldsymbol{\lambda}}$ which satisfies

$$\rho_{\text{mom}} = \rho_{\text{pop}} \leq \rho_{\boldsymbol{\lambda}} = \sum_{j\in\Gamma} \lambda_j \gamma_j \leq \rho_{\text{pop}} + \epsilon = \rho_{\text{mom}} + \epsilon.$$

In particular, we have

$$\sum_{j\in\Gamma} \lambda_j h_j - f \geq 0, \qquad \text{on } \mathbb{K}.$$

Recall that $h_0 > 0$ on $\mathbb{K}$, and consider the new solution $\overline{\boldsymbol{\lambda}}$ defined by $\overline{\lambda}_j = \lambda_j$ for all $j \neq 0$, and $\overline{\lambda}_0 = \lambda_0 + \epsilon$. The solution $\overline{\boldsymbol{\lambda}}$ is feasible in (4.4) because

$$\sum_{j\in\Gamma} \overline{\lambda}_j h_j - f = \sum_{j\in\Gamma} \lambda_j h_j - f + \epsilon h_0 \geq \epsilon h_0 > 0, \qquad \text{on } \mathbb{K},$$

and with associated value $\rho_{\overline{\boldsymbol{\lambda}}} = \rho_{\boldsymbol{\lambda}} + \epsilon\gamma_0$. Invoking Theorem 2.14 we obtain that

$$\sum_{j\in\Gamma} \overline{\lambda}_j h_j - f = f_0 + \sum_{j=1}^{m} f_j g_j,$$

for some s.o.s. polynomials $(f_j) \subset \Sigma[\mathbf{x}]$. Hence, $\overline{\boldsymbol{\lambda}}$ is feasible for the semidefinite relaxation (4.6) as soon as $2i \geq \max_{j=0,\ldots,m} \deg f_j g_j$ (with $g_0 = 1$). But then $\rho_i^* \leq \rho_{\overline{\boldsymbol{\lambda}}} \leq \rho_{\mathrm{mom}} + \epsilon(1 + \gamma_0)$, and as $\epsilon > 0$ is arbitrary, the result follows.

(b) If (4.11) holds for an optimal solution $\mathbf{y}$ of ρ_i, then by Theorem 3.11, $\mathbf{y}$ is the sequence of moments up to order $2s$ of a Borel measure μ finitely supported on $\operatorname{rank} \mathbf{M}_s(\mathbf{y})$ points of $\mathbb{K}$. Hence

$$\rho_i = L_{\mathbf{y}}(f) = \int_{\mathbb{K}} f\, d\mu; \quad \int_{\mathbb{K}} h_j\, d\mu = L_{\mathbf{y}}(h_j) \leqq \gamma_j, \qquad \forall j \in \Gamma,$$

which shows that μ is a feasible solution of the generalized moment problem (4.2), with value $\rho_i \geq \rho_{\mathrm{mom}}$. Therefore μ is an optimal solution of the generalized moment problem, the desired result. □

If (4.11) holds at an optimal solution $\mathbf{y}$ of ρ_i then *finite* convergence takes place, that is the semidefinite relaxation (4.5) is exact, with optimal value $\rho_i = \rho_{\mathrm{mom}}$. Moreover, in this case one may *extract* t ($= \operatorname{rank} \mathbf{M}_s(\mathbf{y})$) points of $\mathbb{K}$, supports of a measure $\mu \in \mathscr{M}(\mathbb{K})_+$, optimal solution of the generalized moment problem (4.2). How to extract those points is the purpose of Algorithm 4.3 detailed in the next section.

Notice that if $\mathbb{K}$ is compact but Assumption 2.1 does not hold, we may still define converging semidefinite relaxations with constraints $\mathbf{M}_i(g_J \mathbf{y}) \succeq \mathbf{0}$, for all $J \subseteq \{1, \ldots, m\}$, where $g_J := \prod_{j \in J} g_j$, and $g_\emptyset = 1$. The only change in the proof is to invoke Theorem 2.13 instead of Theorem 2.14. However, the corresponding semidefinite relaxations have now 2^m linear matrix inequalities instead of m in (4.6), a serious drawback for practical purposes.

4.3 Extraction of Solutions

In this section we describe a procedure that *extracts* an optimal solution of the generalized moment problem (1.1) when the semidefinite relaxation (4.5) is exact at some step i of the hierarchy, i.e., with $\rho_i = \rho_{\mathrm{mom}}$, and the rank condition (4.11) of Theorem 4.1 holds. In this case an optimal solution of (1.1) is a measure supported on $\operatorname{rank} \mathbf{M}_s(\mathbf{y})$ points of $\mathbb{K}$.

Therefore, let $\mathbf{y}$ be an optimal solution of the semidefinite relaxation (4.5) for which (4.11) holds, so that $\rho_i = \rho_{\mathrm{mom}}$. Then the main steps of the extraction algorithm can be sketched as follows.

- **Cholesky factorization.**

As condition (4.11) holds, $\mathbf{y}$ is the vector of a rank $\mathbf{M}_s(\mathbf{y})$-atomic Borel measure μ supported on $\mathbb{K}$. That is, there are r ($=$ rank $\mathbf{M}_s(\mathbf{y})$) points $(\mathbf{x}(k))_{k=1}^r \subset \mathbb{K}$ such that

$$\mu = \sum_{j=1}^{r} \kappa_j^2 \, \delta_{\mathbf{x}(j)}, \quad \kappa_j \neq 0, \ \forall j; \quad \sum_{j=1}^{r} \kappa_j^2 = y_0, \tag{4.12}$$

with $\delta_\bullet$ being the Dirac measure at $\bullet$.

Hence, by construction of the moment matrix $\mathbf{M}_s(\mathbf{y})$,

$$\mathbf{M}_s(\mathbf{y}) = \sum_{j=1}^{r} \kappa_j^2 \, \mathbf{v}_s(\mathbf{x}^*(j))(\mathbf{v}_s(\mathbf{x}^*(j)))' = \mathbf{V}^*\mathbf{D}(\mathbf{V}^*)' \tag{4.13}$$

where $\mathbf{V}^*$ is written columnwise as

$$\mathbf{V}^* = \left[\mathbf{v}_s(\mathbf{x}^*(1)) \ \mathbf{v}_s(\mathbf{x}^*(2)) \ \cdots \ \mathbf{v}_s(\mathbf{x}^*(r)) \right]$$

with $\mathbf{v}_s(\mathbf{x})$ as in (3.9), and $\mathbf{D}$ is a $r\times r$ diagonal matrix with entries $D(i,i) = \kappa_i^2$, $i = 1, \ldots, r$.

In fact, the weights $(\kappa_j)_{j=1}^r$ do not play any role in the sequel. As long as $\kappa_j \neq 0$ for all j, the rank of the moment matrix $\mathbf{M}_s(\mathbf{y})$ associated with the Borel measure μ defined in (4.12) does *not* depend on the weights κ_j, The extraction procedure with another matrix $\mathbf{M}_s(\tilde{\mathbf{y}})$ written as in (4.13) but with different weights $\tilde{\kappa}_j$, would yield the same global minimizers $(\mathbf{x}^*(j))_{j=1}^r$. Of course, the new associated vector $\tilde{\mathbf{y}}$ would also be an optimal solution of the semidefinite relaxation with value $\rho_i = \rho_{\text{mom}}$.

Extract a Cholesky factor $\mathbf{V}$ of the positive semidefinite moment matrix $\mathbf{M}_s(\mathbf{y})$, i.e. a matrix $\mathbf{V}$ with r columns satisfying

$$\mathbf{M}_s(\mathbf{y}) = \mathbf{V}\mathbf{V}'. \tag{4.14}$$

Such a Cholesky factor can be obtained via singular value decomposition, or any cheaper alternative; see e.g. Golub and Loan (1996).

The matrices $\mathbf{V}$ and $\mathbf{V}^*$ span the same linear subspace, so the solution extraction algorithm consists in transforming $\mathbf{V}$ into $\mathbf{V}^*$ by suitable column operations. This is described in the sequel.

- **Column echelon form.**

Reduce $\mathbf{V}$ to column echelon form

$$\mathbf{U} = \begin{bmatrix} 1 & & & & \\ \star & & & & \\ 0 & 1 & & & \\ 0 & 0 & 1 & & \\ \star & \star & \star & & \\ \vdots & & & \ddots & \\ 0 & 0 & 0 & \cdots & 1 \\ \star & \star & \star & \cdots & \star \\ \vdots & & & & \vdots \\ \star & \star & \star & \cdots & \star \end{bmatrix}$$

by Gaussian elimination with column pivoting. By construction of the moment matrix, each row in $\mathbf{U}$ is indexed by a monomial $\mathbf{x}^{\boldsymbol{\alpha}}$ in the basis $\mathbf{v}_s(\mathbf{x})$. Pivot elements in $\mathbf{U}$ (i.e. the first non-zero elements in each column) correspond to monomials $\mathbf{x}^{\beta_j}$, $j = 1, 2, \ldots, r$ of the basis generating the r solutions. In other words, if

$$\mathbf{w}(\mathbf{x}) = \begin{bmatrix} \mathbf{x}^{\beta_1} & \mathbf{x}^{\beta_2} & \ldots & \mathbf{x}^{\beta_r} \end{bmatrix}' \tag{4.15}$$

denotes this generating basis, then

$$\mathbf{v}_s(\mathbf{x}) = \mathbf{U}\mathbf{w}(\mathbf{x}) \tag{4.16}$$

for all solutions $\mathbf{x} = \mathbf{x}^*(j)$, $j = 1, 2, \ldots, r$.

In summary, extracting the solutions amounts to solving the system of polynomial equations (4.16).

- **Solving the system of polynomial equations (4.16).**

Once a generating monomial basis $\mathbf{w}(\mathbf{x})$ is available, it turns out that extracting solutions of the system of polynomial equations (4.16) reduces to solving a linear algebra problem, as described below.

1. Multiplication matrices.

For each degree one monomial x_i, $i = 1, 2, \ldots, n$, extract from $\mathbf{U}$ the $r \times r$ multiplication matrix $\mathbf{N}_i$ containing the coefficients of monomials $x_i \mathbf{x}^{\beta_j}$, $j = 1, 2, \ldots, r$, in the generating basis (4.15), i.e. such that

$$\mathbf{N}_i \mathbf{w}(\mathbf{x}) = x_i \mathbf{w}(\mathbf{x}), \quad i = 1, 2, \ldots, n. \tag{4.17}$$

The entries of global minimizers $\mathbf{x}^*(j)$, $j = 1, 2, \ldots, r$ are all eigenvalues of multiplication matrices $\mathbf{N}_i$, $i = 1, 2, \ldots, n$. That is,

$$\mathbf{N}_i \mathbf{w}(\mathbf{x}^*(j)) = x_i^*(j)\, \mathbf{w}(\mathbf{x}^*(j)), \quad i = 1, \ldots, n; \quad j = 1, \ldots, r.$$

But how to reconstruct the solutions $(\mathbf{x}^*(j))$ from knowledge of the eigenvalues of the $\mathbf{N}_i$s? Indeed, from the n r-uplets of eigenvalues one could build up r^n possible vectors of $\mathbb{R}^n$ whereas we are looking for only r of them.

2. Common eigenspaces.

Observe that for every $j = 1, \ldots, r$, the vector $\mathbf{w}(\mathbf{x}^*(j))$ is an eigenvector common to *all* matrices $\mathbf{N}_i$, $i = 1, \ldots, n$. Therefore, in order to compute $(\mathbf{x}^*(j))$, one builds up a random combination

$$\mathbf{N} = \sum_{i=1}^{n} \lambda_i \mathbf{N}_i$$

of multiplication matrices $\mathbf{N}_i$s, where λ_i, $i = 1, 2, \ldots, n$ are nonnegative real numbers summing up to one. Then with probability 1, the eigenvalues of $\mathbf{N}$ are all distinct and so $\mathbf{N}$ is non-derogatory, i.e., all its eigenspaces are 1-dimensional (and spanned by the vectors $\mathbf{w}(\mathbf{x}^*(j))$, $j = 1, \ldots, r$).

Then, compute the ordered Schur decomposition

$$\mathbf{N} = \mathbf{Q}\,\mathbf{T}\,\mathbf{Q}' \tag{4.18}$$

where

$$\mathbf{Q} = \begin{bmatrix} \mathbf{q}_1 & \mathbf{q}_2 & \cdots & \mathbf{q}_r \end{bmatrix}$$

is an orthogonal matrix (i.e. $\mathbf{q}_i'\mathbf{q}_i = 1$ and $\mathbf{q}_i'\mathbf{q}_j = 0$ for $i \neq j$) and $\mathbf{T}$ is upper-triangular with eigenvalues of $\mathbf{N}$ sorted in increasing order along the diagonal. Finally, the i-th entry $x_i^*(j)$ of $\mathbf{x}^*(j) \in \mathbb{R}^n$ is given by

$$x_i^*(j) = \mathbf{q}_j'\mathbf{N}_i\mathbf{q}_j, \quad i = 1, 2, \ldots, n, \quad j = 1, 2, \ldots, r. \tag{4.19}$$

In summary here is the extraction algorithm:

Algorithm 4.2. (The extraction algorithm)

Input: The moment matrix $\mathbf{M}_s(\mathbf{y})$ of rank r.

Output: The r points $\mathbf{x}^*(i) \in \mathbb{K}$, $i = 1, \ldots, r$, support of an optimal solution of the moment problem.

Algorithm:

1. Get the **Cholesky factorization** $\mathbf{VV}'$ of $\mathbf{M}_s(\mathbf{y})$.
2. Reduce $\mathbf{V}$ to an **echelon form U**
3. Extract from $\mathbf{U}$ the multiplication matrices $\mathbf{N}_i$, $i = 1, \ldots, n$.
4. Compute $\mathbf{N} := \sum_{i=1}^n \lambda_i \mathbf{N}_i$ with randomly generated coefficients λ_i, and the Schur decomposition $\mathbf{N} = \mathbf{QTQ}'$. Compute

$$\mathbf{Q} = \left[\mathbf{q}_1\ \mathbf{q}_2 \cdots \mathbf{q}_r\right], \quad \text{and}$$

$$x_i^*(j) = \mathbf{q}_j' \mathbf{N}_i \mathbf{q}_j, \quad i = 1, 2, \ldots, n, \quad j = 1, 2, \ldots, r.$$

Example 4.1. Consider the moment problem (1.1) in $\mathbb{R}^2$, where

$$\mathbf{x} \mapsto f(\mathbf{x}) = -(x_1 - 1)^2 - (x_1 - x_2)^2 - (x_2 - 3)^2; \; h_1 = 1; \; \gamma_1 = 1,$$

and $\mathbb{K} = \{\mathbf{x} \in \mathbb{R}^2 : 1 - (x_1 - 1)^2 \geq 0; 1 - (x_1 - x_2)^2 \geq 0; 1 - (x_2 - 3)^2 \geq 0\}$.

The first ($i = 1$) semidefinite relaxation yields $\rho_1 = -3$ and $\operatorname{rank} \mathbf{M}_1(\mathbf{y}) = 3$, whereas the second ($i = 2$) semidefinite relaxation yields $\rho_2 = -2$ and $\operatorname{rank} \mathbf{M}_1(\mathbf{y}) = \operatorname{rank} \mathbf{M}_2(\mathbf{y}) = 3$. Hence, the rank condition (4.11) is satisfied, which implies that $-2 = \rho_2 = \rho_{\text{mom}}$. The moment matrix $\mathbf{M}_2(\mathbf{y})$ reads

$$\mathbf{M}_2(\mathbf{y}) = \begin{bmatrix} 1.0000 & 1.5868 & 2.2477 & 2.7603 & 3.6690 & 5.2387 \\ 1.5868 & 2.7603 & 3.6690 & 5.1073 & 6.5115 & 8.8245 \\ 2.2477 & 3.6690 & 5.2387 & 6.5115 & 8.8245 & 12.7072 \\ 2.7603 & 5.1073 & 6.5115 & 9.8013 & 12.1965 & 15.9960 \\ 3.6690 & 6.5115 & 8.8245 & 12.1965 & 15.9960 & 22.1084 \\ 5.2387 & 8.8245 & 12.7072 & 15.9960 & 22.1084 & 32.1036 \end{bmatrix}$$

and the monomial basis is

$$\mathbf{v}_2(\mathbf{x}) = \begin{bmatrix} 1 & x_1 & x_2 & x_1^2 & x_1 x_2 & x_2^2 \end{bmatrix}'.$$

The Cholesky factor (4.14) of $\mathbf{M}_2(\mathbf{y})$ is given by

$$\mathbf{V} = \begin{bmatrix} -0.9384 & -0.0247 & 0.3447 \\ -1.6188 & 0.3036 & 0.2182 \\ -2.2486 & -0.1822 & 0.3864 \\ -2.9796 & 0.9603 & -0.0348 \\ -3.9813 & 0.3417 & -0.1697 \\ -5.6128 & -0.7627 & -0.1365 \end{bmatrix}$$

whose column echelon form reads (after rounding)

$$\mathbf{U} = \begin{bmatrix} 1 & & \\ 0 & 1 & \\ 0 & 0 & 1 \\ -2 & 3 & 0 \\ -4 & 2 & 2 \\ -6 & 0 & 5 \end{bmatrix}.$$

Pivot entries correspond to the following generating basis (4.15)

$$\mathbf{w}(\mathbf{x}) = [1 \quad x_1 \quad x_2]'.$$

From the subsequent rows in matrix $\mathbf{U}$ we deduce from (4.16) that all solutions $\mathbf{x}$ must satisfy the three polynomial equations

$$\begin{aligned} x_1^2 &= -2 + 3x_1 \\ x_1x_2 &= -4 + 2x_1 + 2x_2 \\ x_2^2 &= -6 + 5x_2. \end{aligned}$$

Multiplication matrices (4.17) (by x_1 and x_2) in the basis $\mathbf{w}(\mathbf{x})$ are readily extracted from rows in $\mathbf{U}$:

$$\mathbf{N}_1 = \begin{bmatrix} 0 & 1 & 0 \\ -2 & 3 & 0 \\ -4 & 2 & 2 \end{bmatrix}, \quad \mathbf{N}_2 = \begin{bmatrix} 0 & 0 & 1 \\ -4 & 2 & 2 \\ -6 & 0 & 5 \end{bmatrix}.$$

A randomly chosen convex combination of $\mathbf{N}_1$ and $\mathbf{N}_2$ yields

$$\mathbf{N} = 0.6909\,\mathbf{N}_1 + 0.3091\,\mathbf{N}_2 = \begin{bmatrix} 0 & 0.6909 & 0.3091 \\ -2.6183 & 2.6909 & 0.6183 \\ -4.6183 & 1.3817 & 2.9274 \end{bmatrix}$$

with orthogonal matrix in Schur decomposition (4.18) given by

$$\mathbf{Q} = \begin{bmatrix} 0.4082 & 0.1826 & -0.8944 \\ 0.4082 & -0.9129 & -0.0000 \\ 0.8165 & 0.3651 & 0.4472 \end{bmatrix}.$$

From equations (4.19), we obtain the 3 optimal solutions

$$\mathbf{x}^*(1) = \begin{bmatrix} 1 \\ 2 \end{bmatrix}, \quad \mathbf{x}^*(2) = \begin{bmatrix} 2 \\ 2 \end{bmatrix}, \quad \mathbf{x}^*(3) = \begin{bmatrix} 2 \\ 3 \end{bmatrix}.$$

Numerical stability

All operations of the above solution extraction algorithm are numerically stable, except the Gaussian elimination step with column pivoting. However, practical experiments with GloptiPoly reveal that ill-conditioned problem instances leading to a failure of Gaussian elimination with column pivoting are very scarce. This experimental property of Gaussian elimination was already noticed in Golub and Loan (1996).

Number of extracted solutions

The number of solutions $(\mathbf{x}^*(j))$ extracted by the algorithm is equal to the rank of the moment matrix $\mathbf{M}_s(\mathbf{y})$. Up to our knowledge, when solving a semidefinite relaxation there is no easy way to control the rank of the moment matrix, hence the number of extracted solutions.

4.4 Linear Relaxations

In this section, we derive a hierarchy of linear programming (LP) relaxations whose associated sequence of optimal values converges to the desired value ρ_{mom}. We introduce the following linear optimization problem

$$\begin{aligned} L_i = \sup_{\mathbf{y}} & \sum_{\alpha \in \mathbb{N}^n} L_{\mathbf{y}}(f) \\ \text{s.t. } & L_{\mathbf{y}}(h_j) \leqq \gamma_j, \qquad j \in \Gamma, \\ & L_{\mathbf{y}}(g^{\boldsymbol{\alpha}}(1-g)^{\boldsymbol{\beta}}) \geq 0, \forall \boldsymbol{\alpha}, \boldsymbol{\beta} : \deg \mathbf{g}^{\boldsymbol{\alpha}}(1-\mathbf{g}^{\boldsymbol{\beta}}) \leq 2i, \end{aligned} \tag{4.20}$$

and its dual:

$$
\begin{aligned}
L_i^* = \inf_{\mathbf{c},\boldsymbol{\lambda}} & \sum_{j\in\Gamma} \lambda_j \gamma_j \\
\text{s.t.} & \sum_{j\in\Gamma} \lambda_j h_j - f = \sum_{\boldsymbol{\alpha},\boldsymbol{\beta}} c_{\boldsymbol{\alpha}\boldsymbol{\beta}}\, \mathbf{g}^{\boldsymbol{\alpha}}(1-\mathbf{g})^{\boldsymbol{\beta}}, \\
& c_{\boldsymbol{\alpha}\boldsymbol{\beta}} \geq 0,\ \forall \boldsymbol{\alpha},\boldsymbol{\beta} : \deg \mathbf{g}^{\boldsymbol{\alpha}}(1-\mathbf{g})^{\boldsymbol{\beta}} \leq 2i. \\
& \lambda_j \geq 0,\ \forall j \in \Gamma_+.
\end{aligned}
\tag{4.21}
$$

In (4.21) and possibly after scaling, one assumes that $0 \leq g_j \leq 1$ on $\mathbb{K}$ for all $j = 1, \ldots, m$, and the g_j's generate the algebra $\mathbb{R}[\mathbf{x}]$. The linear relaxation (4.20) is the linear programming (LP) analogue of the semidefinite relaxation (4.5), whereas its dual (4.21) is the LP analogue of the dual semidefinite program (4.6) (or its equivalent form (4.9)).
We next address convergence.

Theorem 4.2. *Let $\{f, (h_j)_{j\in\Gamma}, (g_j)_{j=1}^m\} \subset \mathbb{R}[\mathbf{x}]$, with $h_0 = 1$, and let $\mathbb{K} \subset \mathbb{R}^n$ be as in (4.1). Let Assumption 2.2 hold and let ρ_{mom} be the optimal value of the generalized moment problem defined in (4.2), assumed to be finite. Consider the sequence of primal and dual linear relaxations defined in (4.20) and (4.21), with respective sequences of optimal values $(L_i)_i$ and $(L_i^*)_i$. Then $L_i^* \downarrow \rho_{\text{mom}}$ and $L_i \downarrow \rho_{\text{mom}}$.*

Proof. The proof mimics that of Theorem 4.1, the only change is that we invoke Theorem 2.23 for the representation of the polynomial $\sum_{j\in\Gamma} \lambda_j h_j - f + \epsilon$, positive on $\mathbb{K}$. □

4.5 Extensions

In this section we consider a few extensions of the generalized moment problem. Namely, we consider the case where one has countably many moment constraints, the case of several unknown measures, and the case where some sparsity pattern is present.

4.5.1 *Extensions to countably many moment constraints*

We now consider the moment problem (4.2) with countably many moment constraints

$$\int_{\mathbb{K}} h_j \, d\mu \leqq \gamma_j, \quad j = 0, 1, \ldots,$$

that is, when $\Gamma = \mathbb{N}$. For every $i \in \mathbb{N}$ let $\Gamma_i := \{j \in \{0, 1, \ldots, i\} : \deg h_j \leq 2i\}$ so that Γ_i is finite and $\Gamma_i \uparrow \Gamma$ as $i \to \infty$.

With $i \geq \max_j v_j$, consider the following semidefinite program:

$$\begin{aligned} \rho_i = \sup_{\mathbf{y}} \; & L_{\mathbf{y}}(f) \\ \text{s.t.} \; & L_{\mathbf{y}}(h_j) \leqq \gamma_j, \quad j \in \Gamma_i \\ & \mathbf{M}_i(\mathbf{y}) \succeq \mathbf{0}, \\ & \mathbf{M}_{i-v_j}(g_j \mathbf{y}) \succeq \mathbf{0}, \; j = 1, \ldots, m. \end{aligned} \tag{4.22}$$

Theorem 4.3. *Let Assumption 2.1 hold, and let $\Gamma = \mathbb{N}$. Assume that $h_0 > 0$ on $\mathbb{K}$ and let ρ_{mom} be the optimal value of the generalized moment problem defined in (4.2), assumed to be finite. Consider the sequence of semidefinite relaxations defined in (4.22). Then $\rho_i \downarrow \rho_{\text{mom}}$ as $i \to \infty$.*

The proof is postponed to Section 4.10.

4.5.2 *Extension to several measures*

So far, the moment problem (4.2) we have considered has only one measure as unknown. In some applications developed in later chapters, we need consider an extension of (1.1) that involves several (but finitely many) unknown measures μ_i, $i = 1, \ldots, p$, not necessarily on the same space. Therefore, let Γ, Δ be two finite sets of indices, and for every $i \in \Delta$:

- Let $\{f_i, (h_{ij})_{j \in \Gamma}\} \subset \mathbb{R}[\mathbf{x}^i] = \mathbb{R}[x_1^i, \ldots, x_{n_i}^i]$, for some integers $n_i \in \mathbb{N}$, $i \in \Delta$.
- Let $\mathbb{K}_i \subset \mathbb{R}^{n_i}$ be the basic closed semi-algebraic sets

$$\mathbb{K}_i := \{\mathbf{x}^i \in \mathbb{R}^{n_i} \; : \; g(\mathbf{x}^i) \geq 0, \quad g \in \mathbf{G}_i\}, \tag{4.23}$$

for some finite set of polynomials $\mathbf{G}_i \subset \mathbb{R}[\mathbf{x}^i]$.

Then consider the following problem:

$$\begin{aligned} \rho_{\text{mom}} = \sup_{\mu_i \in \mathscr{M}(\mathbb{K}_i)_+} & \sum_{i \in \Delta} \int_{\mathbb{K}_i} f_i \, d\mu_i \\ \text{s.t.} & \sum_{i \in \Delta} \int_{\mathbb{K}_i} h_{ij} \, d\mu_i \leqq \gamma_j, \quad j \in \Gamma \end{aligned} \tag{4.24}$$

with dual

$$\begin{aligned} \rho_{\text{pop}} = \inf_{\lambda} & \sum_{j \in \Gamma} \gamma_j \lambda_j \\ \text{s.t.} & \sum_{j \in \Gamma} \lambda_j h_{ij}(\mathbf{x}^i) - f_i(\mathbf{x}^i) \geq 0, \quad \forall\, \mathbf{x}^i \in \mathbb{K}_i,\ i \in \Delta. \\ & \lambda_j \geq 0,\ \forall j \in \Gamma_+. \end{aligned} \tag{4.25}$$

Theorem 4.4 below is the analogue of Theorem 1.3 for the multi-measures moment problem (4.24).

Theorem 4.4. *Assume that $\mathbb{K}_i$ is compact for each $i \in \Delta$ and there exists $j \in \Gamma$ such that $h_{ij} > 0$ on $\mathbb{K}_i$ for all $i \in \Delta$. Assume that (4.24) has a feasible solution $(\mu_i)_{i \in \Delta}$. Then $\rho_{\text{mom}} = \rho_{\text{pop}}$ and the moment problem (4.24) has an optimal solution, that is, the* sup *is attained.*
In addition, (4.24) has an optimal solution (μ_i) such that each μ_i is supported on finitely many points of $\mathbb{K}_i$, i.e., μ_i a finite linear combination of Dirac measures, $i \in \Delta$.

As Theorem 1.3 for the single measure moment problem could be extended in Corollary 1.4 to handle countably many moment constraints, Theorem 4.4 can be also extended in a similar way to handle countably many moment constraints.

Semidefinite relaxations

We now describe the semidefinite relaxations associated with (4.24).

Let $v_g := \lceil \deg g/2 \rceil$ for every $g \in \mathbf{G}_i$, $i \in \Delta$, let $d_i := \max\{v_g : g \in \mathbf{G}_i\}$, $i \in \Delta$, and let $D := \max\{\deg h_{ij} \,:\, i \in \Delta, j \in \Gamma\}$. Next, with $k \geq k_0 := \max[D, \deg f, \max_{i \in \Delta} d_i]$, the following semidefinite program

$$
\begin{aligned}
\rho_k = \sup_{\mathbf{y}^i} \; & \sum_{i\in\Delta} L_{\mathbf{y}^i}(f_i) \\
\text{s.t.} \; & \sum_{i\in\Delta} L_{\mathbf{y}^i}(h_{ij}) \lesseqqgtr \gamma_j, \quad j \in \Gamma \\
& \mathbf{M}_k(\mathbf{y}^i) \succeq \mathbf{0}, \quad i \in \Delta \\
& \mathbf{M}_{k-v_g}(g\,\mathbf{y}^i) \succeq \mathbf{0}, \quad g \in \mathbf{G}_i,\, i \in \Delta
\end{aligned}
\tag{4.26}
$$

is the analogue for the multi-measures moment problem (4.24) of (4.5) for the moment problem (4.2). Its dual reads:

$$
\begin{aligned}
\rho_k^* = \inf_{\lambda} \; & \sum_{j\in\Gamma} \lambda_j \gamma_j \\
\text{s.t.} \; & \sum_{j\in\Gamma} \lambda_j h_{ij} - \sigma_i - \sum_{g\in\mathbf{G}_i} \sigma_g g - f_i = 0, \quad \forall i \in \Delta \\
& \sigma_i, \sigma_g \in \Sigma[\mathbf{x}^i];\ \deg \sigma_g\, g \le 2k, \quad \forall g \in \mathbf{G}_i, \forall i \in \Delta \\
& \lambda_j \ge 0,\ \forall j \in \Gamma_+.
\end{aligned}
\tag{4.27}
$$

And we obtain the following multi-measures version of Theorem 4.1.

Theorem 4.5. *Assume that the hypotheses of Theorem 4.4 hold true, and Assumption 2.1 holds for each* $\mathbb{K}_i$, $i \in \Delta$. *Let* ρ_{mom} *be the optimal value of the moment problem (4.24), assumed to be finite. Consider the primal and dual semidefinite relaxations (4.26) and (4.27), with respective optimal values* ρ_k *and* ρ_k^*. *Then:*
(a) $\rho_k^* \downarrow \rho_{\text{mom}}$ *and* $\rho_k \downarrow \rho_{\text{mom}}$ *as* $k\to\infty$.
(b) *If (4.26) has an optimal solution* $(\mathbf{y}^i)_{i\in\Delta}$ *which satisfies*

$$\operatorname{rank} \mathbf{M}_{s_i}(\mathbf{y}^i) = \operatorname{rank} \mathbf{M}_{s_i - d_i}(\mathbf{y}^i), \quad \forall i \in \Delta, \tag{4.28}$$

for some $k_0 \le s_i \le k$, *then* $\rho_k = \rho_{\text{mom}}$ *and the generalized moment problem (4.24) has an optimal solution* $(\mu_i)_{i\in\Delta}$ *with each* μ_i *being finitely supported on* $\operatorname{rank} \mathbf{M}_{s_i}(\mathbf{y}^i)$ *points of* $\mathbb{K}_i$, $i \in \Delta$.

The proof of Theorems 4.4 and 4.5 are verbatim copies of that of Theorems 1.3 and 4.1 with *ad hoc* adjustments.

4.6 Exploiting Sparsity

Consider the moment problem (4.2). Despite their nice properties, the size of the semidefinite relaxations (4.5) grows rapidly with the dimension n. Typically, the moment matrix $\mathbf{M}_i(\mathbf{y})$ is $s(i) \times s(i)$ with $s(i) = \binom{n+i}{n}$, and there are $\binom{n+2i}{n}$ variables $y_{\boldsymbol{\alpha}}$. This makes the applicability of Algorithm 4.1 limited to small to medium size problems only. Fortunately, in many practical applications of large size moment problems, some sparsity pattern is often present and may be exploited.

For instance, under a condition called the *Running Intersection Property*, Theorem 2.27 of Chapter 2 is an example on how a sparsity pattern in the data $(f, \mathbb{K})$ can be exploited to yield a specific "sparse" representation of f, positive on $\mathbb{K}$. In this section we also obtain a sparse version of Algorithm 4.1 when the moment problem (4.2) exhibits some sparsity pattern in its data $\{f, (h_j)_{j\in\Gamma}, \mathbb{K}\}$.

As we have seen in Section 2.7, suppose that there is no *coupling* between some subsets of variables in the polynomials g_k that define the set $\mathbb{K}$, and the polynomials f and h_j, $j \in \Gamma$. By no coupling between two sets of variables we mean that there is *no* monomial involving some variables of the two sets in *any* of the polynomials f, g_k or h_j.

In other words, with $\mathbb{R}[\mathbf{x}] = \mathbb{R}[x_1, \ldots, x_n]$, let $I_0 := \{1, \ldots, n\}$ be the union (with possible overlaps) $\cup_{k=1}^p I_k$ of p subsets I_k, $k = 1, \ldots, p$, with cardinal denoted n_k. For an arbitrary $V \subseteq I_0$, let $\mathbb{R}[\mathbf{x}(V)]$ denote the ring of polynomials in the variables $\mathbf{x}(V) = \{x_i \,:\, i \in V\}$, and so $\mathbb{R}[\mathbf{x}(I_0)] = \mathbb{R}[\mathbf{x}]$. Let Assumptions 2.3 and 2.4 hold, and also assume that for every $j \in \Gamma$,

$$h_j = \sum_{k=1}^{p} h_{jk}, \qquad h_{jk} \in \mathbb{R}[\mathbf{x}(I_k)]. \tag{4.29}$$

So as in Section 2.7, let $m' = m + p$ and $\{1, \ldots, m'\} = \cup_{i=1}^p J_i$ and $\mathbb{K}$ be as in (2.30) after having added the p redundant quadratic constraints (2.29).

For all $i = 1, \ldots, p$, let

$$\mathbb{K}_i := \{\mathbf{x} \in \mathbb{R}^{n_i} \,:\, g_k(\mathbf{x}) \geq 0, \quad k \in J_i\,\}$$

so that the set $\mathbb{K} = \{\mathbf{x} \in \mathbb{R}^n \,:\, g_j(\mathbf{x}) \geq 0, \quad \forall j \in \cup_{i=1}^p J_i\,\}$ also reads

$$\mathbb{K} = \{\,\mathbf{x} \in \mathbb{R}^n \,:\, \mathbf{x}(I_i) \in \mathbb{K}_i, \quad i = 1, \ldots, p\,\}.$$

Let $\Delta = \{1, \ldots, p\}$, and let $I_{jk} := I_j \cap I_k$ with cardinal n_{jk} whenever $I_{jk} \neq \emptyset$. For every $j \in \Delta$, if $I_j \cap I_k \neq \emptyset$, then denote by $\pi_{jk}\mu_j$ the measure on $\mathbb{R}^{n_{jk}}$ which is the projection of μ_j on the variables $\mathbf{x}(I_{jk})$. Define:

$$\Gamma_j := \{ k \,:\, k > j \quad \text{and} \quad I_j \cap I_k \neq \emptyset\} \quad j \in \Delta,\, j < p,$$

and consider the multi-measures generalized moment problem:

$$\begin{aligned} \rho_{\text{mom}}^{\text{sparse}} = \quad & \sup_{\mu_i \in \mathscr{M}(\mathbb{K}_i)_+} \sum_{i \in \Delta} \int f_i \, d\mu_i \\ & \text{s.t.} \sum_{i \in \Delta} h_{ki} \, d\mu_i \leqq \gamma_k, \quad \forall k \in \Gamma \\ & \pi_{jk} \, \mu_j = \pi_{kj} \, \mu_k, \quad \forall k \in \Gamma_j,\, 1 \leq j < p. \end{aligned} \tag{4.30}$$

Notice that in view of Assumptions 2.3 and 2.4, every $\mathbb{K}_i$ is compact, and so, the constraint $\pi_{jk}\mu_j = \pi_{kj}\mu_k$ reduces to the *countably many* moment constraints

$$\int_{\mathbb{K}_j} \mathbf{x}(I_{jk})^{\boldsymbol{\alpha}} \, d\mu_j - \int_{\mathbb{K}_k} \mathbf{x}(I_{jk})^{\boldsymbol{\alpha}} \, d\mu_k = 0, \quad \forall \boldsymbol{\alpha} \in \mathbb{N}^{n_{jk}}. \tag{4.31}$$

Therefore, the moment problem (4.30) is an instance of the multi-measures moment problem (4.24) with the additional feature that it has countably many moment constraints instead of finitely many. Its dual reads:

$$\begin{aligned} \rho_{\text{pop}}^{\text{sparse}} = \quad & \inf_{\lambda, u_{\boldsymbol{\alpha}}^{jk}} \sum_{j \in \Delta} \lambda_j \gamma_j \\ & \text{s.t.} \quad \lambda_k \geq 0, \, \forall k \in \Gamma_+ \\ & \sum_{k \in \Gamma} \lambda_k \, h_{ki} + \sum_{k \in \Gamma_i} \sum_{\boldsymbol{\alpha} \in \mathbb{N}^{n_{ik}}} u_{\boldsymbol{\alpha}}^{ik} \, \mathbf{x}(I_{ik})^{\boldsymbol{\alpha}} \\ & - \sum_{l : i \in \Gamma_l} \sum_{\boldsymbol{\alpha} \in \mathbb{N}^{n_{il}}} u_{\boldsymbol{\alpha}}^{li} \, \mathbf{x}(I_{il})^{\boldsymbol{\alpha}} - f_i \geq 0 \quad \text{on } \mathbb{K}_i, \quad i \in \Delta. \end{aligned} \tag{4.32}$$

Recall from Chapter 2 that the *running intersection property* holds if the collection of sets $\{I_k\}$ satisfies the following condition:

$$\text{For } k = 1, \ldots, p-1, \qquad \left(I_{k+1} \cap \bigcup_{j=1}^{k} I_j \right) \subseteq I_s \quad \text{for some } s \leq k. \tag{4.33}$$

Theorem 4.6. *Let Assumptions 2.3 and 2.4 hold as well as (4.29), and suppose that for some $k \in \Gamma$, $h_{ki} > 0$ on $\mathbb{K}_i$ for every $i = 1, \ldots, p$. Consider the multi-measures moment problem (4.30). Then $\rho_{\text{mom}}^{\text{sparse}} \geq \rho_{\text{mom}}$ and (4.30) has an optimal solution, that is, the* sup *is attained. Moreover, (4.30) has an optimal solution $(\mu_i)_{i\in\Delta}$ such that each μ_i is supported on finitely many points of $\mathbb{K}_i$, i.e., μ_i a finite linear combination of Dirac measures, $i \in \Delta$.*

In addition, if the running intersection property (4.33) holds then $\rho_{\text{mom}}^{\text{sparse}} = \rho_{\text{mom}}$.

Proof. By Theorem 1.3 the moment problem (4.2) is solvable with optimal value ρ_{mom}. Let μ be an optimal solution and for every $i \in \Delta$, let $\mu_i := \pi_i \mu$ be the marginal of μ on $\mathbb{R}^{n_i}$ (projection on the variables $\mathbf{x}(I_i)$). Then the family $(\mu_i)_{i\in\Delta}$ is feasible for the moment problem (4.30) with value

$$\sum_{i\in\Delta} \int_{\mathbb{K}_i} f_i \, d\mu_i = \int_{\mathbb{K}} f \, d\mu = \rho_{\text{mom}},$$

which shows that $\rho_{\text{mom}}^{\text{sparse}} \geq \rho_{\text{mom}}$. We next prove that (4.30) is solvable. As it has a feasible solution, let $(\mu_i^n)_{i\in\Delta}$ be a maximizing sequence of (4.30). Let $k \in \Gamma$ be such that $h_{ki} > 0$ on $\mathbb{K}_i$, for every $i = 1, \ldots, p$. Then $h_{ki} \geq \delta_i$ on $\mathbb{K}_i$ for some $\delta_i > 0$, and so $\sum_{i\in\Delta} \delta_i \, \mu_i(\mathbb{K}_i) \leq \gamma_k$, which shows that all feasible solutions $(\mu_i)_{i\in\Delta}$ are uniformly bounded. Therefore, as every $\mathbb{K}_i$ is compact, and the measures μ_i^n are bounded uniformly in n, there is a subsequence (n_k) and a measure μ_i^* on $\mathbb{K}_i$, such that for every $i = 1, \ldots, p$, $\mu_i^{n_k} \Rightarrow \mu_i^*$ as $k \to \infty$.[1] As $\mathbb{K}_i$ is compact, for every $j \in \Gamma$,

$$\gamma_j = \sum_{i=1}^{p} \int_{\mathbb{K}_i} h_{ji} \, d\mu_i^{n_k} \to \sum_{i=1}^{p} \int_{\mathbb{K}_i} h_{ji} \, d\mu_i^* \quad \text{as } k \to \infty.$$

Similarly, $\pi_{jl} \mu_j^{n_k} \to \pi_{jl} \mu_j^*$ as $k \to \infty$.[2] Hence, $\pi_{jl} \mu_j^* = \pi_{lj} \mu_l^*$ for all (j, l) with

[1]Recall that $\mu_i^{n_k} \Rightarrow \mu_i^*$ is the weak convergence of measures, i.e., $\int_{\mathbb{K}_i} g \, d\mu_i^{n_k} \to \int_{\mathbb{K}_i} g \, d\mu_i^*$ as $k \to \infty$, for all functions g continuous on $\mathbb{K}_i$. See Section B.1

[2]This is because for $g \in \mathbb{R}[\mathbf{x}(I_{jl})]$,

$$\int_{\mathbb{K}_j \cap \mathbb{K}_l} g \, d(\pi_{jl} \mu_j^{n_k}) = \int_{\mathbb{K}_j} \pi_{jl}^* g \, d\mu_j^{n_k} \to \int_{\mathbb{K}_j} \pi_{jl}^* g \, d\mu_j^* = \int_{\mathbb{K}_j} g \, d(\pi_{jl} \mu_j^*),$$

where $\pi_{jl}^* : \mathbb{R}[\mathbf{x}(I_{jl})] \to \mathbb{R}[\mathbf{x}(I_j)]$ is given by $(\pi_{jl}^* g)(\mathbf{x}(I_j)) = g(\mathbf{x}(I_{jl}))$ for all $\mathbf{x}(I_j) \in \mathbb{R}^{n_j}$.

$I_j \cap I_l \neq \emptyset$. And so, $(\mu_i^*)_{i\in\Delta}$ is a feasible solution for (4.30) with value

$$\sum_{i=1}^{p} \int_{\mathbb{K}_i} f_i \, d\mu_i^* \,=\, \lim_{k\to\infty} \sum_{i=1}^{p} \int_{\mathbb{K}_i} f_i \, d\mu_i^{n_k} \,=\, \rho_{\text{mom}}^{\text{sparse}},$$

the desired result.

Finally, to prove that $\rho_{\text{mom}}^{\text{sparse}} = \rho_{\text{mom}}$ if (2.32) holds, and in view of $\rho_{\text{mom}}^{\text{sparse}} \geq \rho_{\text{mom}}$, it suffices to prove that given an optimal solution $(\mu_i)_{i\in\Delta}$ of (4.30), one may construct a measure μ feasible for (4.2), with same value as $\rho_{\text{mom}}^{\text{sparse}}$. By Lemma B.13, there exists a measure μ on $\mathbb{K}$ such that its projection $\pi_k\mu$ on $\mathbb{R}^{n_k}$ (i.e., on the variables $\mathbf{x}(I_k)$) is the measure μ_k, $k \in \Delta$. Therefore, as $h_{ki} \in \mathbb{R}[\mathbf{x}(I_i)]$ for every $k \in \Gamma, i \in \Delta$, we obtain

$$\int_{\mathbb{K}} h_k \, d\mu \,=\, \sum_{i\in\Delta} \int_{\mathbb{K}} h_{ki} \, d\mu \,=\, \sum_{i\in\Delta} \int_{\mathbb{K}_i} h_{ki} \, d\mu_i \,\leqq\, \gamma_k, \quad \forall\, k \in \Gamma,$$

which proves that μ is feasible for the moment problem (4.2). Similarly, as $f = \sum_i f_i$ with $f_i \in \mathbb{R}[\mathbf{x}(I_i)]$, we obtain

$$\rho_{\text{mom}}^{\text{sparse}} = \sum_{i=1}^{p} \int_{\mathbb{K}_i} f_i \, d\mu_i \,=\, \sum_{i=1}^{p} \int_{\mathbb{K}} f_i \, d\mu \,=\, \int_{\mathbb{K}} f \, d\mu,$$

and so, μ is an optimal solution of (4.2) with value $\rho_{\text{mom}} = \rho_{\text{mom}}^{\text{sparse}}$. □

4.6.1 *Sparse semidefinite relaxations*

Recall that the moment and localizing matrices $\mathbf{M}_i(\mathbf{y})$ and $\mathbf{M}_i(g_j\mathbf{y})$ used in the semidefinite relaxations (4.5) are indexed in the canonical basis of $\mathbb{R}[\mathbf{x}]$. For every $\boldsymbol{\alpha} \in \mathbb{N}^n$, let $\text{supp}\,(\alpha) \in I$ be the support of α, i.e.,

$$\text{supp}\,(\boldsymbol{\alpha}) \,:=\, \{\, i \in \{1,\ldots,n\} \,:\, \quad \alpha_i \neq 0\,\}, \qquad \alpha \in \mathbb{N}^n.$$

For instance, with $n = 6$ and $\boldsymbol{\alpha} := (004020)$, $\text{supp}\,(\alpha) = \{3,5\}$. For each $k = 1,\ldots,p$, let $\mathscr{I}_k$ be the set of all subsets of I_k. Next, define

$$S_k := \{\, \boldsymbol{\alpha} \in \mathbb{N}^n \,:\, \quad \text{supp}\,(\boldsymbol{\alpha}) \,\in\, \mathscr{I}_k\,\}, \qquad k = 1,\ldots,p. \tag{4.34}$$

A polynomial $h \in \mathbb{R}[\mathbf{x}(I_k)]$ can be viewed as a member of $\mathbb{R}[\mathbf{x}]$, and is written

$$h(\mathbf{x}) \,=\, h(\mathbf{x}(I_k)) \,=\, \sum_{\alpha\in S_k} h_\alpha \, \mathbf{x}^{\boldsymbol{\alpha}}. \tag{4.35}$$

Instead of defining moments variables $\mathbf{y}^i = (y^i_{\boldsymbol{\alpha}})$ (with $\boldsymbol{\alpha} \in \mathbb{N}^{n_i}$) for each measure μ_i, $i = 1, \ldots, p$, and relating them via the equality constraints (4.31), we find better to use a single $\mathbf{y} = (y_{\boldsymbol{\alpha}})$ with $\boldsymbol{\alpha} \in \mathbb{N}^n$ but use only those $y_{\boldsymbol{\alpha}}$ with $\mathrm{supp}(\boldsymbol{\alpha}) \in \cup_{i=1}^p \mathscr{I}_k$.

With $k \in \{1, \ldots, p\}$ fixed, and $g \in \mathbb{R}[\mathbf{x}(I_k)]$, let $\mathbf{M}_i(\mathbf{y}, I_k)$ (resp. $\mathbf{M}_i(g\,\mathbf{y}, I_k)$) be the moment (resp. localizing) submatrix obtained from $\mathbf{M}_i(\mathbf{y})$ (resp. $\mathbf{M}_i(g\,\mathbf{y})$) by retaining only those rows (and columns) $\alpha \in \mathbb{N}^n$ of $\mathbf{M}_i(\mathbf{y})$ (resp. $\mathbf{M}_i(g\,\mathbf{y})$) such that $\mathrm{supp}(\boldsymbol{\alpha}) \in \mathscr{I}_k$.

In doing so, $\mathbf{M}_i(\mathbf{y}, I_k)$ and $\mathbf{M}_i(g\,\mathbf{y}, I_k)$ can be viewed as moment and localizing matrices with rows and columns indexed in the canonical basis of $\mathbb{R}[\mathbf{x}(I_k)]_i$. Indeed, $\mathbf{M}_i(\mathbf{y}, I_k)$ contain only variables y_α with $\mathrm{supp}\,(\alpha) \in \mathscr{I}_k$, and so does $\mathbf{M}_i(g\,\mathbf{y}, I_k)$ because $g \in \mathbb{R}[\mathbf{x}(I_k)]$. And for every polynomial $u \in \mathbb{R}[\mathbf{x}(I_k)]_i$, with coefficient vector $\mathbf{u}$ in the canonical basis, we also have

$$
\begin{aligned}
\langle \mathbf{u}, \mathbf{M}_i(\mathbf{y}, I_k)\mathbf{u}\rangle &= L_{\mathbf{y}}(u^2), \qquad &\forall\, u \in \mathbb{R}[\mathbf{x}(I_k)]_i \\
\langle \mathbf{u}, \mathbf{M}_i(g\,\mathbf{y}, I_k)\mathbf{u}\rangle &= L_{\mathbf{y}}(g\,u^2), \qquad &\forall\, u \in \mathbb{R}[\mathbf{x}(I_k)]_i,
\end{aligned}
$$

and therefore,

$$
\begin{aligned}
\mathbf{M}_i(\mathbf{y}, I_k) \succeq 0 &\Leftrightarrow L_{\mathbf{y}}(u^2) \geq 0, \qquad &\forall\, u \in \mathbb{R}[\mathbf{x}(I_k)]_i \\
\mathbf{M}_i(g\,\mathbf{y}, I_k) \succeq 0 &\Leftrightarrow L_{\mathbf{y}}(g\,u^2) \geq 0, \qquad &\forall\, u \in \mathbb{R}[\mathbf{x}(I_k)]_i.
\end{aligned}
$$

The sparse analogue of the semidefinite relaxations (4.5) read:

$$
\begin{aligned}
\rho_i^{\text{sparse}} = \ & \sup_{\mathbf{y}}\ L_{\mathbf{y}}(f) \\
& \text{s.t. } L_{\mathbf{y}}(h_j) \leqq \gamma_j, \qquad \forall\, j \in \Gamma \\
& \quad\ \ \mathbf{M}_i(\mathbf{y}, I_k) \succeq \mathbf{0}, \qquad k = 1, \ldots, p \\
& \quad\ \ \mathbf{M}_{i-v_j}(g_j\mathbf{y}, I_k) \succeq \mathbf{0},\ \forall\, j \in J_k,\ k = 1, \ldots, p.
\end{aligned}
\tag{4.36}
$$

The dual of (4.36) is the semidefinite program:

$$
\begin{aligned}
(\rho_i^{\text{sparse}})^* = \; & \inf_{\lambda, \sigma_{kj}} \sum_{j \in \Gamma} \lambda_j \gamma_j \\
\text{s.t.} \; & \sum_{j \in \Gamma} \lambda_j h_j - \sum_{k=1}^{p} \left(\sigma_{k0} + \sum_{l \in J_k} \sigma_{kl}\, g_l \right) - f = 0 \\
& \lambda_j \geq 0, \; j \in \Gamma_+ \\
& \sigma_{k0}, \sigma_{kl} \in \Sigma[\mathbf{x}(I_k)], \quad l \in J_k, \; k = 1, \ldots, p \\
& \deg \sigma_{k0}, \deg \sigma_{kl} g_l \leq 2i, \quad l \in J_k, \; k = 1, \ldots, p.
\end{aligned}
\tag{4.37}
$$

Theorem 4.7. *Assume that the moment problem (4.2) has optimal value $\rho_{\text{mom}} > -\infty$, let Assumptions 2.3 and 2.4 hold as well as (4.29), and suppose that for some $q \in \Gamma$, $h_{qi} > 0$ on $\mathbb{K}_i$ for every $i = 1, \ldots, p$. Consider the sparse semidefinite relaxations (4.36) and (4.37).*

If the running intersection property (4.33) holds then $\rho_i^{\text{sparse}} \downarrow \rho_{\text{mom}}$ as $i \to \infty$.

The proof of Theorem 4.7 being tedious, it is postponed to Section 4.10.

4.6.2 *Computational complexity*

The number of variables for the sparse semidefinite relaxation (4.36) (with optimal value ρ_i^{sparse}) is bounded by $\sum_{k=1}^{p} \binom{n_k+2i}{n_k}$, and so, if all n_k's are *close* to each other, say $n_k \approx n/p$ for all k, then one has at most $O(p(\frac{n}{p})^{2i})$ variables, a big saving when compared with $O(n^{2i})$ in the original semidefinite relaxation (4.5).

In addition, one also has p moment matrices of size $O((\frac{n}{p})^i)$ and $m + p$ localizing matrices of size $O((\frac{n}{p})^{i-v})$ (where $v = \max_j \lceil \deg g_j / 2 \rceil$) to be compared with a single moment matrix of size $O(n^i)$ and m localizing matrices of size at most $O(n^{i-v})$ in (4.5).

When using an interior point method to solve (4.36), it is definitely better to handle p matrices, each of size $(n/p)^i$, rather than a single one of size n^i.

4.7 Summary

We have presented two general numerical approximation schemes for computing (or approximating) the optimal value ρ_{mom} of the generalized moment problem (4.2) with polynomial data. An important observation is that convexity of the underlying sets $\mathbb{K}$ does not influence the approach. The underlying (linear or semidefinite) relaxations are convex for arbitrary basic semi-algebraic sets $\mathbb{K}$. In fact, the set $\mathbb{K}$ might have a very complicated geometry. This observation underlies the generality of the approach. Of course, the price to pay is that the size of the relaxations might be very large.

The first numerical scheme consists of solving a hierarchy of semidefinite relaxations of increasing size, where the increase in size is due to allowing larger and larger (polynomial) degrees in some s.o.s. representation of polynomials, positive on $\mathbb{K}$. Thanks to some powerful Positivstellensatz, convergence of the sequence of optimal values to the exact optimal value ρ_{mom} is guaranteed. In addition, if some rank condition holds at an optimal solution of some semidefinite relaxation in the hierarchy, finite convergence takes place and one may extract finitely many points which are the support of a measure, optimal solution of the generalized moment problem.

The second numerical scheme consists of solving an analogous hierarchy of linear relaxations of increasing size. Again, the increase in size is due to allowing larger and larger (polynomial) degrees in some other representation of polynomials, positive on $\mathbb{K}$. Thanks to some other *ad hoc* Positivstellensatz, convergence of the sequence of optimal values to the exact optimal value ρ_{mom} is also guaranteed.

Importantly, the above numerical scheme can also be applied for variations of the generalized moment problem (4.2), provided the problem is described with polynomials, or to some extent, piecewise polynomials. Convergence is proved here for the special (but still relatively general) cases considered in this section, but sometimes, convergence in more complicated moment problems may also be proved using tailored arguments, depending on the problem on hand.

The methodology we outlined leads to semidefinite and linear relaxations whose solution approximates (and sometimes computes exactly) the optimal value ρ_{mom} of problem (4.2). We have also considered variants the moment problem (4.2), namely:

- (4.2) with countably many moment constraints.

- (4.24) with several measures on (possibly) different spaces (as in e.g. Exercise 4.3), with their support contained on basic semi-algebraic sets, explicitly defined by polynomial inequalities.

For those variants, we have proved convergence of the corresponding semidefinite relaxations.

Finally, we have also considered sparse semidefinite relaxations to treat the moment problem (4.2) when its data $(f, \mathbb{K}, h_j)$ exhibit a certain sparsity pattern (often present in many large scale problems). When the sparsity patterns satisfies some condition (well-known in graph theory), then convergence is also proved. This opens up the door to the possibility of solving moment problems of large dimensions.

Of course, one may also think of several other variants of (4.2) for which the methodology we outlined can be applied but with no general convergence result available. However, the semidefinite relaxations provide tighter and tighter upper bounds on the optimal value of these variants of (4.2).

4.8 Exercises

Exercise 4.1. Suppose $\mathbb{K} = \mathbb{R}^n$. Provide an example such that the sequence of relaxations in Theorem 4.1 does not converge to ρ_{mom}.

Exercise 4.2. Prove that Theorem 4.4 can be extended to handle countably many moment constraints, i.e., if the set Γ is countable instead of being finite.

Exercise 4.3. Let Γ be a finite set of indices, $f_1 \in \mathbb{R}[\mathbf{x}]$, $f_2 \in \mathbb{R}[z]$. Let $(h_{1j})_{j\in\Gamma} \subset \mathbb{R}[\mathbf{x}]$ and $(h_{2j})_{j\in\Gamma} \subset \mathbb{R}[z]$, and let $\mathbb{K}_1 \subset \mathbb{R}^n$, $\mathbb{K}_2 \subset \mathbb{R}$ be basic semi-algebraic sets. Consider the problem

$$\begin{aligned} \rho_{\text{mom}} = \sup_{\mu_1 \in \mathscr{M}(\mathbb{K}_1), \mu_2 \in \mathscr{M}(\mathbb{K}_2)} & \int_{\mathbb{K}_1} f_1 d\mu_1 + \int_{\mathbb{K}_2} f_2 d\mu_2 \\ \text{s.t.} & \int_{\mathbb{K}_1} h_{1j} d\mu_1 + \int_{\mathbb{K}_2} h_{2j} d\mu_2 = \gamma_j, \quad j \in \Gamma \end{aligned} \tag{4.38}$$

involving two measures μ_1 and μ_2. Let $\mathbb{K}_2 \subset \mathbb{R}$ be the interval $[0, +\infty)$. Let Assumption 2.1 hold for $\mathbb{K}_1$ and $h_{ki} > 0$, $i = 1, 2$ for some $k \in \Gamma$.
(a) Can we extend Theorem 4.4?
(b) Write the semidefinite relaxations (4.26) for Problem (4.38).

(c) Can we prove that the sequence of relaxations converges to the value ρ_{mom}?

Exercise 4.4. Assume that Theorem 1.3 holds. Give a simple proof of Theorem 4.7. (*Hint:* Use Theorem 2.28.)

4.9 Notes and Sources

The approach in this chapter was outlined in Lasserre (2008b). Theorem 4.7 was proved in Lasserre (2006a) for polynomial optimization problems with sparsity, i.e., when $\Gamma = \{0\}$ and $h_0 = 1$.

4.10 Proofs

4.10.1 *Proof of Theorem 4.3*

Call $\mathbf{Q}_i$ the semidefinite program (4.22). We first prove that $\mathbf{Q}_i$ has a feasible solution. Let μ be a feasible solution of the moment problem (4.2), and let $\mathbf{y} = (y_{\boldsymbol{\alpha}})$ be the sequence of moments of μ (well-defined because μ is supported on $\mathbb{K}$). Recalling the definition of $\mathbf{M}_i(\mathbf{y})$ and $\mathbf{M}_{i-v_j}(g_j\,\mathbf{y})$, one has $\mathbf{M}_i(\mathbf{y}) \succeq 0$ and $\mathbf{M}_{i-v_j}(g_j\,\mathbf{y}) \succeq 0$. In addition,

$$L_{\mathbf{y}}(h_k) = \int_{\mathbb{K}} h_k d\mu \leqq \gamma_k, \qquad \forall k \in \Gamma,$$

which proves that $\mathbf{y}$ is a feasible solution of $\mathbf{Q}_i$. Therefore $\rho_{\text{mom}} \leq \rho_i$ for every i. We next prove that $\rho_i < \infty$ for all sufficiently large i.

Let $2i_0 \geq \max[\deg h_0, \deg f, \max_j v_j]$, and let $\mathbb{N}^n_i := \{\boldsymbol{\alpha} \in \mathbb{N}^n : |\boldsymbol{\alpha}| \leq i\}$, $i \in \mathbb{N}$. As $\mathbb{K}$ is compact, there exists N such that $N \pm \mathbf{x}^{\boldsymbol{\alpha}} > 0$ on $\mathbb{K}$ for all $\boldsymbol{\alpha} \in \mathbb{N}^n_{i_0}$, and in view of Assumption 2.1, the polynomial $\mathbf{x} \mapsto N \pm \mathbf{x}^{\boldsymbol{\alpha}}$ belongs to the quadratic module $Q \subset \mathbb{R}[\mathbf{x}]$ generated by $\{g_j\} \subset \mathbb{R}[\mathbf{x}]$, i.e.,

$$Q := \left\{ \sigma_0 + \sum_{j=1}^{m} \sigma_j\, g_j \;:\; \sigma_j \in \Sigma[\mathbf{x}] \quad \forall j \in \{0,\ldots,m\} \right\}.$$

Similarly, as $h_0 > 0$ on $\mathbb{K}$ and $\mathbb{K}$ is compact, $h_0 - \delta > 0$ on $\mathbb{K}$ for some $\delta > 0$, and so $h_0 - \delta \in Q$.

But there is even some integer $l(i_0)$ such that $h_0 - \delta \in Q_{l(i_0)}$ and $N \pm x^{\boldsymbol{\alpha}} \in Q(l(i_0))$ for all $\boldsymbol{\alpha} \in \mathbb{N}_{i_0}$, where $Q(t) \subset Q$ is the set of elements of Q which have a representation $\sigma_0 + \sum_{j=1}^m \sigma_j\, g_j$ for some $\{\sigma_j\} \subset \Sigma[\mathbf{x}]$

with $\deg \sigma_0 \leq 2t$ and $\deg \sigma_j g_j \leq 2t$ for all $j = 1, \ldots, m$. Of course, we also have $h_0 - \delta \in Q(l)$ and $N \pm \mathbf{x}^{\boldsymbol{\alpha}} \in Q(l)$ for all $\boldsymbol{\alpha} \in \mathbb{N}^n_{i_0}$, whenever $l \geq l(i_0)$. Therefore, consider $i = l(i_0)$ $(\geq i_0)$.

For every feasible solution $\mathbf{y}$ of $\mathbf{Q}_{l(i_0)}$, the constraint $L_{\mathbf{y}}(h_0) \leqq \gamma_0$ yields that $y_0 \leq M := \gamma_0/\delta$. Indeed, as $h_0 - \delta \in Q(l(i_0))$,

$$\gamma_0 - \delta y_0 \geqq L_{\mathbf{y}}(h_0 - \delta) \,=\, L_{\mathbf{y}}(\sigma_0) + \sum_{j=1}^{m} L_{\mathbf{y}}(\sigma_j \, g_j)$$

for some $(\sigma_j) \subset \Sigma[\mathbf{x}]$. Therefore, the semidefinite constraints $M_{l(i_0)}(\mathbf{y}) \succeq 0$ and $M_{l(i_0)-v_j}(g_j \, \mathbf{y}) \succeq 0$ yield $L_{\mathbf{y}}(h_0 - \delta) \geq 0$. Similarly, as $N \pm \mathbf{x}^{\boldsymbol{\alpha}} \in Q(l(i_0))$,

$$N y_0 \pm L_{\mathbf{y}}(\mathbf{x}^{\boldsymbol{\alpha}}) \,=\, L_{\mathbf{y}}(N \pm \mathbf{x}^{\boldsymbol{\alpha}}) \,=\, L_{\mathbf{y}}(\sigma_0) + \sum_{j=1}^{m} L_{\mathbf{y}}(\sigma_j \, g_j)$$

for some $(\sigma_j) \subset \Sigma[\mathbf{x}]$. Hence, for same reasons, $L_{\mathbf{y}}(N \pm \mathbf{x}^{\boldsymbol{\alpha}}) \geq 0$, which yields

$$|\, L_{\mathbf{y}}(\mathbf{x}^{\boldsymbol{\alpha}}) \,| \leq NM, \qquad \boldsymbol{\alpha} \in \mathbb{N}^n_{i_0}.$$

Next, $L_{\mathbf{y}}(f) \leq NM \sum_{\boldsymbol{\alpha}} |f_{\boldsymbol{\alpha}}|$ because $2i_0 \geq \deg f$ and so $\rho_{l(i_0)} < +\infty$.

From what precedes, and with $s \in \mathbb{N}$ arbitrary, let $l(s) \geq s$ be such that

$$N_s \pm \mathbf{x}^{\boldsymbol{\alpha}} \in Q(l(s)) \qquad \forall \boldsymbol{\alpha} \in \mathbb{N}^n_s, \tag{4.39}$$

for some N_s. Next, let $r \geq l(i_0)$ (so that $\rho_r < +\infty$), and let $\mathbf{y}^r$ be a nearly optimal solution of $\mathbf{Q}_r$ with value

$$\rho_r \,\geq\, L_{\mathbf{y}^r}(f) \,\geq\, \rho_r - \frac{1}{r} \quad \left(\geq \rho_{\mathrm{mom}} - \frac{1}{r}\right). \tag{4.40}$$

Fix $s \in \mathbb{N}$. Notice that from (4.39), for all $r \geq l(s)$, one has

$$|\, L_{\mathbf{y}^r}(\mathbf{x}^{\boldsymbol{\alpha}}) \,| \,\leq\, N_s M, \qquad \forall \boldsymbol{\alpha} \in \mathbb{N}^n_s.$$

Therefore, for all $r \geq l(i_0)$,

$$|y^r_{\boldsymbol{\alpha}}| \,=\, |\, L_{\mathbf{y}^r}(\mathbf{x}^{\boldsymbol{\alpha}}) \,| \,\leq\, N'_s, \quad \forall \boldsymbol{\alpha} \in \mathbb{N}^n_s, \tag{4.41}$$

where $N'_s = \max[N_s M, V_s]$, with

$$V_s := \max\,\{|y^r_{\boldsymbol{\alpha}}| \,:\, \boldsymbol{\alpha} \in \mathbb{N}_s;\; l(i_0) \leq r \,<\, l(s)\}.$$

Complete each $\mathbf{y}^r$ with zeros to make it an infinite vector in l_∞, indexed in the canonical basis of $\mathbb{R}[\mathbf{x}]$. In view of (4.41), one has

$$|y^r_{\boldsymbol{\alpha}}| \leq N'_s, \qquad \forall\, \boldsymbol{\alpha} \in \mathbb{N}; \quad 2s-1 \leq |\boldsymbol{\alpha}| \leq 2s,$$

for all $s = 1, 2, \ldots$.

Hence, define the new sequence $\widehat{\mathbf{y}}^r \in l_\infty$ defined by $\widehat{y}^r_0 := y^r_0/M$, and

$$\widehat{y}^r_{\boldsymbol{\alpha}} := \frac{y^r_{\boldsymbol{\alpha}}}{N'_s} \qquad \forall\, \boldsymbol{\alpha} \in \mathbb{N}^n, \quad 2s-1 \leq |\boldsymbol{\alpha}| \leq 2s$$

for all $s = 1, 2, \ldots$, and in l_∞, consider the sequence $(\widehat{\mathbf{y}}^r)_r$ as $r \to \infty$. Obviously, the sequence $(\widehat{\mathbf{y}}^r)_r$ is in the unit ball B_1 of l_∞, and so, by the Banach–Alaoglu theorem (see, e.g., (Ash, 1972, Theorem. 3.5.16)), there exists $\widehat{\mathbf{y}} \in B_1$ and a subsequence $\{r_i\}$, such that $\widehat{\mathbf{y}}^{r_i} \to \widehat{\mathbf{y}}$ as $i \to \infty$ for the weak $\star$ topology $\sigma(l_\infty, l_1)$ of l_∞. In particular, pointwise convergence holds, that is,

$$\lim_{i\to\infty} \widehat{y}^{r_i}_{\boldsymbol{\alpha}} \to \widehat{y}_{\boldsymbol{\alpha}}, \qquad \forall\, \boldsymbol{\alpha} \in \mathbb{N}^n.$$

Next, define $y_0 := M\widehat{y}_0$ and

$$y_{\boldsymbol{\alpha}} := \widehat{y}_{\boldsymbol{\alpha}} \times N'_s, \quad 2s-1 \leq |\boldsymbol{\alpha}| \leq 2s, \quad s = 1, 2, \ldots.$$

The pointwise convergence $\widehat{\mathbf{y}}^{r_i} \to \widehat{\mathbf{y}}$ implies the pointwise convergence $\mathbf{y}^{r_i} \to \mathbf{y}$, i.e.,

$$\lim_{i\to\infty} y^{r_i}_{\boldsymbol{\alpha}} \to y_{\boldsymbol{\alpha}}, \qquad \forall\, \boldsymbol{\alpha} \in \mathbb{N}^n. \tag{4.42}$$

Let $s \in \mathbb{N}$ be fixed. From the pointwise convergence (4.42), $\lim_{i\to\infty} \mathbf{M}_s(\mathbf{y}^{r_i}) = \mathbf{M}_s(\mathbf{y}) \succeq 0$. Similarly, $\lim_{i\to\infty} \mathbf{M}_s(g_j\, \mathbf{y}^{r_i}) = \mathbf{M}_s(g_j\, \mathbf{y}) \succeq 0$, for all $j = 1, \ldots, m$. And as s was arbitrary, we obtain

$$\mathbf{M}_r(\mathbf{y}) \succeq 0; \quad \mathbf{M}_r(g_j\, \mathbf{y}) \succeq 0, \quad j = 1, \ldots, m; \quad r = 0, 1, 2, \ldots,$$

which, by Theorem 3.8(b), implies that $\mathbf{y}$ has a finite Borel representing measure μ with support contained in $\mathbb{K}$. In addition, fix $k \in \Gamma$ arbitrary. The pointwise convergence (4.42) yields $\lim_{i\to\infty} L_{\mathbf{y}^{r_i}}(h_k) = L_{\mathbf{y}}(h_k)$, and as $k \in \Gamma_r$ for all sufficiently large r, we obtain $L_{\mathbf{y}}(h_k) \leqq \gamma_k$. And so, as $k \in \Gamma$ was arbitrary,

$$\int_{\mathbb{K}} h_k \, d\mu = L_{\mathbf{y}}(h_k) = \lim_{i\to\infty} L_{\mathbf{y}^{r_i}}(h_k) \leqq \gamma_k, \qquad \forall k \in \Gamma,$$

which proves that μ is a feasible solution of the moment problem (4.2).

On the other hand, (4.40) and (4.42) also yield

$$\rho_{\text{mom}} \leq \lim_{i\to\infty} \rho_{r_i} = \lim_{i\to\infty} L_{\mathbf{y}^{r_i}}(f) = L_{\mathbf{y}}(f) = \int_{\mathbb{K}} f\, d\mu,$$

and so μ is an optimal solution of (4.2).

4.10.2 *Proof of Theorem 4.7*

With $\rho_{\text{mom}}^{\text{sparse}}$ being the optimal value of (4.30), it is sufficient to prove that $\rho_i^{\text{sparse}} \downarrow \rho_{\text{mom}}^{\text{sparse}}$ because by Theorem 4.6 we have $\rho_{\text{mom}} = \rho_{\text{mom}}^{\text{sparse}}$. Define the sets:

$$\begin{aligned}
\Theta_{ki} &:= \{\, \boldsymbol{\alpha} \in \mathbb{N}^n \;:\; \text{supp}\,(\boldsymbol{\alpha}) \in \mathscr{I}_k; \quad |\boldsymbol{\alpha}| \leq 2i \,\}, \quad k = 1, \ldots, p, \\
\Theta_i &:= \bigcup_{k=1}^{p} \Theta_{ki} = \left\{ \boldsymbol{\alpha} \in \mathbb{N}^n \;:\; \text{supp}\,(\boldsymbol{\alpha}) \in \bigcup_{k=1}^{p} \mathscr{I}_k; \quad |\boldsymbol{\alpha}| \leq 2r \right\}, \\
\Theta &:= \bigcup_{i\in\mathbb{N}} \Theta_i = \left\{ \boldsymbol{\alpha} \in \mathbb{N}^n \;:\; \text{supp}\,(\boldsymbol{\alpha}) \in \bigcup_{k=1}^{p} \mathscr{I}_k \right\},
\end{aligned}$$

and call $\mathbf{Q}_i$ the semidefinite program (4.36).

We first prove that $\mathbf{Q}_i$ has a feasible solution. Let ν be a feasible solution of the moment problem (4.2), and let

$$y_{\boldsymbol{\alpha}} = \int \mathbf{x}^{\boldsymbol{\alpha}}\, d\nu \qquad \forall\, \boldsymbol{\alpha} \in \Theta_i.$$

Recalling the definition of $\mathbf{M}_i(\mathbf{y}, I_k)$ and $\mathbf{M}_{i-v_j}(g_j\, \mathbf{y}, I_k)$, one has $\mathbf{M}_i(\mathbf{y}, I_k) \succeq 0$ and $\mathbf{M}_{i-v_j}(g_j\, \mathbf{y}, I_k) \succeq 0$; therefore, $\mathbf{y}$ is an obvious feasible solution of $\mathbf{Q}_i$. Next we prove that $\rho_i^{\text{sparse}} < +\infty$ for all sufficiently large i.

Let $2i_0 \geq \max[\deg f, \max_j \deg g_j, \max_k \deg h_k]$. In view of Assumption 2.3 and as $h_{qk} > 0$ on $\mathbb{K}_k$ for every $k = 1, \ldots, p$, $h_{qk} - \delta_k > 0$ on $\mathbb{K}_k$ for some $\delta_k > 0$, $k = 1, \ldots, p$. Next, again by compactness of $\mathbb{K}_k$, there exists N such that $N \pm \mathbf{x}^{\boldsymbol{\alpha}} > 0$ on $\mathbb{K}_k$ for all $\boldsymbol{\alpha} \in \Theta_{ki_0}$, and all $k = 1, \ldots, p$. Therefore, for every $k = 1, \ldots, p$ and $\boldsymbol{\alpha} \in \Theta_{ki_0}$, the polynomials $\mathbf{x} \mapsto h_{qk}(\mathbf{x}) - \delta_k$ and $\mathbf{x} \mapsto N \pm \mathbf{x}^{\boldsymbol{\alpha}}$ belong to the quadratic module $Q_k \subset \mathbb{R}[\mathbf{x}(I_k)]$ generated by $\{g_j\}_{j\in J_k} \subset \mathbb{R}[\mathbf{x}(I_k)]$, i.e.,

$$Q_k := \left\{ \sigma_0 + \sum_{j\in J_k} \sigma_j\, g_j \;:\quad \sigma_j \in \Sigma[\mathbf{x}(I_k)] \quad \forall j \in \{0\} \cup J_k \right\}.$$

But there is even some $l(i_0)$ such that $h_{qk} - \delta_k$ and $N \pm \mathbf{x}^{\boldsymbol{\alpha}}$ belong to $Q_k(l(i_0))$ for all $\boldsymbol{\alpha} \in \Theta_{ki_0}$ and $k = 1, \ldots, p$, where $Q_k(t) \subset Q_k$ is the set of elements of Q_k which have a representation $\sigma_0 + \sum_{j \in J_k} \sigma_j \, g_j$ for some $\{\sigma_j\} \subset \Sigma[\mathbf{x}(I_k)]$ with $\deg \sigma_0 \leq 2t$ and $\deg \sigma_j g_j \leq 2t$ for all $j \in J_k$. Of course we also have $h_{qk} - \delta_k \in Q_k(l)$ and $N \pm \mathbf{x}^{\boldsymbol{\alpha}} \in Q_k(l)$ for all $\boldsymbol{\alpha} \in \Theta_{ki_0}$, whenever $l \geq l(i_0)$. Therefore, consider $i = l(i_0)$ $(\geq i_0)$.

For every feasible solution $\mathbf{y}$ of $\mathbf{Q}_{l(i_0)}$,

$$\gamma_q - y_0 \sum_{k=1}^{p} \delta_k \;\geq\; \sum_{k=1}^{p} L_{\mathbf{y}}(h_{qk} - \delta_k) \geq 0.$$

Indeed, for every $k = 1, \ldots, p$,

$$L_{\mathbf{y}}(h_{qk} - \delta_k) \;=\; L_{\mathbf{y}}(\sigma_0) + \sum_{j \in J_k} L_{\mathbf{y}}(\sigma_j \, g_j)$$

for some $(\sigma_j) \subset \Sigma[\mathbf{x}(I_k)]$, and so the semidefinite constraints $\mathbf{M}_{l(i_0)}(\mathbf{y}, I_k) \succeq 0$ and $\mathbf{M}_{l(i_0)-v_j}(g_j \, \mathbf{y}, I_k) \succeq 0$, $j \in J_k$, yield $L_{\mathbf{y}}(h_{qk} - \delta_k) \geq 0$, $k = 1, \ldots, p$. Therefore $y_0 \leq M := \gamma_q (\sum_k \delta_k)^{-1}$. Similarly,

$$N y_0 \pm L_{\mathbf{y}}(\mathbf{x}^{\boldsymbol{\alpha}}) \;=\; L_{\mathbf{y}}(N \pm \mathbf{x}^{\boldsymbol{\alpha}}) \;=\; L_{\mathbf{y}}(\sigma_0) + \sum_{j \in J_k} L_{\mathbf{y}}(\sigma_j \, g_j)$$

for some $(\sigma_j) \subset \Sigma[\mathbf{x}(I_k)]$. And so for same reasons, $L_{\mathbf{y}}(N \pm \mathbf{x}^{\boldsymbol{\alpha}}) \geq 0$ which in turn implies

$$|\, L_{\mathbf{y}}(\mathbf{x}^{\boldsymbol{\alpha}}) \,| \leq NM, \qquad \boldsymbol{\alpha} \in \Theta_{ki_0}; \quad k = 1, \ldots, p.$$

As $2i_0 \geq \deg f$, it follows that $L_{\mathbf{y}}(f) \leq NM \sum_{\boldsymbol{\alpha}} |f_{\boldsymbol{\alpha}}|$. This is because by Assumption 2.4, $f_{\boldsymbol{\alpha}} \neq 0 \Rightarrow \boldsymbol{\alpha} \in \Theta_{i_0}$. Hence $\rho_{l(i_0)}^{\text{sparse}} < +\infty$.

From what precedes, and with $s \in \mathbb{N}$ arbitrary, let $l(s) \geq s$ be such that

$$N_s \pm \mathbf{x}^{\boldsymbol{\alpha}} \in Q_k(l(s)) \qquad \forall \boldsymbol{\alpha} \in \Theta_{ks}; \quad k = 1, \ldots, p, \tag{4.43}$$

for some N_s. Next, let $r \geq l(i_0)$ (so that $\rho_r^{\text{sparse}} < +\infty$), and let $\mathbf{y}^r$ be a nearly optimal solution of $\mathbf{Q}_r$ with value

$$\rho_r^{\text{sparse}} \geq L_{\mathbf{y}^r}(f) \geq \rho_r^{\text{sparse}} - \frac{1}{r} \quad \left(\geq \rho_{\text{mom}} - \frac{1}{r} \right). \tag{4.44}$$

Fix $s \in \mathbb{N}$. Notice that from (4.43), for all $r \geq l(s)$, one has

$$|\, L_{\mathbf{y}^r}(\mathbf{x}^{\boldsymbol{\alpha}}) \,| \;\leq\; N_s M \qquad \forall \boldsymbol{\alpha} \in \Theta_s.$$

Therefore, for all $r \geq i_0$,

$$|y_{\alpha}^r| = |L_{\mathbf{y}^r}(\mathbf{x}^{\alpha})| \leq N_s' \quad \forall \boldsymbol{\alpha} \in \Theta_s, \tag{4.45}$$

where $N_s' = \max[N_s M, V_s]$, with

$$V_s := \max\{|y_{\boldsymbol{\alpha}}^r| \,:\, \boldsymbol{\alpha} \in \Theta_s;\ l(i_0) \leq r < l(s)\}.$$

Complete each $\mathbf{y}^r$ with zeros to make it an infinite vector in l_∞, indexed in the canonical basis of $\mathbb{R}[\mathbf{x}]$. Notice that $\mathbf{y}_{\boldsymbol{\alpha}}^r \neq 0$ only if $\boldsymbol{\alpha} \in \Theta$.

In view of (4.45), for every $s = 1, 2, \ldots$, one has:

$$|y_{\boldsymbol{\alpha}}^r| \leq N_s' \qquad \forall \boldsymbol{\alpha} \in \Theta; \quad 2s-1 \leq |\boldsymbol{\alpha}| \leq 2s. \tag{4.46}$$

Hence, define the new sequence $\widehat{\mathbf{y}}^r \in l_\infty$ defined by $\widehat{y}_0^r := y_0^r/M$, and

$$\widehat{y}_{\boldsymbol{\alpha}}^r := \frac{y_{\boldsymbol{\alpha}}^r}{N_s'} \qquad \forall \alpha \in \Theta, \quad 2s-1 \leq |\boldsymbol{\alpha}| \leq 2s,$$

for all $s = 1, 2, \ldots$, and in l_∞, consider the sequence $(\widehat{\mathbf{y}}^r)_r$ as $r \to \infty$. Obviously, the sequence $(\widehat{\mathbf{y}}^r)_r$ is in the unit ball B_1 of l_∞, and so, by the Banach–Alaoglu theorem (see, e.g., (Ash, 1972, Theorem. 3.5.16)), there exists $\widehat{\mathbf{y}} \in B_1$ and a subsequence $\{r_i\}$, such that $\widehat{\mathbf{y}}^{r_i} \to \widehat{\mathbf{y}}$ as $i \to \infty$ for the weak $\star$ topology $\sigma(l_\infty, l_1)$ of l_∞. In particular, pointwise convergence holds, that is,

$$\lim_{i\to\infty} \widehat{y}_{\boldsymbol{\alpha}}^{r_i} \to \widehat{y}_{\boldsymbol{\alpha}}, \qquad \forall \boldsymbol{\alpha} \in \mathbb{N}^n.$$

Notice that $\widehat{y}_{\boldsymbol{\alpha}} \neq 0$ only if $\boldsymbol{\alpha} \in \Theta$. Next, define $y_0 := M\widehat{y}_0$ and

$$y_{\boldsymbol{\alpha}} := N_s' \widehat{y}_{\boldsymbol{\alpha}}, \quad 2s-1 \leq |\boldsymbol{\alpha}| \leq 2s, \quad s = 1, 2, \ldots.$$

The pointwise convergence $\widehat{\mathbf{y}}^{r_i} \to \widehat{\mathbf{y}}$ implies the pointwise convergence $\mathbf{y}^{r_i} \to \mathbf{y}$:

$$\lim_{i\to\infty} y_{\boldsymbol{\alpha}}^{r_i} \to y_{\boldsymbol{\alpha}} \qquad \forall \boldsymbol{\alpha} \in \Theta. \tag{4.47}$$

Let $s \in \mathbb{N}$ be fixed. From the pointwise convergence (4.47),

$$\lim_{i\to\infty} \mathbf{M}_s(\mathbf{y}^{r_i}, I_k) = \mathbf{M}_s(\mathbf{y}, I_k) \succeq 0, \qquad k = 1, \ldots, p.$$

Similarly

$$\lim_{i\to\infty} \mathbf{M}_s(g_j\,\mathbf{y}^{r_i}, I_k) = \mathbf{M}_s(g_j\,\mathbf{y}, I_k) \succeq 0, \quad j \in J_k, \qquad k = 1, \ldots, p.$$

As s was arbitrary, we obtain that for all $k = 1, \ldots, p$,

$$\mathbf{M}_r(\mathbf{y}, I_k) \succeq 0; \quad \mathbf{M}_r(g_j\,\mathbf{y}, I_k) \succeq 0, \quad j \in J_k; \quad r = 0, 1, 2, \ldots. \tag{4.48}$$

Introduce the subsequence $\mathbf{y}^k$ obtained from $\mathbf{y}$ by

$$\mathbf{y}^k := \{\, y_{\boldsymbol{\alpha}} : \mathrm{supp}\,(\boldsymbol{\alpha}) \in \mathscr{I}_k \,\} \qquad \forall\, k = 1, \ldots, p. \tag{4.49}$$

Recall that $\mathbf{M}_r(\mathbf{y}, I_k)$ (resp., $\mathbf{M}_r(g_j\,\mathbf{y}, I_k)$) is also the moment matrix $\mathbf{M}_r(\mathbf{y}^k)$ (resp., the localizing matrix $\mathbf{M}_r(g_j\,\mathbf{y}^k)$) for the sequence $\mathbf{y}^k$ indexed in the canonical basis of $\mathbb{R}[\mathbf{x}(I_k)]$.

By Theorem 3.8(b), (4.48) implies that $\mathbf{y}^k$ has a representing measure μ_k with support contained in $\mathbb{K}_k$, $k = 1, \ldots, p$.

Next, let j, k be such that I_{jk} $(= I_j \cap I_k) \neq \emptyset$, and recall that $\mathscr{I}_{jk}$ is the set of all subsets of I_{jk}. Observe that from the definition (4.49) of $\mathbf{y}^j$ and $\mathbf{y}^k$, one has

$$y^j_{\boldsymbol{\alpha}} = y^k_{\boldsymbol{\alpha}} \qquad \forall\, \boldsymbol{\alpha} \text{ with } \mathrm{supp}\,(\boldsymbol{\alpha}) \in \mathscr{I}_{jk},$$

and as measures on compact sets are moment determinate, it follows that the marginal probability measures of μ_j and μ_k on the variables $\mathbf{x}(I_{jk})$ are the *same* probability measure, denoted μ_{jk}. That is,

$$y^k_{\boldsymbol{\alpha}} = y^j_{\boldsymbol{\alpha}} = \int_{\mathbb{K}_{jk}} \mathbf{x}^{\boldsymbol{\alpha}}\, d\mu_{jk} \qquad \forall\, \boldsymbol{\alpha} \text{ with } \mathrm{supp}\,(\boldsymbol{\alpha}) \in \mathscr{I}_{jk},$$

for some measure μ_{jk} supported on the set

$$\mathbb{K}_{jk} = \{\, \mathbf{x} \in \mathbb{R}^{n_{jk}} : \quad \mathbf{x}(I_j) \in \mathbb{K}_j\,; \quad \mathbf{x}(I_k) \in \mathbb{K}_k \,\}.$$

But then, by Lemma B.13 (to be adapted to measures instead of probability measures), there exists a measure μ on $\mathbb{R}^n$ with support contained in $\mathbb{K}$, such that its projection $\pi_k\mu$ on $\mathbb{R}^{n_k}$ (on the variables $\mathbf{x}(I_k)$) is just μ_k, for every $k = 1, \ldots, p$.

Next, from the pointwise convergence (4.47),

$$\int_{\mathbb{K}} h_k\, d\mu = L_{\mathbf{y}}(h_k) = \lim_{i\to\infty} L_{\mathbf{y}^{r_i}}(h_k) \leqq \gamma_k \quad \forall k \in \Gamma,$$

which proves that μ is a feasible solution of the moment problem (4.2).

On the other hand, (4.44) and (4.47) again, also yield

$$\rho_{\mathrm{mom}} \leq \lim_{i\to\infty} \rho^{\mathrm{sparse}}_{r_i} = \lim_{i\to\infty} L_{\mathbf{y}^{r_i}}(f) = L_{\mathbf{y}}(f) = \int_{\mathbb{K}} f\, d\mu,$$

from which we conclude that μ is in fact an optimal solution of (4.2).

Part II

Applications

Chapter 5

Global Optimization over Polynomials

This chapter is on global optimization, probably the simplest instance of the generalized moment problem. We detail the semidefinite relaxations of Chapter 4 for minimizing a polynomial on $\mathbb{R}^n$ and on a compact basic semi-algebraic set. We also discuss the linear relaxations and several particular cases, e.g. when $\mathbb{K}$ is a polytope or a finite variety. In particular, this latter case encompasses all $0-1$ discrete optimization problems.

In this chapter, we address the following two fundamendal optimization problems:

(a) The unconstrained polynomial optimization problem:

$$f^* := \min \{ f(\mathbf{x}) \,:\, \mathbf{x} \in \mathbb{R}^n \} \tag{5.1}$$

(b) The constrained polynomial optimization problem:

$$f^*_{\mathbb{K}} := \min \{ f(\mathbf{x}) \,:\, \mathbf{x} \in \mathbb{K} \} \tag{5.2}$$

where $f \in \mathbb{R}[\mathbf{x}]$ is a real-valued polynomial and

$$\mathbb{K} = \{\mathbf{x} \in \mathbb{R}^n \,:\, g_i(\mathbf{x}) \geq 0,\ i = 1, \ldots, m\}, \tag{5.3}$$

and $f, g_i \in \mathbb{R}[\mathbf{x}]$, $i = 1, \ldots, m$.

While the framework of this chapter is a special case of the previous chapter, because of the more special structure we will be able to develop a deeper theory. The basic semi-algebraic set $\mathbb{K}$ is assumed to be a compact subset of $\mathbb{R}^n$. We do not assume that $\mathbb{K}$ is convex or even connected. As observed earlier, this is a rather rich modeling framework that includes linear, quadratic, $0/1$, mixed $0/1$ optimization problems as special cases. In particular, constraints of the type $x_i \in \{0,1\}$ can be written as $x_i^2 - x_i \geq 0$ and $x_i - x_i^2 \geq 0$.

When $n = 1$, we have seen in Chapter 2 that a nonnegative univariate polynomial can be written as a sum of squares of polynomials. We will see that this condition naturally leads to reformulating Problems (5.1)-(5.2) as semidefinite optimization problems, for which efficient algorithms and software packages are available. It is quite remarkable that a nonconvex problem can be reformulated as a convex one and underscores the importance of the representation theorems from Chapter 2.

On the other hand, the multivariate case radically differs from the univariate case, because not every nonnegative polynomial can be written as a sum of squares of polynomials. Moreover, Problem (5.1) involving a polynomial f of degree greater than or equal to four on n-variables is $\mathscr{N}P$-hard. However, we will see that Problem (5.1) can be approximated as closely as desired (and often can be obtained exactly) by solving a finite sequence of semidefinite optimization problems of the same type as in the one-dimensional case. A similar conclusion also holds for Problem (5.2).

5.1 The Primal and Dual Perspectives

Let $\mathbb{K} \subset \mathbb{R}^n$ be as in (5.3) and recall that $\mathscr{M}(\mathbb{R}^n)_+$ and $\mathscr{M}(\mathbb{K})_+$ are the spaces of finite Borel measures on $\mathbb{R}^n$ and $\mathbb{K}$, respectively. Following Chapter 1, we reformulate Problems (5.1), (5.2) as follows:

$$\begin{aligned} \rho_{\text{mom}} := \min_{\mu \in \mathscr{M}(\Omega)_+} & \int_\Omega f \, d\mu \\ \text{s.t. } & \mu(\Omega) = 1 \end{aligned} \tag{5.4}$$

i.e., (5.4) is just (1.1) (but now minimizing) with $\Gamma = \{1\}$, $h_1 = 1$, $\gamma_1 = 1$, and with $\Omega = \mathbb{R}^n$ or $\Omega = \mathbb{K}$ with $\mathbb{K}$ as in (5.3). Therefore by Theorem 1.1,

Theorem 5.1.
(a) *Problem (5.1) is equivalent to problem (5.4) with* $\Omega = \mathbb{R}^n$*, and so* $f^* = \rho_{\text{mom}}$.
(b) *Problem (5.2) is equivalent to problem (5.4) with* $\Omega = \mathbb{K}$*, and so,* $f^*_{\mathbb{K}} = \rho_{\text{mom}}$.

Taking the dual perspective developed in Chapter 1, we also rewrite Problem (5.1) (resp. Problem (5.2)) as

$$\begin{aligned} \rho_{\text{pop}} &= \sup \lambda \\ &\quad \text{s.t. } f(\mathbf{x}) - \lambda \geq 0, \quad \forall \mathbf{x} \in \Omega, \end{aligned} \tag{5.5}$$

with $\Omega = \mathbb{R}^n$ (resp. $\Omega = \mathbb{K}$). It is clear that $\rho_{\text{pop}} = \rho_{\text{mom}} = f^*$ if $\Omega = \mathbb{R}^n$ and $\rho_{\text{pop}} = \rho_{\text{mom}} = f^*_{\mathbb{K}}$ if $\Omega = \mathbb{K}$.

We now consider linear and semidefinite relaxations for Problems (5.1) and (5.2), as special cases of the numerical scheme developed in Chapter 4 for the generalized moment problem. In doing so, for Problem (5.1), we obtain a single semidefinite relaxation, which is exact if and only if the polynomial $\mathbf{x} \mapsto f(\mathbf{x}) - f^*$ is a sum of squares. For Problem (5.2), we obtain a hierarchy of semidefinite relaxations whose sizes depend on the degree d permitted for the s.o.s. weights in the representation (2.13) in Theorem 2.14. The larger the degree the better the optimal value λ. We may also consider the alternative representation of $f(\mathbf{x}) - \lambda$ as an element of the cone C_G in Eq. (2.18), and invoke Theorem 2.23. In doing so we obtain a hierarchy of linear relaxations instead of semidefinite relaxations.

5.2 Unconstrained Polynomial Optimization

In the unconstrained case $\Omega = \mathbb{R}^n$, the only interesting case is when $\deg f$ is even, otherwise necessarily $f^* = -\infty$. Therefore, let $2r$ be the degree of $f \in \mathbb{R}[\mathbf{x}]$, and let $i \geq r$. The semidefinite relaxation (4.5) of Chapter 4 reads:

$$\rho_i = \inf_{\mathbf{y}} \{ L_{\mathbf{y}}(f) \, : \, \mathbf{M}_i(\mathbf{y}) \succeq \mathbf{0}, \, y_0 = 1 \} \tag{5.6}$$

with associated dual:

$$\rho_i^* = \sup_{\lambda,\mathbf{X}} \{ \lambda : \mathbf{X} \succeq \mathbf{0}; \quad \langle \mathbf{X}, \mathbf{B}_{\boldsymbol{\alpha}} \rangle = f_{\boldsymbol{\alpha}} - \lambda \delta_{\boldsymbol{\alpha}=0}, \quad \forall |\boldsymbol{\alpha}| \leq 2i \} \tag{5.7}$$

where $\mathbf{M}_i(\mathbf{y})$, the moment matrix associated with the sequence $\mathbf{y}$, is written $\mathbf{M}_i(\mathbf{y}) = \sum_{|\boldsymbol{\alpha}|\leq 2i} y_{\boldsymbol{\alpha}} \mathbf{B}_{\boldsymbol{\alpha}}$ for appropriate symmetric matrices $\{\mathbf{B}_{\boldsymbol{\alpha}}\}$, and $\delta_{\boldsymbol{\alpha}=0}$ is the Kronecker symbol. In the present context, the interpretation (4.9) reads:

$$\rho_i^* = \sup_{\lambda} \{ \lambda : f - \lambda \in \Sigma(\mathbf{x}) \}. \tag{5.8}$$

Therefore, one has to consider only one semidefinite relaxation, namely that with $i = r$, because obviously, if f has degree $2r$ then $f - \lambda$ cannot be a sum of squares of polynomials with degree larger than r. That is, $\rho_i^* = \rho_r^*$ for all $i > r$.

Proposition 5.2. *There is no duality gap, that is, $\rho_r = \rho_r^*$. And, if $\rho_r > -\infty$ then (5.8) has an optimal solution.*

Proof. The result follows from the duality theory of semidefinite optimization if we can prove that there is a strictly feasible solution $\mathbf{y}$ of Problem (5.6), i.e., such that $\mathbf{M}_r(\mathbf{y}) \succ \mathbf{0}$ (Slater condition); see Theorem C.17. Let μ be a probability measure on $\mathbb{R}^n$ with a strictly positive density f with respect to the Lebesgue measure and with all its moments finite; that is, μ is such that

$$y_{\boldsymbol{\alpha}} = \int_{\mathbb{R}^n} \mathbf{x}^{\boldsymbol{\alpha}} \, d\mu = \int_{\mathbb{R}^n} \mathbf{x}^{\boldsymbol{\alpha}} f(\mathbf{x}) \, d\mathbf{x} < \infty, \quad \forall \boldsymbol{\alpha} \in \mathbb{N}^n.$$

Then $\mathbf{M}_r(\mathbf{y})$, with $\mathbf{y}$ as above, is such that $\mathbf{M}_r(\mathbf{y}) \succ \mathbf{0}$. To see this, recall that for every polynomial $q \in \mathbb{R}[\mathbf{x}]$ of degree at most r, and vector of coefficients $\mathbf{q} \in \mathbb{R}^{s(r)}$,

$$\begin{aligned} \langle q, q \rangle_{\mathbf{y}} = \langle \mathbf{q}, \mathbf{M}_r(\mathbf{y}) \mathbf{q} \rangle &= \int_{\mathbb{R}^n} q^2 \, d\mu \qquad \text{(from (3.11))} \\ &= \int_{\mathbb{R}^n} q(\mathbf{x})^2 f(\mathbf{x}) \, d\mathbf{x} \\ &> 0, \text{ whenever } q \neq 0 \ \text{ (as } f > 0). \end{aligned}$$

Therefore, $\mathbf{y}$ is strictly feasible for Problem (5.6), i.e., $\mathbf{M}_r(\mathbf{y}) \succ \mathbf{0}$, the desired result. □

We next prove the central result of this section.

Theorem 5.3. *Let $f \in \mathbb{R}[\mathbf{x}]$ be a $2r$-degree polynomial with global minimum $f^* > -\infty$.*
(a) *If the nonnegative polynomial $f - f^*$ is s.o.s., then Problem (5.1) is equivalent to the semidefinite optimization problem (5.6), i.e., $f^* = \rho_r^* = \rho_r$ and if $\mathbf{x}^* \in \mathbb{R}^n$ is a global minimizer of (5.1), then the moment vector*

$$\mathbf{y}^* := (x_1^*, \ldots, x_n^*, (x_1^*)^2, x_1^* x_2^*, \ldots, (x_1^*)^{2r}, \ldots, (x_n^*)^{2r}) \tag{5.9}$$

is a minimizer of Problem (5.6).
(b) *If Problem (5.7) has a feasible solution and $f^* = \rho_r^*$, then $f - f^*$ is s.o.s.*

Proof. (a) Let $f - f^*$ be s.o.s., that is,

$$f(\mathbf{x}) - f^* = \sum_{i=1}^{k} q_i(\mathbf{x})^2, \qquad \mathbf{x} \in \mathbb{R}^n, \tag{5.10}$$

for some polynomials $\{q_i\}_{i=1}^k \subset \mathbb{R}[\mathbf{x}]_r$ with coefficient vector $\mathbf{q}_i \in \mathbb{R}^{s(r)}$, $i = 1, 2, \ldots, k$, with $s(r) = \binom{n+r}{n}$. Equivalently, with $\mathbf{v}_r(\mathbf{x})$ as in (3.9),

$$f(\mathbf{x}) - f^* = \langle \mathbf{X}, \mathbf{M}_r(\mathbf{y}) \rangle, \qquad \mathbf{x} \in \mathbb{R}^n, \tag{5.11}$$

with $\mathbf{X} = \sum_{i=1}^k \mathbf{q}_i \mathbf{q}_i'$ and $\mathbf{y} = \mathbf{v}_{2r}(\mathbf{x})$. From Eq. (5.11) it follows that

$$\langle \mathbf{X}, \mathbf{B}_\mathbf{0} \rangle = f_\mathbf{0} - f^*; \quad \langle \mathbf{X}, \mathbf{B}_\alpha \rangle = f_\alpha, \text{ for all } 0 \neq \alpha, \ |\alpha| \leq 2r,$$

so that (as $\mathbf{X} \succeq \mathbf{0}$) $\mathbf{X}$ is feasible for problem (5.7) with value $\lambda := f^*$. Since $\mathbf{y}^*$ in (5.9) is feasible for Problem (5.6) with value f^* and $\rho_r^* = \rho_r$, it follows that $\mathbf{y}^*$ and $\mathbf{X}$ are optimal solutions to Problems (5.6) and (5.7), respectively.

(b) Suppose that Problem (5.7) has a feasible solution and $f^* = \rho_r^*$. Then, from Proposition 5.2, Problem (5.7) has an optimal solution $(\mathbf{X}^*, f^*)$, with $\mathbf{X}^* \succeq \mathbf{0}$, and there is no duality gap, that is, $\rho_r = \rho_r^*$. As $\mathbf{X}^* \succeq \mathbf{0}$, we use its spectral decomposition to write $\mathbf{X}^* = \sum_{i=1}^k \mathbf{q}_i \mathbf{q}_i'$ for some vector $\{\mathbf{q}_i\}$

in $\mathbb{R}^{s(r)}$. Using the feasibility of $(\mathbf{X}^*, f^*)$ in (5.7), we obtain

$$\left\langle \mathbf{X}^*, \sum_{\alpha} \mathbf{B}_{\alpha} \mathbf{x}^{\alpha} \right\rangle = f(\mathbf{x}) - f^*,$$

which, using $\mathbf{X}^* = \sum_{i=1}^{k} \mathbf{q}_i \mathbf{q}_i'$ and

$$\sum_{\alpha} \mathbf{B}_{\alpha} \mathbf{x}^{\alpha} = \mathbf{M}_r(\mathbf{v}_{2r}(\mathbf{x})) = \mathbf{v}_r(\mathbf{x}) \mathbf{v}_r(\mathbf{x})',$$

yields the desired sum of squares

$$f(\mathbf{x}) - f^* = \sum_{i=1}^{k} \langle \mathbf{q}_i \mathbf{q}_i', \mathbf{v}_r(\mathbf{x}) \mathbf{v}_r(\mathbf{x})' \rangle = \sum_{i=1}^{k} \langle \mathbf{q}_i, \mathbf{v}_r(\mathbf{x}) \rangle^2 = \sum_{i=1}^{k} q_i^2(\mathbf{x}),$$

with the polynomials $\mathbf{x} \mapsto q_i(\mathbf{x}) := \langle \mathbf{q}_i, \mathbf{v}_r(\mathbf{x}) \rangle$, for all $i = 1, \ldots, k$. □

From the proof of Theorem 5.3, it is obvious that if $f^* = \rho_r^*$, then any global minimizer $\mathbf{x}^*$ of f is a zero of each polynomial q_i, where $\mathbf{X}^* = \sum_{i=1}^{k} \mathbf{q}_i \mathbf{q}_i'$ at an optimal solution $\mathbf{X}^*$ of problem (5.7). When $f - f^*$ is s.o.s., solving Problem (5.7) provides the polynomials q_i of such a decomposition. As a corollary, we obtain the following.

Corollary 5.4. *Let $f \in \mathbb{R}[\mathbf{x}]$ be of degree $2r$. Assume that Problem (5.7) has a feasible solution. Then,*

$$f(\mathbf{x}) - f^* = \sum_{i=1}^{k} q_i(\mathbf{x})^2 - [f^* - \rho_r^*], \quad \mathbf{x} \in \mathbb{R}^n, \tag{5.12}$$

for some real-valued polynomials $q_i \in \mathbb{R}[\mathbf{x}]$ of degree at most r, $i = 1, 2, \ldots, k$.

The proof is the same as that of Theorem 5.3(b), except now we may not have $f^* = \rho_r^*$, but instead $\rho_r^* \leq f^*$. Hence, ρ_r^* always provides a lower bound on f^*.

Corollary 5.4 states that, up to some constant, one may always write $f - f^*$ as a s.o.s. whenever Problem (5.7) has a feasible solution. The previous development leads to the following algorithm for either solving Problem (5.1) or providing a lower bound on its optimal value f^*.

Algorithm 5.1. **(Unconstrained polynomial optimization)**

Input: A polynomial $\mathbf{x} \mapsto f(\mathbf{x}) = \sum_{\alpha \in \mathbb{N}^n} f_\alpha \mathbf{x}^\alpha$ of degree $2r$.

Output: The value $f^* = \min_{\mathbf{x} \in \mathbb{R}^n} f(\mathbf{x})$ or a lower bound ρ_r on f^*.

Algorithm:
1. Solve the semidefinite optimization problem (5.6) with optimal value ρ_r and optimal solution $\mathbf{y}^*$ (if $\mathbf{y}^*$ exists).
2. If $\operatorname{rank} \mathbf{M}_{r-1}(\mathbf{y}^*) = \operatorname{rank} \mathbf{M}_r(\mathbf{y}^*)$, then $f^* = \rho_r$ and there are at least $\operatorname{rank} \mathbf{M}_r(\mathbf{y}^*)$ global minimizers, which are extracted by applying Algorithm 4.2.
3. Otherwise, ρ_r only provides a lower bound $\rho_r \leq f^*$.

We next show that Algorithm 5.1 correctly determines whether ρ_r is the exact solution value f^* or a lower bound.

Theorem 5.5. *Let* $f \in \mathbb{R}[\mathbf{x}]$ *with degree* $2r$*, and suppose that the optimal value* ρ_r *of Problem (5.6) is attained at some optimal solution* $\mathbf{y}^*$*. If* $\operatorname{rank} \mathbf{M}_{r-1}(\mathbf{y}^*) = \operatorname{rank} \mathbf{M}_r(\mathbf{y}^*)$*, then* $f^* = \rho_r$*, and there are at least* $\operatorname{rank} \mathbf{M}_r(\mathbf{y}^*)$ *global minimizers.*

Proof. We have already shown that $\rho_r \leq f^*$. If $\operatorname{rank} \mathbf{M}_{r-1}(\mathbf{y}^*) = \operatorname{rank} \mathbf{M}_r(\mathbf{y}^*)$, we let $s = \operatorname{rank} \mathbf{M}_r(\mathbf{y}^*)$. Then $\mathbf{M}_r(\mathbf{y}^*)$ is a flat extension of $\mathbf{M}_{r-1}(\mathbf{y}^*)$, and so, by Theorem 3.7, $\mathbf{y}^*$ is the vector of moments up to order $2r$, of some s-atomic probability measure μ^* on $\mathbb{R}^n$. Therefore, $\rho_r = L_{\mathbf{y}}(f) = \int f \, d\mu^*$, which proves that μ^* is an optimal solution of (5.4), and $\rho_r = \rho_{\text{mom}} = f^*$, because we always have $\rho_r \leq \rho_{\text{mom}}$. We next show that each of the s atoms of μ^* is a global minimizer of f. Indeed, being s-atomic, there is a family $(\mathbf{x}(k))_{k=1}^s \subset \mathbb{R}^n$ and a family $(\beta_k)_{k=1}^s \subset \mathbb{R}$, such that

$$\mu^* = \sum_{k=1}^s \beta_k \, \delta_{\mathbf{x}(k)}, \quad \beta_k > 0, \quad \forall k = 1, \ldots, s; \quad \sum_{k=1}^s \beta_k = 1,$$

Hence,

$$f^* = \rho_r = \int_{\mathbb{R}^n} f\, d\mu^* = \sum_{k=1}^{s} \beta_k\, f(\mathbf{x}(k)),$$

which, in view of $f(\mathbf{x}(k)) \geq f^*$, for all $k = 1, \ldots, s$, implies the desired result $f(\mathbf{x}(k)) = f^*$, for all $k = 1, \ldots, s$. □

As an illustration, suppose that $\mathbf{y}^*$ satisfies $\operatorname{rank} \mathbf{M}_r(\mathbf{y}^*) = 1$. Therefore, $\mathbf{M}_r(\mathbf{y}^*) = \mathbf{v}_r(\mathbf{x}^*)\mathbf{v}_r(\mathbf{x}^*)'$ for some $\mathbf{x}^* \in \mathbb{R}^n$, that is, $\mathbf{y}^*$ is the vector of moments up to order $2r$ of the Dirac measure at $\mathbf{x}^*$, and one **reads** an optimal solution $\mathbf{x}^*$, from the subvector of first "moments" $y^*_{\boldsymbol{\alpha}}$ with $|\boldsymbol{\alpha}| = 1$. Note also that when $n = 1$, $f - f^*$ is always a s.o.s., so that we expect that $\rho_r = f^*$, where $2r$ is the degree of f.

In other words, the global minimization of a univariate polynomial is a convex optimization problem, the semidefinite program (5.6).

Therefore, in view of Section 2.4, we also expect that $\rho_r = f^*$ for quadratic polynomials, and bivariate polynomials of degree 4. Let us illustrate these properties with an example.

Example 5.1. We consider the polynomial f on two variables:

$$\mathbf{x} \mapsto f(\mathbf{x}) = (x_1^2 + 1)^2 + (x_2^2 + 1)^2 + (x_1 + x_2 + 1)^2.$$

Note that in this case $r = 2$ and $f - f^*$ is a bivariate polynomial of degree 4, and therefore it is a sum of squares. We thus expect that $\rho_2 = f^*$. Solving Problem (5.6) for $i = 2$, yields a minimum value of $\rho_2 = -0.4926$. In this case, it turns out that $\mathbf{M}_2(\mathbf{y}^*)$ has rank one, and from the optimal solution $\mathbf{y}^*$, we check that

$$\mathbf{y} = (1, x_1^*, x_2^*, (x_1^*)^2, x_1^* x_2^*, (x_2^*)^2, \ldots, (x_1^*)^4, \ldots, (x_2^*)^4),$$

with $x_1^* = x_2^* \approx -0.2428$. We observe that the solution $\mathbf{x}^*$ is a good approximation of a global minimizer of problem (5.1), since the gradient vector verifies

$$\left.\frac{\partial f}{\partial x_1}\right|_{\mathbf{x}=\mathbf{x}^*} = \left.\frac{\partial f}{\partial x_2}\right|_{\mathbf{x}=\mathbf{x}^*} = 4 \cdot 10^{-9}.$$

In this example, it follows that the semidefinite relaxation is exact. The reason we have $\nabla f(\mathbf{x}^*) \approx \mathbf{0}$, but not exactly $\mathbf{0}$, is likely due anavoidable numerical inaccuracies when using an SDP solver.

The next example discusses when the approach does not work and the effect of perturbations.

Example 5.2. Consider the bivariate polynomial $\mathbf{x} \mapsto f(\mathbf{x}) = x_1^2 x_2^2 (x_1^2 + x_2^2 - 1)$. In Exercise 2.4(a) we have shown that $f + 1$ is positive, but not a sum of squares. Note that the global optimum is $f^* = -1/27$ and there are four global minimizers $\mathbf{x}^* = (\pm\sqrt{3}/3, \pm\sqrt{3}/3)$. If we consider the problem $f^* = \min f(x_1, x_2)$ and apply Algorithm 5.1 for $r = 3$, we obtain that $\rho_3 = -\infty$, that is the bound is uninformative.

However consider the perturbed problem of minimizing the polynomial $f_\epsilon := f(\mathbf{x}) + \epsilon(x_1^{10} + x_2^{10})$ with $\epsilon = 0.001$. Applying Algorithm 5.1 using the software GloptiPoly we find that $\rho_5 \approx f^*$ and four optimal solutions extracted $(x_1, x_2) \approx \mathbf{x}^*$. In other words, while the approach fails when applied to the original polynomial f, it succeeds when applied to the perturbed polynomial f_ϵ.

5.3 Constrained Polynomial Optimization: Semidefinite Relaxations

In this section, we address Problem (5.2) with $f \in \mathbb{R}[\mathbf{x}]$ being a polynomial of degree $2v_0$ or $2v_0 - 1$. The set $\mathbb{K}$ defined in (5.3) is assumed to be a compact basic semi-algebraic subset of $\mathbb{R}^n$. For all $j = 1, \ldots, m$, and depending on its parity, let $2v_j$, or $2v_j - 1$, denote the degree of the polynomials g_j in the definition (5.3) of the set $\mathbb{K}$.

We assume that Assumption 2.1 holds, which as we discussed in Section 2.5, is verified in many cases, for example if there is one polynomial g_k such that $\{\mathbf{x} \in \mathbb{R}^n : g_k(\mathbf{x}) \geq 0\}$ is compact; see the discussion following Theorem 2.14 and Theorem 2.15.

Under Assumption 2.1, from Theorem 2.14, every polynomial $f \in \mathbb{R}[\mathbf{x}]$ which is strictly positive on $\mathbb{K}$, can be written

$$f(\mathbf{x}) = \sigma_0(\mathbf{x}) + \sum_{j=1}^{m} g_j(\mathbf{x})\sigma_j(\mathbf{x}), \quad \forall \mathbf{x} \in \mathbb{R}^n, \tag{5.13}$$

for some polynomials $\sigma_j \in \Sigma[\mathbf{x}]$, for all $j = 0, 1, \ldots, m$.

From the general scheme described in Chapter 4, the semidefinite relaxations (4.5) are as follows: For $i \geq \max_{j=0,\ldots,m} v_j$,

$$\begin{aligned} \rho_i = \inf_{\mathbf{y}} \; & L_{\mathbf{y}}(f) \\ \text{s.t. } & \mathbf{M}_i(\mathbf{y}) \succeq \mathbf{0}, \\ & \mathbf{M}_{i-v_j}(g_j\mathbf{y}) \succeq \mathbf{0}, \quad j = 1, \ldots, m \\ & y_{\mathbf{0}} \; = 1. \end{aligned} \tag{5.14}$$

The dual of problem (5.14) is the semidefinite program:

$$\begin{aligned} \rho_i^* = & \sup_{\lambda, \{\sigma_j\}} \; \lambda \\ & \text{s.t. } f - \lambda = \sigma_0 + \sum_{j=1}^{m} \sigma_j \, g_j \\ & \deg \sigma_0 \leq 2i; \; \deg \sigma_j \leq 2i - 2v_j, \; j = 1, \ldots, m. \end{aligned} \tag{5.15}$$

Note that if in the definition of $\mathbb{K}$ there is an equality constraint $g_j(\mathbf{x}) = 0$ (i.e. two opposite inequality constraints $g_j(\mathbf{x}) \geq 0$ and $-g_i(\mathbf{x}) \geq 0$), then one has the equality constraint $\mathbf{M}_{i-v_j}(g_j\,\mathbf{y}) = 0$ in (5.14) and accordingly, in (5.15) the weight $\sigma_j \in \mathbb{R}[\mathbf{x}]$ is not required to be a s.o.s.

The overall algorithm is as follows:

Algorithm 5.2. (Constrained polynomial optimization)

Input: A polynomial $\mathbf{x} \mapsto f(\mathbf{x})$ of degree $2v_0$ or $2v_0 - 1$; a set $\mathbb{K} = \{\mathbf{x} \in \mathbb{R}^n : g_j(\mathbf{x}) \geq 0, \; j = 1, \ldots, m\}$, where the polynomials g_j are of degree $2v_j$ or $2v_j - 1$, $j = 1, \ldots, m$; a number k of the highest relaxation.

Output: The value $f^*_{\mathbb{K}} = \min_{\mathbf{x} \in \mathbb{K}} f(\mathbf{x})$ and a list of global minimizers or a lower bound ρ_k on $f^*_{\mathbb{K}}$.

Algorithm:

1. Solve the semidefinite optimization problem (5.14) with optimal value ρ_i, and optimal solution $\mathbf{y}^*$ (if it exists).

2. If there is no optimal solution $\mathbf{y}^*$ then ρ_k only provides a lower bound $\rho_k \leq f^*_{\mathbb{K}}$. If $i < k$ then increase i by one and go to Step 1; otherwise stop and output ρ_k.

3. If $\operatorname{rank} \mathbf{M}_{i-v}(\mathbf{y}^*) = \operatorname{rank} \mathbf{M}_i(\mathbf{y}^*)$ (with $v := \max_j v_j$), then $\rho_i = f^*_{\mathbb{K}}$ and there are at least $\operatorname{rank} \mathbf{M}_i(\mathbf{y}^*)$ global minimizers, which are extracted by applying Algorithm 4.2.

4. If $\operatorname{rank} \mathbf{M}_{i-v}(\mathbf{y}^*) \neq \operatorname{rank} \mathbf{M}_i(\mathbf{y}^*)$, and $i < k$, then increase i by one and go to Step 1; otherwise, stop and output ρ_k only provides a lower bound $\rho_k \leq f^*_{\mathbb{K}}$.

And we have the following convergence result:

Theorem 5.6. *Let Assumption 2.1 hold and consider the semidefinite relaxation (5.14) with optimal value* ρ_i.
(a) $\rho_i \uparrow f^*_{\mathbb{K}}$ *as* $i \to \infty$.
(b) *Assume that (5.2) has a unique optimal solution* $\mathbf{x}^* \in \mathbb{K}$ *and let* $\mathbf{y}^i$ *be a nearly optimal solution of (5.14) with value* $L_{\mathbf{y}}(f) \leq \rho_i + 1/i$. *Then as* $i \to \infty$, $L_{\mathbf{y}^i}(x_j) \to x^*_j$ *for every* $j = 1, \ldots, n$.

Theorem 5.6(a) is a direct consequence of Theorem 4.1. For Theorem 5.6(b) one uses same arguments as in the proof of Theorem 4.3. A subsequence $\mathbf{y}^{i_k}$ converges pointwise to a sequence $\mathbf{y}$ which is shown to be the moment sequence of a measure μ supported on $\mathbb{K}$, and with value $L_{\mathbf{y}}(f) = \lim_{k\to\infty} L_{\mathbf{y}^{i_k}}(f) \leq f^*_{\mathbb{K}}$. Hence μ is the Dirac measure $\delta_{\mathbf{x}^*}$ at the unique global minimizer $\mathbf{x}^* \in \mathbb{K}$. Therefore the whole sequence $\mathbf{y}^i$ converges pointwise and in particular, for every $j = 1, \ldots, n$, $L_{\mathbf{y}^i}(x_j) \to x^*_j$ as $i \to \infty$.

5.3.1 *Obtaining global minimizers*

As for the unconstrained case, after solving the semidefinite relaxation (5.14) for some value of $i \in \mathbb{N}$, we are left with two issues:
(a) How do we know that $\rho_i < f^*_{\mathbb{K}}$, or $\rho_i = f^*_{\mathbb{K}}$?
(b) If $\rho_i = f^*_{\mathbb{K}}$, can we get at least one global minimizer $\mathbf{x}^* \in \mathbb{K}$?

Again, an easy case is when (5.14) has an optimal solution $\mathbf{y}^*$ which satisfies $\operatorname{rank} \mathbf{M}_i(\mathbf{y}^*) = 1$, and so, necessarily, $\mathbf{M}_i(\mathbf{y}^*) = \mathbf{v}_i(\mathbf{x}^*)\mathbf{v}_i(\mathbf{x}^*)'$ for some $\mathbf{x}^* \in \mathbb{R}^n$. In addition, the constraints $\mathbf{M}_{i-v_j}(g_j\mathbf{y}^*) \succeq 0$ imply that $\mathbf{x}^* \in \mathbb{K}$. That is, $\mathbf{y}^*$ is the vector of moments up to order $2i$ of the Dirac measure at $\mathbf{x}^* \in \mathbb{K}$, and one reads the optimal solution $\mathbf{x}^*$, from the subvector of first "moments" y^*_{α} with $|\alpha| = 1$.

For the case of multiple global minimizers, we have the following sufficient condition which is implemented at step 3 of Algorithm 5.2:

Theorem 5.7. *Let* $f \in \mathbb{R}[\mathbf{x}]$, *and suppose that the optimal value* ρ_i *of Problem (5.14) is attained at some optimal solution* $\mathbf{y}^*$.
Let $v := \max_{j=1,\ldots,m} v_j$. *If* $\operatorname{rank} \mathbf{M}_{i-v}(\mathbf{y}^*) = \operatorname{rank} \mathbf{M}_i(\mathbf{y}^*)$, *then* $f^*_{\mathbb{K}} = \rho_i$ *and there are at least* $s := \operatorname{rank} \mathbf{M}_i(\mathbf{y}^*)$ *global minimizers, and they can be extracted by Algorithm 4.2.*

Proof. We already know that $\rho_i \leq f_{\mathbb{K}}^*$ for all i. If $s = \operatorname{rank} \mathbf{M}_{i-v}(\mathbf{y}^*) = \operatorname{rank} \mathbf{M}_i(\mathbf{y}^*)$, then by Theorem 3.11, $\mathbf{y}^*$ is the moment vector of some s-atomic probability measure μ^* on $\mathbb{K}$. We then argue that each of the s atoms of μ^* is a global minimizer of f on $\mathbb{K}$. Indeed, being s-atomic, there is a family $(\mathbf{x}(k))_{k=1}^s \subset \mathbb{K}$ and a family $(\beta_k)_{k=1}^s \subset \mathbb{R}$, such that

$$\mu^* = \sum_{k=1}^{s} \beta_k \, \delta_{\mathbf{x}(k)}, \quad \beta_k > 0, \quad \forall k = 1, \ldots, s; \quad \sum_{k=1}^{s} \beta_k = 1,$$

Hence,

$$f_{\mathbb{K}}^* \geq \rho_i = L_{\mathbf{y}^*}(f) = \int_{\mathbb{K}} f \, d\mu^* = \sum_{k=1}^{s} \beta_k \, f(\mathbf{x}(k)) \geq f_{\mathbb{K}}^*,$$

which clearly implies $f_{\mathbb{K}}^* = \rho_i$, and $f(\mathbf{x}(k)) = f_{\mathbb{K}}^*$, for all $k = 1, \ldots, s$, the desired result. □

The rank-test of Theorem 5.7 implies that if it is satisfied, we can conclude that there are $\operatorname{rank} \mathbf{M}_i(\mathbf{y}^*)$ global minimizers encoded in the optimal solution $\mathbf{y}^*$ of (5.14). In order to extract these solutions from $\mathbf{y}^*$, we apply Algorithm 4.2. This extraction algorithm is implemented in the GloptiPoly software described in Appendix D.

Example 5.3. Consider the optimization problem

$$\begin{aligned} f_{\mathbb{K}}^* = \min\ & x_1^2 x_2^2 (x_1^2 + x_2^2 - 1) \\ \text{s.t. } & x_1^2 + x_2^2 \leq 4, \end{aligned}$$

which is the same as in Example 5.2, but with the additional ball constraint $\|\mathbf{x}\|^2 \leq 4$. Therefore the optimal value is $f_{\mathbb{K}}^* = -1/27$ with global minimizers $\mathbf{x}^* = (\pm\sqrt{3}/3, \pm\sqrt{3}/3)$. Applying Algorithm 5.2 for $k = 4$ and using GloptyPoly, the optimal value is obtained with $\rho_4 \approx -1/27$ and the four optimal solutions $(\pm 0.5774, \pm 0.5774)' \approx \mathbf{x}^*$ are extracted.

When we add the additional nonconvex constraint $x_1 x_2 \geq 1$, we find that the optimal value is obtained with $\rho_3 \approx 1$ and the two optimal solutions $(x_1^*, x_2^*)' = (-1, -1)'$ and $(x_1^*, x_2^*)' = (1, 1)'$ are extracted. In both cases, the rank-test is satisfied for an optimal solution $\mathbf{y}^*$ and a global optimality certificate is provided thanks to Theorem 5.7.

Example 5.4. Let $n = 2$ and $f \in \mathbb{R}[\mathbf{x}]$ be the concave polynomial

$$f(\mathbf{x}) := -(x_1 - 1)^2 - (x_1 - x_2)^2 - (x_2 - 3)^2$$

and let $\mathbb{K} \subset \mathbb{R}^2$ be the set:

$$\mathbb{K} := \{\mathbf{x} \in \mathbb{R}^2 \,:\, 1 - (x_1 - 1)^2 \geq 0,\, 1 - (x_1 - x_2)^2 \geq 0,\, 1 - (x_2 - 3)^2 \geq 0\}.$$

The optimal value is $f^*_{\mathbb{K}} = -2$. Solving problem (5.14) for $i = 1$, yields $\rho_1 = -3$ instead of the desired value -2. On the other hand, solving problem (5.14) for $i = 2$ yields $\rho_2 \approx -2$ and the three optimal solutions $(x_1^*, x_2^*) = (1,2), (2,2), (2,3)$ are extracted. Hence, with polynomials of degree 4 instead of 2, we obtain (a good approximation of) the correct value. Note that there exist scalars $\lambda_j = 1 \geq 0$ such that

$$f(\mathbf{x}) + 3 = 0 + \sum_{j=1}^{3} \lambda_j g_j(\mathbf{x}),$$

but $f(\mathbf{x}) - f^*_{\mathbb{K}}$ $(= f(\mathbf{x}) + 2)$ cannot be written in this way (otherwise ρ_1 would be the optimal value -2).

For quadratically constrained nonconvex quadratic problems, the semidefinite program (5.14) with $i = 1$ is a well-known relaxation. But ρ_1 which sometimes provides directly the exact global minimum value, is only a lower bound in general.

5.3.2 *The univariate case*

If we consider the univariate case $n = 1$, and with $\mathbb{K} = [a, b]$ or $\mathbb{K} = [0, \infty)$, the corresponding sequence of semidefinite relaxations (5.14) simply reduces to a *single* relaxation.

> In other words, the minimization of a univariate polynomial on an interval of the real line (bounded or not) is a convex optimization problem and reduces to solving a semidefinite program.

Indeed, consider for instance, the case where f has degree $2r$ and $\mathbb{K} \subset \mathbb{R}$ is the interval $[-1, 1]$. By Theorem 2.6(b)

$$f(x) - f^*_{\mathbb{K}} = f_0(x) + f_3(x)(1 - x^2), \qquad x \in \mathbb{R}, \tag{5.16}$$

for some s.o.s. polynomials $f_0, f_3 \in \Sigma[x]$ such that the degree of the summands is less than $2r$.

Theorem 5.8. *Let $f \in \mathbb{R}[x]$ be a univariate polynomial of degree $2r$. The semidefinite relaxation (5.14) for $i = r$ of the problem*

$$f^*_{\mathbb{K}} = \min_x \{f(x) : \; 1 - x^2 \geq 0\},$$

is exact, i.e., $\rho_r = \rho^*_r = f^*_{\mathbb{K}}$. In addition, both (5.14) and (5.15) have an optimal solution.

Proof. From (5.16), $(f^*_{\mathbb{K}}, f_0, f_3)$ is a feasible solution of (5.15) with $i = r$, and so, optimal, because $\rho^*_r = f^*_{\mathbb{K}}$. Therefore, from $\rho^*_r \leq \rho_r \leq f^*_{\mathbb{K}}$, we also obtain $\rho_r = f^*_{\mathbb{K}}$, which in turn implies that $\mathbf{y}^* := \mathbf{v}_{2r}(\mathbf{x}^*)$ is an optimal solution, for any global minimizer $x^* \in \mathbb{K}$. □

A similar argument holds if $f \in \mathbb{R}[x]$ has odd degree $2r - 1$, in which case

$$f(x) - f^*_{\mathbb{K}} = f_1(x)(1 + x) + f_2(x)(1 - x), \quad x \in \mathbb{R},$$

for some s.o.s. polynomials $f_1, f_2 \in \Sigma[x]$ such that the degree of the summands is less than $2r - 1$. Again, for the problem

$$f^*_{\mathbb{K}} = \min_x \{f(x) : \; 1 - x \geq 0, \; 1 + x \geq 0\},$$

the relaxation (5.14) with $i = r$ is exact.

5.3.3 *Numerical experiments*

In this section, we report on the performance of Algorithm 5.2 using its implementation in the software package GloptiPoly (see Chapter D) on a series of benchmark non-convex continuous optimization problems.

In Table 5.1, we record the problem name, the source of the problem, the number of decision variables (**var**), the number of inequality or equality constraints (**cstr**), and the maximum degree arising in the polynomial expressions (**deg**), the CPU time in seconds (**CPU**) and the order of the relaxation (**order**).

At the time of the experiment, Gloptipoly was using the semidefinite optimization solver SeDuMi; see Sturm (1999). As indicated by the label **dim** in the rightmost column, quadratic problems 2.8, 2.9 and 2.11 in Floudas *et al.* (1999) involve more than 19 variables and could not be handled by the current version of GloptiPoly. Except for problems 2.4 and

Table 5.1 Continuous optimization problems. CPU times and semidefinite relaxation orders required to reach global optimality.

problem	var	cstr	deg	CPU	order
(Lasserre, 2001, Ex. 1)	2	0	4	0.13	2
(Lasserre, 2001, Ex. 2)	2	0	4	0.13	2
(Lasserre, 2001, Ex. 3)	2	0	6	1.13	8
(Lasserre, 2001, Ex. 5)	2	3	2	0.22	2
(Floudas *et al.*, 1999, Pb. 2.2)	5	11	2	11.8	3
(Floudas *et al.*, 1999, Pb. 2.3)	6	13	2	1.86	2
(Floudas *et al.*, 1999, Pb. 2.4)	13	35	2	1012	2
(Floudas *et al.*, 1999, Pb. 2.5)	6	15	2	1.58	2
(Floudas *et al.*, 1999, Pb. 2.6)	10	31	2	67.7	2
(Floudas *et al.*, 1999, Pb. 2.7)	10	25	2	75.3	2
(Floudas *et al.*, 1999, Pb. 2.8)	20	10	2	-	dim
(Floudas *et al.*, 1999, Pb. 2.9)	24	10	2	-	dim
(Floudas *et al.*, 1999, Pb. 2.10)	10	11	2	45.3	2
(Floudas *et al.*, 1999, Pb. 2.11)	20	10	2	-	dim
(Floudas *et al.*, 1999, Pb. 3.2)	8	22	2	3032	3
(Floudas *et al.*, 1999, Pb. 3.3)	5	16	2	1.20	2
(Floudas *et al.*, 1999, Pb. 3.4)	6	16	2	1.50	2
(Floudas *et al.*, 1999, Pb. 3.5)	3	8	2	2.42	4
(Floudas *et al.*, 1999, Pb. 4.2)	1	2	6	0.17	3
(Floudas *et al.*, 1999, Pb. 4.3)	1	2	50	0.94	25
(Floudas *et al.*, 1999, Pb. 4.4)	1	2	5	0.25	3
(Floudas *et al.*, 1999, Pb. 4.5)	1	2	4	0.14	2
(Floudas *et al.*, 1999, Pb. 4.6)	2	2	6	0.41	3
(Floudas *et al.*, 1999, Pb. 4.7)	1	2	6	0.20	3
(Floudas *et al.*, 1999, Pb. 4.8)	1	2	4	0.16	2
(Floudas *et al.*, 1999, Pb. 4.9)	2	5	4	0.31	2
(Floudas *et al.*, 1999, Pb. 4.10)	2	6	4	0.58	4

3.2, the computational load is moderate. In almost all reported instances the global optimum was reached exactly by a semidefinite relaxation of small order.

5.3.4 *Exploiting sparsity*

As already mentioned in Chapter 4, despite their nice properties, the size of the semidefinite relaxations (5.14) grows rapidly with the dimension n. Typically, the moment matrix $\mathbf{M}_i(\mathbf{y})$ is $s(i) \times s(i)$ with $s(i) = \binom{n+i}{n}$, and there are $\binom{n+2i}{n}$ variables $y_{\boldsymbol{\alpha}}$. This makes the applicability of Algorithm 5.2 limited to small to medium size problems only. Fortunately, in many practical applications of large size moment problems, some sparsity pattern is often present and may be exploited as outlined in Chapter 4.

Suppose that there is no *coupling* between some subsets of variables in the polynomials g_k that define the set $\mathbb{K}$, and the polynomial f. That is, by no coupling between two sets of variables we mean that there is *no* monomial involving some variables of such subsets in any of the polynomials f, (g_k).

Recalling the notation of Section 4.6, let $I_0 := \{1,\ldots,n\}$ be the union $\cup_{k=1}^p I_k$ of p subsets I_k, $k = 1,\ldots,p$, with cardinal denoted n_k (with possible overlaps). For an arbitrary $J \subseteq I_0$, let $\mathbb{R}[\mathbf{x}(J)]$ denote the ring of polynomials in the variables $\mathbf{x}(J) = \{x_i \,:\, i \in J\}$, and so $\mathbb{R}[\mathbf{x}(I_0)] = \mathbb{R}[\mathbf{x}]$. So let Assumption 2.3, 2.4 hold and as in section 2.7, let $m' = m + p$ and $\{1,\ldots,m'\} = \cup_{i=1}^p J_i$ and $\mathbb{K}$ be as in (2.30) after having added in its definition the p redundant quadratic constraints (2.29).

With $k \in \{1,\ldots,p\}$ fixed, and $g \in \mathbb{R}[\mathbf{x}(I_k)]$, let $\mathbf{M}_i(\mathbf{y}, I_k)$ (resp. $\mathbf{M}_i(g\,\mathbf{y}, I_k)$) be the moment (resp. localizing) submatrix obtained from $\mathbf{M}_i(\mathbf{y})$ (resp. $\mathbf{M}_i(g\,\mathbf{y})$) by retaining only those rows (and columns) $\alpha \in \mathbb{N}^n$ of $\mathbf{M}_i(\mathbf{y})$ (resp. $\mathbf{M}_i(g\,\mathbf{y})$) such that $\mathrm{supp}(\boldsymbol{\alpha}) \in \mathscr{I}_k$. Hence in the present context, the sparse semidefinite relaxation (4.36) reads:

$$\begin{aligned} \rho_i^{\text{sparse}} = \; & \inf_{\mathbf{y}} \; L_{\mathbf{y}}(f) \\ & \text{s.t. } \mathbf{M}_i(\mathbf{y}, I_k) \succeq \mathbf{0}, \qquad k = 1,\ldots,p \\ & \quad\;\; \mathbf{M}_{i-v_j}(g_j\mathbf{y}, I_k) \succeq \mathbf{0}, \; j \in J_k, \; k = 1,\ldots,p \\ & \quad\;\; y_{\mathbf{0}} = 1 \end{aligned} \tag{5.17}$$

whereas its dual is the semidefinite program

$$\begin{aligned} (\rho_i^{\text{sparse}})^* = \; & \sup_{\lambda,\sigma_{kj}} \; \lambda \\ & \text{s.t. } f - \lambda = \sum_{k=1}^{p} \left(\sigma_{k0} + \sum_{j \in J_k} \sigma_{kj}\, g_j \right) \\ & \qquad \sigma_{k0}, \sigma_{kj} \in \Sigma[\mathbf{x}(I_k)], \quad k = 1,\ldots,p \\ & \qquad \deg \sigma_{k0}, \deg \sigma_{kj} g_j \leq 2i, \quad k = 1,\ldots,p. \end{aligned} \tag{5.18}$$

Theorem 5.9. *Let Assumptions 2.3 and 2.4 hold. Consider the sparse semidefinite relaxations (5.17) and (5.18).*

If the running intersection property (4.33) holds then $\rho_i^{\text{sparse}} \uparrow f_{\mathbb{K}}^*$ *and* $(\rho_i^{\text{sparse}})^* \uparrow f_{\mathbb{K}}^*$ *as* $i \to \infty$.

Theorem 5.9 is just Theorem 4.7 adapted to the present context of global optimization. To see the gain in terms of number of variables and size of the moment and localizing matrices, the reader is referred to the comment just after Theorem 4.7. For instance, if $\tau := \sup_k |I(k)|$ is relatively small (say e.g. $6, 7$) then one may solve optimization problems with $n = 1000$ variables whereas with $n = 1000$, one may not even implement the first standard semidefinite relaxation (5.14) with $i = 1$!

5.4 Linear Programming Relaxations

In this section, and still with $\mathbb{K}$ as in (5.3), we derive linear programming relaxations, the specialized version for problem (5.2) of those described in Chapter 4 for the generalized moment problem with polynomial data.

Let $\widehat{g}_j$ be the normalized version associated with g_j, $j = 1, \ldots, m$, and defined in (2.17). Let $i \in \mathbb{N}$ be fixed and consider the following linear optimization problem:

$$L_i = \inf_{\mathbf{y}} \{L_{\mathbf{y}}(f) : y_0 = 1; \quad L_{\mathbf{y}}(\widehat{\mathbf{g}}^{\boldsymbol{\alpha}}(1-\widehat{\mathbf{g}})^{\boldsymbol{\beta}}) \geq 0, \quad \forall\, |\boldsymbol{\alpha}+\boldsymbol{\beta}| \leq i\}. \tag{5.19}$$

with $L_{\mathbf{y}}$ being as in (3.3). Indeed, Problem (5.19) is a linear optimization problem, because since $|\boldsymbol{\alpha}+\boldsymbol{\beta}| \leq i$, the conditions $L_{\mathbf{y}}(\widehat{g}^{\boldsymbol{\alpha}}(1-\widehat{g})^{\boldsymbol{\beta}}) \geq 0$ yield finitely many linear inequality constraints on finitely many coefficients $y_{\boldsymbol{\alpha}}$ of the infinite sequence $\mathbf{y}$. The dual of Problem (5.19) is

$$L_i^* = \sup_{\lambda, \mathbf{u} \geq 0} \{\lambda : f - \lambda = \sum_{\boldsymbol{\alpha},\boldsymbol{\beta} \in \mathbb{N}^m,\, |\boldsymbol{\alpha}+\boldsymbol{\beta}| \leq i} u_{\boldsymbol{\alpha},\boldsymbol{\beta}}\, \widehat{\mathbf{g}}^{\boldsymbol{\alpha}}\, (1-\widehat{\mathbf{g}})^{\boldsymbol{\beta}}\}. \tag{5.20}$$

From Theorem 4.2 we obtain

Theorem 5.10. *The sequence $(L_i)_i$ is monotone nondecreasing, and under Assumption 2.2,*

$$\lim_{i\to\infty} L_i^* = \lim_{i\to\infty} L_i = f_{\mathbb{K}}^*. \tag{5.21}$$

The proof which uses Theorem 4.2 is left as an exercise. So the relaxations (5.19)-(5.20) are linear programs, a good news because in principle, with current LP packages, one is able to solve linear programs with million variables and constraints! However, as we will see, the linear programs (5.19)-(5.20) suffer from some serious drawbacks.

5.4.1 The case of a convex polytope

In the particular case where the basic semi-algebraic set $\mathbb{K} \subset \mathbb{R}^n$ is compact and all the polynomials g_i in the definition (5.3) of $\mathbb{K}$ are affine (and so $\mathbb{K}$ is a polytope), one may specialize the linear relaxation (5.19) to:

$$L_i = \inf_{\mathbf{y}} \{ L_{\mathbf{y}}(f) \,:\, y_0 = 1; \quad L_{\mathbf{y}}(\mathbf{g}^{\boldsymbol{\alpha}}) \geq 0, \quad \forall \, |\boldsymbol{\alpha}| \leq i \}, \tag{5.22}$$

and its associated dual:

$$L_i^* = \max_{\lambda, \mathbf{u} \geq 0} \{ \lambda \,:\, f - \lambda = \sum_{\boldsymbol{\alpha} \in \mathbb{N}^m;\, |\boldsymbol{\alpha}| \leq i} u_{\boldsymbol{\alpha}} \, \mathbf{g}^{\boldsymbol{\alpha}} \}. \tag{5.23}$$

Theorem 5.11. *If all g_j's in (5.3) are affine and $\mathbb{K}$ is a convex polytope with nonempty interior, then:*

$$\lim_{i \to \infty} L_i^* = \lim_{i \to \infty} L_i = f_{\mathbb{K}}^*. \tag{5.24}$$

The proof is analogous to that of Theorem 5.10 and is left as an exercise.

5.4.2 Contrasting LP and semidefinite relaxations

Theorem 5.11 implies that we can approach the global optimal value $f_{\mathbb{K}}^*$ as closely as desired by solving linear optimization problems (5.19) of increasing size. This should be interesting because very powerful linear optimization software packages are available, in contrast with semidefinite optimization software packages that have not yet reached the level of maturity of the linear optimization packages.

Unfortunately we next show that in general the LP-relaxations (5.19) cannot be exact, that is, the convergence in (5.21) (or in (5.24) when $\mathbb{K}$ is a convex polyope) is only asymptotic, not finite. Indeed, assume that the convergence is finite, i.e., $L_i^* = f_{\mathbb{K}}^*$ for some $i \in \mathbb{N}$. Suppose that the interior of $\mathbb{K}$ (int $\mathbb{K}$) is given by

$$\operatorname{int} \mathbb{K} = \{ \mathbf{x} \in \mathbb{R}^n \,:\, g_j(\mathbf{x}) > 0, \quad j = 1, \ldots, m \}.$$

Then, if there exists a global minimizer $\mathbf{x}^* \in \operatorname{int} \mathbb{K}$, and $1 - \widehat{g}_j(\mathbf{x}^*) > 0$ for all $j = 1, \ldots, m$, we would get the contradiction

$$0 = f(\mathbf{x}^*) - L_i^* = \sum_{\boldsymbol{\alpha}, \boldsymbol{\beta} \in \mathbb{N}^m} \mathbf{u}_{\boldsymbol{\alpha}, \boldsymbol{\beta}} \, \widehat{\mathbf{g}}(\mathbf{x}^*)^{\boldsymbol{\alpha}} (1 - \widehat{\mathbf{g}}(\mathbf{x}^*))^{\boldsymbol{\beta}} > 0.$$

Similarly, if $\mathbf{x}^* \in \partial\mathbb{K}$, let $J^* := \{j : g_j(\mathbf{x}^*) = 0\}$. For the same reasons, if there exists $\mathbf{x} \in \mathbb{K}$ with $f(\mathbf{x}) > f(\mathbf{x}^*)$ and $g_j(\mathbf{x}) = 0$ for all $j \in J^*$, then finite convergence cannot take place.

Example 5.5. Let $\mathbb{K} := \{x \in \mathbb{R} : 0 \leq x \leq 1\}$ and $x \mapsto f(x) := x(x-1)$. This is a convex optimization problem with global minimizer $x^* = 1/2$ in the interior of $\mathbb{K}$, and optimal value $-1/4$. The optimal values $(L_i)_i$ of the linear relaxations (5.19) are reported in Table 5.2. The example shows the rather slow monotone convergence of $L_i \rightarrow -0.25$, despite the original problem is convex.

Table 5.2 The slow convergence of the linear relaxation (5.19).

i	2	4	6	10	15
L_i	-1/3	-1/3	-0.3	-0.27	-0.2695

On the other hand, with $-f$ instead of f, the problem becomes a harder concave minimization problem. But this time the second relaxation is exact! Indeed, $f^*_{\mathbb{K}} = 0$ and we have

$$f(x) - f^*_{\mathbb{K}} = x - x^2 = x(1-x) = g_1(x)g_2(x),$$

with $g_1(x) = x$ and $g_2(x) = 1 - x$.

Example 5.5 illustrates that the convergence $L_i \uparrow f^*_{\mathbb{K}}$ as $i\rightarrow\infty$, is in general asymptotic, not finite; as underlined, the global minimizer being in the interior of $\mathbb{K}$, convergence *cannot* be finite. In addition, Example 5.5 exhibits an annoying paradox, namely that LP-relaxations may perform better for the concave minimization problem than for the *a priori* easier convex minimization problem. Finally, notice that for large values of i, the constraints of the LP-relaxations (5.20) should contain very large coefficients due to the presence of binomial coefficients in the terms $(1 - \widehat{\mathbf{g}})^\beta$, a source of numerical instability and ill-conditioned problems.

5.5 Global Optimality Conditions

In this section, we derive global optimality conditions for polynomial optimization generalizing the local optimality conditions due to Karush-Kuhn-Tucker (KKT) for nonlinear optimization.

A vector $(\mathbf{x}^*, \boldsymbol{\lambda}^*) \in \mathbb{R}^n \times \mathbb{R}^m$ satisfies the KKT conditions associated with problem (5.2) (and is called a KKT pair) if

$$\begin{aligned} \nabla f(\mathbf{x}^*) &= \sum_{j=1}^{m} \boldsymbol{\lambda}_j^* \nabla g_j(\mathbf{x}^*), \\ \boldsymbol{\lambda}_j^* g_j(\mathbf{x}^*) &= 0, \quad j = 1, \ldots, m, \\ g_j(\mathbf{x}^*),\ \lambda_j^* &\geq 0, \quad j = 1, \ldots, m. \end{aligned} \tag{5.25}$$

• The nonnegative dual variables $\boldsymbol{\lambda}^* \in \mathbb{R}^m$ are called the Lagrange Karush-Kuhn-Tucker multipliers; See Section C.
• In fact, most local optimization algorithms try to find a pair of vectors $(\mathbf{x}^*, \boldsymbol{\lambda}^*) \in \mathbb{R}^n \times \mathbb{R}_+^m$ that satisfies (5.25).
• In general, $\mathbf{x}^*$ is *not* a global minimizer of the Lagrangian polynomial

$$\mathbf{x} \mapsto \quad L_f(\mathbf{x}) := f(\mathbf{x}) - f_{\mathbb{K}}^* - \sum_{j=1}^{m} \lambda_j^* g_j(\mathbf{x}). \tag{5.26}$$

• However, if f is convex, the g_i's are concave, and the interior of $\mathbb{K}$ is nonempty (i.e., Slater's condition holds), then (5.25) are necessary and sufficient optimality conditions for $\mathbf{x}^*$ to be an optimal solution of (5.2). Moreover, $\mathbf{x}^*$ is a global minimizer of the Lagrangian polynomial L_f which is nonnegative on $\mathbb{R}^n$, with $L_f(\mathbf{x}^*) = 0$.

The developments in the earlier section lead to the following global optimality conditions.

Theorem 5.12. *Let $\mathbf{x}^* \in \mathbb{K}$ be a global minimizer for problem (5.2), with global optimum $f_{\mathbb{K}}^*$, and assume that $f - f_{\mathbb{K}}^*$ has the representation (5.13), i.e.,*

$$f(\mathbf{x}) - f_{\mathbb{K}}^* = \sigma_0(\mathbf{x}) + \sum_{j=1}^{m} \sigma_j(\mathbf{x})\, g_j(\mathbf{x}), \quad \mathbf{x} \in \mathbb{R}^n, \tag{5.27}$$

for some s.o.s. polynomials $\{\sigma_j\}_{i=0}^m \subset \Sigma[\mathbf{x}]$. Then:
(a) *$\sigma_j(\mathbf{x}^*), g_j(\mathbf{x}^*) \geq 0$, for all $j = 1, \ldots, m$.*
(b) *$\sigma_j(\mathbf{x}^*) g_j(\mathbf{x}^*) = 0$, for all $j = 1, \ldots, m$.*
(c) *$\nabla f(\mathbf{x}^*) = \sum_{i=1}^m \sigma_j(\mathbf{x}^*)\, \nabla g_j(\mathbf{x}^*)$, that is $(\mathbf{x}^*, \boldsymbol{\lambda}^*)$ is a KKT pair, with $\lambda_i^* := \sigma_j(\mathbf{x}^*)$, $\forall j = 1, \ldots, m$.*
(d) *$\mathbf{x}^*$ is a global minimizer of the polynomial $f - f_{\mathbb{K}}^* - \sum_{j=1}^m \sigma_j g_j$.*

Proof. **(a)** Part (a) is obvious from $\mathbf{x}^* \in \mathbb{K}$, and the polynomials σ_j's are s.o.s.

(b) From (5.27) and the fact that $\mathbf{x}^*$ is a global minimizer, we obtain

$$f(\mathbf{x}^*) - f^*_{\mathbb{K}} = 0 = \sigma_0(\mathbf{x}^*) + \sum_{j=1}^{m} \sigma_j(\mathbf{x}^*) g_j(\mathbf{x}^*),$$

which in turn implies part (b) because $g_j(\mathbf{x}^*) \geq 0$ for all $j = 1, \ldots, m$, and the polynomials (σ_j) are all s.o.s., hence nonnegative. This also implies $\sigma_0(\mathbf{x}^*) = 0$.
(c) Differentiating and using the fact that the polynomials (σ_j) are s.o.s., and using part (b), yields part (c).

(d) From (5.27) we obtain,

$$f - f^*_{\mathbb{K}} - \sum_{j=1}^{m} \sigma_j\, g_j = \sigma_0 \geq 0,$$

because $\sigma_0 \in \mathbb{R}[\mathbf{x}]$ is s.o.s., and using property **(b)**,

$$f(\mathbf{x}^*) - f^*_{\mathbb{K}} - \sum_{j=1}^{m} \sigma_j(\mathbf{x}^*)\, g_j(\mathbf{x}^*) = 0,$$

which shows that $\mathbf{x}^*$ is a global minimizer of $f - f^*_{\mathbb{K}} - \sum_j \sigma_j g_j$. □

> Theorem 5.12 implies that when $f - f^*_{\mathbb{K}}$ has the representation (5.27), then:
>
> **(a)** (5.27) should be interpreted as a **global optimality condition**.
>
> **(b)** The s.o.s. polynomial coefficients $\{\sigma_j\} \subset \mathbb{R}[\mathbf{x}]$ should be interpreted as generalized Karush-Kuhn-Tucker Lagrange multipliers.
>
> **(c)** The polynomial $f - f^*_{\mathbb{K}} - \sum_j \sigma_j g_j$ is a generalized Lagrangian, with s.o.s. polynomial multipliers instead of nonnegative scalars. It is s.o.s. (hence nonnegative on $\mathbb{R}^n$), vanishes at every global minimizer $\mathbf{x}^* \in \mathbb{K}$, and so $\mathbf{x}^*$ is also a global minimizer of this Lagrangian.

Note that in the local KKT optimality conditions (5.25), only the constraints $g_j(\mathbf{x}) \geq 0$ binding at $\mathbf{x}^*$, have a possibly nontrivial associated Lagrange (scalar) multiplier λ_j^*. In contrast, in the global optimality condition (5.27), every constraint $g_j(\mathbf{x}) \geq 0$ has a possibly nontrivial s.o.s. polynomial Lagrange multiplier $\mathbf{x} \mapsto \sigma_j(\mathbf{x})$. Note that if $g_j(\mathbf{x}^*) > 0$, then necessarily $\sigma_j(\mathbf{x}^*) = 0 = \lambda_j^*$, as in the local KKT optimality conditions.

> In non convex optimization, a constraint $g_j(\mathbf{x}) \geq 0$ that is not binding at a global minimizer $\mathbf{x}^* \in \mathbb{K}$ may still be important, i.e., if removed from the definition of $\mathbb{K}$, then the global minimum $f_{\mathbb{K}}^*$ may decrease strictly. In this case, and in contrast to local KKT optimality conditions (5.25), g_j is necessarily involved in the representation (5.13) of $f - f_{\mathbb{K}}^*$ (when the latter exists), hence with a nontrivial s.o.s. multiplier σ_j which vanishes at $\mathbf{x}^*$.

To see this consider the following trivial example.

Example 5.6. Let $n = 1$ and consider the following problem:

$$f_{\mathbb{K}}^* = \min_x \{-x : \; 1/2 - x \geq 0, \; x^2 - 1 = 0\},$$

with optimal value $f_{\mathbb{K}}^* = 1$, and global minimizer $x^* = -1$. The constraint $1/2 - x \geq 0$ is not binding at $x^* = -1$, but if removed, the global minimum jumps to -1 with new global minimizer $x^* = 1$. In fact, we have the representation

$$f(x) - f_{\mathbb{K}}^* = -(x+1) = (x^2 - 1)(x + 3/2) + (1/2 - x)(x+1)^2,$$

which shows the important role of the constraint $1/2 - x \geq 0$ in the representation of $f - f_{\mathbb{K}}^*$, via its nontrivial multiplier $x \mapsto \sigma_1(x) := (x+1)^2$. Note also that $\sigma_1(x^*) = 0 = \lambda_1$ and $\sigma_2(x^*) = x^* + 3/2 = -1/2 = \lambda_2$ are the KKT Lagrange multipliers $(\lambda_1, \lambda_2) \in \mathbb{R}_+ \times \mathbb{R}$ in the local optimality conditions (5.25).

5.6 Convex Polynomial Programs

If practice seems to reveal that convergence of the semidefinite relaxations is often fast and even finite, we have seen that their size grows rapidly with the rank i in the hierarchy. And so, if sparsity in the original problem data is not exploited, the approach is limited to small to medium size problems only. On the other hand, it is well-known that a large class of convex optimization problems can be solved efficiently. Therefore, as the moment approach is dedicated to solving difficult non convex (most of the time $\mathcal{NP}$-hard) problems, it should have the highly desirable feature to somehow *recognize* "easy" problems like convex ones. That is, when applied to such easy problems it should show some significant improvement or a particular nice behavior not necessarily valid in the general case. This is the issue that we investigate in this section.

5.6.1 *An extension of Jensen's inequality*

Recall that if μ is a probability measure on $\mathbb{R}^n$ with $E_\mu(\mathbf{x}) < \infty$, Jensen's inequality states that if $f \in L_1(\mu)$ and f is convex, then

$$E_\mu(f) \; (= \int_{\mathbb{R}^n} f \, d\mu) \; \geq \; f(E_\mu(\mathbf{x})),$$

a very useful property in many applications. We next provide an extension of Jensen's inequality when one restricts its application to the class of s.o.s.-convex polynomials. Recall that $\mathbb{N}^n_{2d} = \{\boldsymbol{\alpha} \in \mathbb{N}^n \, : \, |\boldsymbol{\alpha}| \leq 2d\}$.

Theorem 5.13. *Let $f \in \mathbb{R}[\mathbf{x}]_{2d}$ be s.o.s.-convex; see Definition 2.3. Let $\mathbf{y} = (y_\alpha)_{\alpha \in \mathbb{N}^n_{2d}}$ satisfy $y_0 = 1$ and $\mathbf{M}_d(\mathbf{y}) \succeq 0$. Then:*

$$L_{\mathbf{y}}(f) \; \geq \; f(L_{\mathbf{y}}(\mathbf{x})), \qquad (5.28)$$

where $L_{\mathbf{y}}(\mathbf{x}) = (L_{\mathbf{y}}(x_1), \ldots, L_{\mathbf{y}}(x_n))$.

Proof. Let $\mathbf{z} \in \mathbb{R}^n$ be fixed, arbitrary, and consider the polynomial $\mathbf{x} \mapsto f(\mathbf{x}) - f(\mathbf{z})$. Then from Lemma 2.30,

$$f(\mathbf{x}) - f(\mathbf{z}) \; = \; \langle \nabla f(\mathbf{z}), \mathbf{x} - \mathbf{z} \rangle + \langle (\mathbf{x} - \mathbf{z}), \mathbf{F}(\mathbf{x})(\mathbf{x} - \mathbf{z}) \rangle, \qquad (5.29)$$

with $\mathbf{F} : \mathbb{R}^n \to \mathbb{R}[\mathbf{x}]^{n \times n}$ being the matrix polynomial

$$\mathbf{x} \mapsto \quad \mathbf{F}(\mathbf{x}) := \int_0^1 \int_0^t \nabla^2 f(\mathbf{z} + s(\mathbf{x} - \mathbf{z})) \, ds \, dt.$$

As f is s.o.s.-convex, by Lemma 2.29, $\mathbf{F}$ is a s.o.s. matrix polynomial and so the polynomial $\mathbf{x} \mapsto g(\mathbf{x}) := \langle (\mathbf{x} - \mathbf{z}), \mathbf{F}(\mathbf{x})(\mathbf{x} - \mathbf{z}) \rangle$ is s.o.s., i.e., $g \in \Sigma[\mathbf{x}]$.

Then applying $L_{\mathbf{y}}$ to the polynomial $\mathbf{x} \mapsto f(\mathbf{x}) - f(\mathbf{z})$ and using (5.29) yields (recall that $y_{\mathbf{0}} = 1$)

$$\begin{aligned} L_{\mathbf{y}}(f) - f(\mathbf{z}) &= \langle \nabla f(\mathbf{z}), L_{\mathbf{y}}(\mathbf{x}) - \mathbf{z} \rangle + L_{\mathbf{y}}(g) \\ &\geq \langle \nabla f(\mathbf{z}), L_{\mathbf{y}}(\mathbf{x}) - \mathbf{z} \rangle \quad [\text{because } L_{\mathbf{y}}(g) \geq 0]. \end{aligned}$$

As $\mathbf{z} \in \mathbb{R}^n$ was arbitrary, taking $\mathbf{z} := L_{\mathbf{y}}(\mathbf{x}) \,(= (L_{\mathbf{y}}(x_1), \ldots, L_{\mathbf{y}}(x_n))$ yields the desired result. □

Hence (5.28) is Jensen's inequality extended to linear functionals $L_{\mathbf{y}} : \mathbb{R}[\mathbf{x}]_{2d} \to \mathbb{R}$ in the dual cone of $\Sigma[\mathbf{x}]_d$, that is, vectors $\mathbf{y} = (y_{\alpha})$ such that $\mathbf{M}_d(\mathbf{y}) \succeq 0$ and $y_{\mathbf{0}} = L_{\mathbf{y}}(1) = 1$; hence $\mathbf{y}$ is *not* necessarily the (truncated) moment sequence of some probability measure μ. As a consequence we also get:

Corollary 5.14. *Let f be a convex univariate polynomial, $g \in \mathbb{R}[\mathbf{x}]$ (and so $f \circ g \in \mathbb{R}[\mathbf{x}]$). Let $d := \lceil (\deg f \circ g)/2 \rceil$, and let $\mathbf{y} = (y_{\alpha})_{\alpha \in \mathbb{N}^n_{2d}}$ be such that $y_{\mathbf{0}} = 1$ and $\mathbf{M}_d(\mathbf{y}) \succeq 0$. Then:*

$$L_{\mathbf{y}}(f \circ g) \geq f(L_{\mathbf{y}}(g)). \tag{5.30}$$

5.6.2 *The s.o.s.-convex case*

Next, with $f \in \mathbb{R}[\mathbf{x}]_d$ and $2i \geq \max[\deg f, \max_j \deg g_j]$, consider the semidefinite program:

$$\rho_i = \inf_{\mathbf{y}} \{ L_{\mathbf{y}}(f) \,:\, \mathbf{M}_i(\mathbf{y}) \succeq 0;\, L_{\mathbf{y}}(g_j\, \mathbf{y}) \geq 0,\, j = 1, \ldots, m;\, y_{\mathbf{0}} = 1 \} \tag{5.31}$$

and its dual

$$\rho_i^* = \sup_{\gamma, \boldsymbol{\lambda} \in \mathbb{R}^m_+, \sigma_0 \in \Sigma[\mathbf{x}]_i} \{ \gamma \,:\, f - \gamma = \sigma_0 + \sum_{j=1}^m \lambda_j\, g_j \}. \tag{5.32}$$

Theorem 5.15. *Let $\mathbb{K}$ be as in (2.10) and Slater's condition hold. Let $f \in \mathbb{R}[\mathbf{x}]$ be such that $f^*_{\mathbb{K}} := \inf_{\mathbf{x}} \{ f(\mathbf{x}) \,:\, \mathbf{x} \in \mathbb{K} \} = f(\mathbf{z})$ for some $\mathbf{z} \in \mathbb{K}$. Assume that f and $-g_j$ are s.o.s.-convex, $j = 1, \ldots, m$.*

Then $f^*_{\mathbb{K}} = \rho_i = \rho_i^*$. Moreover, if $\mathbf{y}$ is an optimal solution of (5.31) then $\mathbf{x}^* := (L_{\mathbf{y}}(x_i)) \in \mathbb{K}$ is a global minimizer of f on $\mathbb{K}$.

Proof. Recall the definition of $Q_c(g)$ in (2.35). By Theorem 2.32, $f - f^*_{\mathbb{K}} \in Q_c(g)$, i.e., $f - f^*_{\mathbb{K}} = \sigma_0 + \sum_j \lambda_j g_j$ for some $\boldsymbol{\lambda} \in \mathbb{R}^m_+$ and some $\sigma_0 \in \Sigma[\mathbf{x}]$. Therefore, $(f^*_{\mathbb{K}}, \boldsymbol{\lambda}, \sigma_0)$ is a feasible solution of (5.32), which yields $\rho^*_i \geq f^*_{\mathbb{K}}$, and which combined with $\rho^*_i \leq \rho_i \leq f^*_{\mathbb{K}}$ yields $\rho^*_i = \rho_i = f^*_{\mathbb{K}}$. Obviously (5.31) is solvable and so, let $\mathbf{y}$ be an optimal solution. By Theorem 5.13, $f^*_{\mathbb{K}} = \rho^*_i = L_{\mathbf{y}}(f) \geq f(L_{\mathbf{y}}(\mathbf{x}))$, and similarly, $0 \geq L_{\mathbf{y}}(-g_j) \geq -g_j(L_{\mathbf{y}}(\mathbf{x}))$, $j = 1, \ldots, m$, which shows that $\mathbf{x}^* := L_{\mathbf{y}}(\mathbf{x}) \, (= L_{\mathbf{y}}(x_i)) \in \mathbb{K}$ is a global minimizer of f on $\mathbb{K}$. □

Therefore, when the polynomials f and $(-g_j)$ are s.o.s., the first semidefinite program in the hierarchy of semidefinite programs (5.14) is exact as it is either identical to (5.31) or more constrained, hence with optimal value $\rho_i = f^*_{\mathbb{K}}$. In other words, the methodology recognizes s.o.s.-convexity.

5.6.3 *The strictly convex case*

If f or some of the $-g_j$'s is not s.o.s.-convex but $\nabla^2 f \succ 0$ (so that f is strictly convex) and $-g_j$ is convex for every $j = 1, \ldots, m$, then one obtains the following result.

Theorem 5.16. *Let $\mathbb{K}$ be as in (2.10) and let Assumption 2.1 and Slater's condition hold. Assume that for all $\mathbf{x} \in \mathbb{R}^n$, $\nabla^2 f(\mathbf{x}) \succ 0$ and $-\nabla^2 g_j(\mathbf{x}) \succeq 0$, $j = 1, \ldots, m$.*

Then the hierarchy of semidefinite relaxations (5.14) -(5.15) has finite convergence. That is, $f^*_{\mathbb{K}} = \rho^*_i = \rho_i$ for some index i. In addition, both primal and dual relaxations are solvable.

Proof. Let $f^*_{\mathbb{K}} := \min_{\mathbf{x} \in \mathbb{K}} f(\mathbf{x})$. By Theorem 2.33,

$$f - f^*_{\mathbb{K}} = \sigma_0 + \sum_{j=1}^m \sigma_j \, g_j,$$

for some s.o.s. polynomials $(\sigma_j) \subset \Sigma[\mathbf{x}]$. Let $2i_0 \geq \max_j \deg \sigma_j + \deg g_j$ (with $g_0 = 1$). Then $(f^*_{\mathbb{K}}, (\sigma_j))$ is a feasible solution of the semidefinite program (5.15). Hence $f^*_{\mathbb{K}} \leq \rho^*_i \leq \rho_i \leq f^*_{\mathbb{K}}$, which yields the desired result $f^*_{\mathbb{K}} = \rho_i = \rho^*_i$. □

When compared to Theorem 5.15 for the s.o.s.-convex case, in the strictly convex case the simplified semidefinite relaxation (5.31) is not guaranteed to be exact. However, finite convergence still occurs for the standard semidefinite relaxations (5.14). Hence the hierarchy of semidefinite relaxations exhibits a particularly nice behavior for convex problems, a highly desirable property since convex optimization problems are easier to solve. In contrast, and using arguments from Section 5.4.2, such a nice behavior cannot be expected in general for the LP-relaxations (5.19).

5.7 Discrete Optimization

In this section, we consider problem (5.2) with $\mathbb{K}$ a finite variety. More precisely, in the definition (5.3) of $\mathbb{K}$, the g_i's are such that $\mathbb{K}$ can be rewritten

$$\mathbb{K} = \{\mathbf{x} \in \mathbb{R}^n \,:\, g_i(\mathbf{x}) = 0,\; i = 1, \ldots, m; \quad h_j(\mathbf{x}) \geq 0,\; j = 1, \ldots, s\}. \tag{5.33}$$

Indeed, (5.33) is a particular case of (5.3) where some inequality constraints $g_i(\mathbf{x}) \geq 0$ and $-g_i(\mathbf{x}) \geq 0$ are present.

The polynomials $(g_i)_{i=1}^m \subset \mathbb{R}[\mathbf{x}]$ define an ideal $I := \langle g_1, \ldots, g_m \rangle$ of $\mathbb{R}[\mathbf{x}]$, and we will consider the case where I is a zero-dimensional *radical ideal*, that is, I is radical, and the variety

$$V_{\mathbb{C}}(I) := \{\mathbf{x} \in \mathbb{C}^n \,:\, g_i(\mathbf{x}) = 0,\; i = 1, \ldots, m\},$$

is a finite set (see Sections 2.6 and A.2).

This is an important special case as it covers all basic 0/1 optimization problems. For instance, if we let $g_i(\mathbf{x}) := x_i^2 - x_i$, for all $i = 1, \ldots, n$, then we recover boolean optimization, in which case, $V_{\mathbb{C}}(I) = \{0, 1\}^n$ and $I = \langle x_1^2 - x_1, \ldots, x_n^2 - x_n \rangle$ is radical. Similarly, given $(r_i)_{i=1}^n \subset \mathbb{N}$, and a finite set of points $(x_{ij})_{j=1}^{r_i} \subset \mathbb{R}$, $i = 1, \ldots, n$, let

$$g_i(\mathbf{x}) := \prod_{j=1}^{r_i} (x_i - x_{ij}), \quad i = 1, \ldots, n.$$

Then we recover (bounded) integer optimization problems, in which case $V_{\mathbb{C}}(I)$ is the **grid** $\{(x_{1j_1}, x_{2j_2}, \cdots, x_{nj_n})\}$, where $j_k \in \{1, \ldots, r_k\}$ for all $k = 1, \ldots, n$.

Theorem 5.17. *Let $f \in \mathbb{R}[\mathbf{x}]$, $\mathbb{K} \subset \mathbb{R}^n$ be as in (5.33), and let the ideal $I = \langle g_1, \ldots, g_m \rangle \subset \mathbb{R}[\mathbf{x}]$ be radical and zero-dimensional. Let (ρ_i) be the optimal values of the semidefinite relaxations defined in (5.14). Then, there is some $i_0 \in \mathbb{N}$ such that $\rho_i = f^*_{\mathbb{K}}$ for all $i \geq i_0$.*

Proof. The polynomial $f - f^*_{\mathbb{K}}$ is nonnegative on $\mathbb{K}$. Therefore, by Theorem 2.26 there exist polynomials $(\sigma_k)_{k=1}^m \subset \mathbb{R}[\mathbf{x}]$, and s.o.s. polynomials $(\sigma_0, (v_j)_{j=1}^s) \subset \Sigma[\mathbf{x}]$, such that:

$$f - f^*_{\mathbb{K}} = \sigma_0 + \sum_{k=1}^{m} \sigma_k \, g_k + \sum_{j=1}^{s} v_j \, h_j.$$

Let d_1, d_2, d_3 be the degree of σ_0, the maximum degree of the polynomials $(\sigma_k \, g_k)_{i=1}^m$, and $(v_j \, h_j)_{j=1}^s$ respectively, and let $2i_0 \geq \max[d_1, d_2, d_3]$.

Then, $(f^*_{\mathbb{K}}, (\sigma_k), (v_j))$ is a feasible solution for of the relaxation (5.15) for $i = i_0$, and with value $f^*_{\mathbb{K}}$, so that $\rho^*_{i_0} \geq f^*_{\mathbb{K}}$. As we also have $\rho^*_i \leq \rho_i \leq f^*_{\mathbb{K}}$ whenever the semidefinite relaxations are well-defined, we conclude that $\rho^*_i = \rho_i = f^*_{\mathbb{K}}$ for all $i \geq i_0$. Finally, let μ be the Dirac probability measure at some global minimizer $\mathbf{x}^* \in \mathbb{K}$ of problem (5.2), and let $\mathbf{y} \in \mathbb{R}^\infty$ be the vector of its moments. Then $\mathbf{y}$ is feasible for all the semidefinite relaxations (5.14), with value $f^*_{\mathbb{K}}$, which completes the proof. □

In fact, more will be said in the next chapter. In particular, as soon as the real variety $V_{\mathbb{R}}(I)$ is finite (and even if I is not real radical), the semidefinite relaxation (5.14) is exact for some index i in the hierarchy, and one may extract *all* optimal solutions. See Theorem 6.1.

5.7.1 *Boolean optimization*

It is worth noting that in the semidefinite relaxations (5.14), the constraints $\mathbf{M}_r(g_i\mathbf{y}) = 0$ translate into simplifications via elimination of variables in the moment matrix $\mathbf{M}_r(\mathbf{y})$ and the localizing matrices $\mathbf{M}_r(h_j\mathbf{y})$. Indeed, consider for instance the boolean optimization case, i.e., when $g_i(\mathbf{x}) = x_i^2 - x_i$ for all $i = 1, \ldots, n$. Then the constraints $\mathbf{M}_r(g_i\mathbf{y}) = \mathbf{0}$ for all $i = 1, \ldots, n$, simply state that whenever $|\boldsymbol{\alpha}| \leq 2r$, one replaces every variable $y_{\boldsymbol{\alpha}}$ with the variable $y_{\boldsymbol{\beta}}$, where for $k = 1, \ldots, n$:

$$\beta_k = \begin{cases} 0, \text{ if } \alpha_k = 0, \\ 1, \text{ otherwise.} \end{cases}$$

Indeed, with $x_i^2 = x_i$ for all $i = 1, \ldots, n$, one has $\mathbf{x}^{\boldsymbol{\alpha}} = \mathbf{x}^{\boldsymbol{\beta}}$, with $\boldsymbol{\beta}$ as above. For instance, with $n = 2$, we obtain

$$\mathbf{M}_2(\mathbf{y}) = \begin{bmatrix} y_{00} & y_{10} & y_{01} & y_{10} & y_{11} & y_{01} \\ y_{10} & y_{10} & y_{11} & y_{10} & y_{11} & y_{11} \\ y_{01} & y_{11} & y_{01} & y_{11} & y_{11} & y_{01} \\ y_{10} & y_{10} & y_{11} & y_{10} & y_{11} & y_{11} \\ y_{11} & y_{11} & y_{11} & y_{11} & y_{11} & y_{11} \\ y_{01} & y_{11} & y_{01} & y_{11} & y_{11} & y_{01} \end{bmatrix}.$$

In addition, every column (row) of $\mathbf{M}_r(\mathbf{y})$ corresponding to a monomial $\mathbf{x}^{\boldsymbol{\alpha}}$, with $\alpha_k > 1$ for some $k \in \{1, \ldots, n\}$, is identical to the column corresponding to the monomial $\mathbf{x}^{\boldsymbol{\beta}}$, with $\boldsymbol{\beta}$ as above. Hence, the constraint $\mathbf{M}_r(\mathbf{y}) \succeq \mathbf{0}$ reduces to the new constraint $\widehat{\mathbf{M}}_r(\mathbf{y}) \succeq \mathbf{0}$, with the new simplified moment matrix

$$\widehat{\mathbf{M}}_2(\mathbf{y}) = \begin{bmatrix} y_{00} & y_{10} & y_{01} & y_{11} \\ y_{10} & y_{10} & y_{11} & y_{11} \\ y_{01} & y_{11} & y_{01} & y_{11} \\ y_{11} & y_{11} & y_{11} & y_{11} \end{bmatrix}.$$

Theorem 5.17 has little practical value. For instance, in the case of boolean optimization, we may easily show that $\rho_i = \rho_{i_0}$ for all $i \geq i_0 := n$. But in this case the simplified matrix $\widehat{\mathbf{M}}_{i_0}(\mathbf{y})$ has size $2^n \times 2^n$, and solving problem (5.2) by simple enumeration would be as efficient! However, in general, one obtains the exact global optimum at some earlier relaxation with value ρ_i, i.e., with $i \ll i_0$.

Numerical experiments.

One also reports the performance of GloptiPoly on a series of small-size combinatorial optimization problems (in particular, the MAXCUT problem). In Table 5.3 we first let GloptiPoly converge to the global optimum, in general extracting several solutions. The number of extracted solutions is reported in the column labelled *sol.*

Then, we slightly perturbed the criterion to be optimized in order to destroy the problem symmetry. Proceeding this way, the optimum solution

Table 5.3 Discrete optimization problems. CPU times and semidefinite relaxation order required to reach global optimum and extract several solutions.

problem	var	cstr	deg	CPU	order	sol
QP (Floudas *et al.*, 1999, Pb. 13.2.1.1)	4	4	2	0.10	1	1
QP (Floudas *et al.*, 1999, Pb. 13.2.1.2)	10	0	2	3.61	2	1
Max-Cut P_1 Floudas *et al.* (1999)	10	0	2	38.1	3	10
Max-Cut P_2 Floudas *et al.* (1999)	10	0	2	2.7	2	2
Max-Cut P_3 Floudas *et al.* (1999)	10	0	2	2.6	2	2
Max-Cut P_4 Floudas *et al.* (1999)	10	0	2	2.6	2	2
Max-Cut P_5 Floudas *et al.* (1999)	10	0	2	-	4	dim
Max-Cut P_6 Floudas *et al.* (1999)	10	0	2	2.6	2	2
Max-Cut P_7 Floudas *et al.* (1999)	10	0	2	44.3	3	4
Max-Cut P_8 Floudas *et al.* (1999)	10	0	2	2.6	2	2
Max-Cut P_9 Floudas *et al.* (1999)	10	0	2	49.3	3	6
Max-Cut cycle C_5 Anjos (2001)	5	0	2	0.19	3	10
Max-Cut complete K_5 Anjos (2001)	5	0	2	0.19	4	20
Max-Cut 5-node Anjos (2001)	5	0	2	0.24	3	6
Max-Cut antiweb AW_9^2 Anjos (2001)	9	0	2	-	4	dim
Max-Cut 10-node Petersen Anjos (2001)	10	0	2	39.6	3	10
Max-Cut 12-node Anjos (2001)	12	0	2	-	3	dim

is generically unique and convergence to the global optimum is easier.

Of course, the size of the combinatorial problems is relatively small and GloptiPoly cannot compete with *ad hoc* heuristics, which may solve problems with many more variables. But these numerical experiments are reported only to show the potential of the method, as in most cases the global optimum is reached at the second semidefinite relaxation in the hierarchy.

5.7.2 *Back to unconstrained optimization*

We have seen in Section 5.2 that for the unconstrained optimization problem (5.1), the semidefinite relaxations (5.6) reduce to a single one, and with a 0-1 answer, depending on whether or not, the polynomial $f - f^*$ is a sum of squares. Therefore, in general, according to Theorem 5.3, the SDP (5.6) provides only a lower bound on f^*.

However, if one knows *a priori* some bound M on the euclidean norm $\|\mathbf{x}^*\|$ of a global minimizer $\mathbf{x}^* \in \mathbb{R}^n$, then it suffices to replace the original unconstrained problem (5.1) with the constrained problem (5.2), with $\mathbb{K}$ being the basic semi-algebraic set

$$\mathbb{K} := \{\mathbf{x} \in \mathbb{R}^n : \quad M^2 - \|\mathbf{x}\|^2 \geq 0\}.$$

It is immediate to verify that Assumption 2.1 holds, and therefore, the machinery described in Section 5.3 applies, and the semidefinite relaxations (5.14) with $\mathbb{K}$ as above, converge to $f^* = f^*_{\mathbb{K}}$.

Another approach which avoids the *a priori* knowledge of this bound M, consists of taking

$$\mathbb{K} := \{\mathbf{x} \in \mathbb{R}^n : \quad \nabla f(\mathbf{x}) = 0\}, \tag{5.34}$$

since if a global minimizer $\mathbf{x}^* \in \mathbb{R}^n$ exists, then necessarily $\nabla f(\mathbf{x}^*) = 0$, and in addition, $\mathbf{x}^*$ is also a global minimizer of f on $\mathbb{K}$ defined in (5.34). However, convergence of the relaxations (5.14) has been proved for a compact set $\mathbb{K}$. Fortunately, the set $\mathbb{K}$ in (5.34) has nice properties. Indeed:

Proposition 5.18. *For almost all polynomials $f \in \mathbb{R}[\mathbf{x}]_d$, the ideal $I_f = \langle \partial f/\partial x_1, \ldots, \partial f/\partial x_n \rangle$ is zero-dimensional and radical.*

And so we get the following result:

Theorem 5.19. *With $f \in \mathbb{R}[\mathbf{x}]_d$, and $\mathbb{K} \subset \mathbb{R}^n$ as in (5.34), consider the semidefinite relaxation defined in (5.14) with optimal value ρ_i. Let*

$$\mathscr{F}_d := \{f \in \mathbb{R}[\mathbf{x}]_d \,:\, \exists \mathbf{x}^* \in \mathbb{R}^n \text{ s.t. } f(\mathbf{x}^*) = f^* = \min\{f(\mathbf{x}) \,:\, \mathbf{x} \in \mathbb{R}^n\}\}.$$

Then for almost all $f \in \mathscr{F}_d$, $\rho_i = f^* = f^*_{\mathbb{K}}$, for some index i, i.e. finite convergence takes place.

Theorem 5.19 is a direct consequence of Proposition 5.18 and Theorem 5.17.

5.8 Global Minimization of a Rational Function

In this section, we consider the problem:

$$r_{\mathbb{K}} := \inf \left\{ \frac{p(\mathbf{x})}{q(\mathbf{x})} \,:\, \mathbf{x} \in \mathbb{K} \right\}, \tag{5.35}$$

where $p, q \in \mathbb{R}[\mathbf{x}]$ and $\mathbb{K}$ is defined as in (5.3). Note that if p, q have no common zero on $\mathbb{K}$, then $r_{\mathbb{K}} > -\infty$ only if q does not change sign on $\mathbb{K}$. Therefore, we will assume that q is strictly positive on $\mathbb{K}$.

Proposition 5.20. *Let $r_{\mathbb{K}}$ be the optimal value of Problem (5.35) and consider the following instance of the generalized problem of moments*

$$\begin{aligned} \rho_{\text{mom}} = \inf_{\mu \in \mathscr{M}(\mathbb{K})_+} & \int_{\mathbb{K}} p\, d\mu \\ \text{s.t.} & \int_{\mathbb{K}} q\, d\mu = 1. \end{aligned} \tag{5.36}$$

Then $r_{\mathbb{K}} = \rho_{\text{mom}}$.

Proof. Assume first that $r_{\mathbb{K}} > -\infty$, so that $p(\mathbf{x}) \geq r_{\mathbb{K}}\, q(\mathbf{x})$ for all $\mathbf{x} \in \mathbb{K}$, and let $\mu \in \mathscr{M}(\mathbb{K})_+$ be a feasible measure for problem (5.36). Then $\int_{\mathbb{K}} p\, d\mu \geq r_k \int_{\mathbb{K}} q\, d\mu$, leading to $\rho_{\text{mom}} \geq r_{\mathbb{K}}$.

Conversely, let $\mathbf{x} \in \mathbb{K}$ be fixed arbitrary, and let $\mu \in \mathscr{M}(\mathbb{K})_+$ be the measure $q(\mathbf{x})^{-1}\delta_{\mathbf{x}}$, where $\delta_{\mathbf{x}}$ is the Dirac measure at the point $\mathbf{x} \in \mathbb{K}$. Then, $\int_{\mathbb{K}} q\, d\mu = 1$, so that μ is a feasible measure for problem (5.36). Moreover, its value satisfies $\int_{\mathbb{K}} p\, d\mu = p(\mathbf{x})/q(\mathbf{x})$. As $\mathbf{x} \in \mathbb{K}$ was arbitrary, $\rho_{\text{mom}} \leq r_{\mathbb{K}}$, and the result follows. Finally if $r_{\mathbb{K}} = -\infty$ then from what precedes we also have $\rho_{\text{mom}} = -\infty$. □

One recognize in (5.36) a particular instance of the generalized moment problem (1.1). Let $2v_k$ or $2v_k - 1$ be the degree of the polynomial $g_k \in \mathbb{R}[\mathbf{x}]$ in the definition (5.3) of $\mathbb{K}$, for all $k = 1, \ldots, m$. Proceeding as in Section 5.3, we obtain the semidefinite relaxation for $i \geq \max\{\deg p, \deg q, \max_k v_k\}$, which is analogous to (5.14):

$$\begin{aligned} \rho_i = \inf_{\mathbf{y}}\ & L_{\mathbf{y}}(p) \\ \text{s.t. } & \mathbf{M}_i(\mathbf{y}),\ \mathbf{M}_{i-v_k}(g_k\, \mathbf{y}) \succeq \mathbf{0}, \quad k = 1, \ldots, m \\ & L_{\mathbf{y}}(q) = 1. \end{aligned} \tag{5.37}$$

Note that in contrast to (5.14) where $y_0 = 1$, in general $y_0 \neq 1$ in (5.37). In fact, the last constraint $L_{\mathbf{y}}(q) = 1$ in (5.37) yields $y_0 = 1$ whenever $q = 1$, that is, problem (5.36) reduces to problem (5.2). The dual of problem (5.37) reads:

$$\begin{aligned}\rho_i^* = &\sup_{\sigma_k,\lambda} \lambda \\ &\text{s.t. } p - \lambda q = \sigma_0 + \sum_{k=1}^m \sigma_k\, g_k \\ &\sigma_k \in \Sigma[\mathbf{x}]; \quad \deg \sigma_k \leq i - v_k,\, k = 0, 1, \ldots, m.\end{aligned} \tag{5.38}$$

(with $v_0 = 1$). Recall that $q > 0$ on $\mathbb{K}$.

Theorem 5.21. *Let $\mathbb{K}$ be as in (5.3), and let Assumption 2.1 hold. Consider the semidefinite relaxations (5.37) and (5.38). Then,*
(a) *$r_{\mathbb{K}}$ is finite and $\rho_i^* \uparrow r_{\mathbb{K}}$, $\rho_i \uparrow r_{\mathbb{K}}$ as $i \to \infty$.*
(b) *In addition, if $\mathbb{K} \subset \mathbb{R}^n$ has nonempty interior then $\rho_i^* = \rho_i$ for every i.*
(c) *Let $\mathbf{x}^* \in \mathbb{K}$ be a global minimizer of p/q on $\mathbb{K}$. If the polynomial $p - q\, r_{\mathbb{K}} \in \mathbb{R}[\mathbf{x}]$, nonnegative on $\mathbb{K}$, has the representation (5.13), then both problems (5.37) and (5.38) have an optimal solution, and $\rho_i^* = \rho_i = r_{\mathbb{K}}$, for all $i \geq i_0$, for some $i_0 \in \mathbb{N}$.*

Proof. **(a)** follows directly from Theorem 4.1 and $r_{\mathbb{K}} = \rho_{\text{mom}}$.
(b) $\rho_i^* = \rho_i$ follows from Slater's condition which is satisfied for problem (5.37). Indeed, let $\mu \in \mathscr{M}(\mathbb{K})_+$ be a measure with uniform distribution on $\mathbb{K}$, and scaled to ensure $\int q\, d\mu = 1$. (As $\mathbb{K}$ is compact, this is always possible.) Then, the vector $\mathbf{y}$ of its moments, is a strictly feasible solution of problem (5.37) ($\mathbf{M}_i(\mathbf{y}) \succ \mathbf{0}$, and $\mathbb{K}$ having a nonempty interior implies $\mathbf{M}_{i-v_k}(g_k\, \mathbf{y}) \succ \mathbf{0}$ for all $k = 1, \ldots, m$). Thus, there is no duality gap, i.e., $\rho_i^* = \rho_i$, and the result follows from (a).
(c) If $p - q r_{\mathbb{K}}$ has the representation (5.13), then as we did for part (a), we can find a feasible solution to problem (5.38) with value $r_{\mathbb{K}}$, for all $i \geq i_0$, for some $i_0 \in \mathbb{N}$. Hence,

$$\rho_i^* = \rho_i = r_{\mathbb{K}}, \qquad \forall i \geq i_0.$$

Finally, (5.37) has an optimal solution $\mathbf{y}$. It suffices to take for $\mathbf{y}$ the vector of moments of the measure $q(\mathbf{x}^*)^{-1} \delta_{\mathbf{x}^*}$, i.e.,

$$y_{\boldsymbol{\alpha}} = q(\mathbf{x}^*)^{-1} (\mathbf{x}^*)^{\boldsymbol{\alpha}}, \qquad \boldsymbol{\alpha} \in \mathbb{N}^n.$$

□

5.9 Exploiting Symmetry

In this section we briefly describe how symmetry can be exploited to replace a semidefinite program invariant under the action of some group of permutations, with a much simpler one. In particular, it can be applied to the semidefinite relaxations (5.14) when $f \in \mathbb{R}[\mathbf{x}]$ and the polynomials $(g_j) \subset \mathbb{R}[\mathbf{x}]$ that define $\mathbb{K}$ are all invariant under some group of permutations.

Let S_n be the space of $n \times n$ real symmetric matrices and let $\mathrm{Aut}(S_n)$ be the group of automorphisms on S_n. Let $\mathscr{G}$ be a finite group acting on $\mathbb{R}^n$ via $\rho_0 : \mathscr{G} \to \mathrm{GL}(\mathbb{R}^n)$, which in turn induces an action $\rho : \mathscr{G} \to \mathrm{Aut}(S_n)$ on S_n by $\rho(g)(\mathbf{X}) = \rho_0(g)^T \mathbf{X}\, \rho_0(g)$ for every $g \in \mathscr{G}$, $\mathbf{X} \in S_n$.

Assume that $\rho_0(g)$ is orthonormal for every $g \in \mathscr{G}$. A matrix $\mathbf{X} \in \mathbb{R}^{n\times n}$ is said to be invariant under the action of $\mathscr{G}$ if $\rho(g)(\mathbf{X}) = \mathbf{X}$ for every $g \in \mathscr{G}$, and $\mathbf{X}$ is invariant if and only if $\mathbf{X}$ is an element of the *commutant algebra*

$$A^{\mathscr{G}} := \{\mathbf{X} \in \mathbb{R}^{n\times n} \,:\, \rho_0(g)\,\mathbf{X} = \mathbf{X}\,\rho_0(g), \quad \forall\, g \in \mathscr{G}\}. \tag{5.39}$$

Of particular interest is when $\mathscr{G}$ is a subgroup of the group $\mathscr{P}_n$ of permutations of $\{1,\ldots,n\}$, in which case $\rho_0(g)(\mathbf{x}) = (x_{g(i)})$ for every $\mathbf{x} \in \mathbb{R}^n$, and $\rho(g)(\mathbf{X})_{ij} = X_{g(i),g(j)}$ for every $1 \le i,j \le n$.

For every $(i,j) \in \{1,\ldots,n\}\times\{1,\ldots,n\}$, the orbit $O_{\mathscr{G}}(i,j)$ under action of $\mathscr{G}$, is the set of couples $\{(g(i),g(j)) \,:\, g \in \mathscr{G}\}$. With ω the number of orbits, and $1 \le l \le \omega$, define the $n \times n$ matrix $\tilde{\mathbf{D}}_l$ by $(\tilde{\mathbf{D}}_l)_{ij} := 1$ if (i,j) belongs to orbit l, and 0 otherwise. Normalize to $\mathbf{D}_l := \tilde{\mathbf{D}}_l / \sqrt{\langle \tilde{\mathbf{D}}_l, \tilde{\mathbf{D}}_l\rangle}$, for every $1 \le l \le \omega$, and define:

- The multiplication table

$$\mathbf{D}_i\,\mathbf{D}_j = \sum_{l=1}^{\omega} \gamma_{ij}^l\,\mathbf{D}_l, \qquad i,j = 1,\ldots,\omega.$$

- The $\omega \times \omega$ matrices $\mathbf{L}_1,\ldots,\mathbf{L}_\omega$ by:

$$(\mathbf{L}_k)_{ij} := \gamma_{kj}^i, \qquad i,j,k = 1,\ldots,\omega.$$

Then the commutant algebra (5.39) reads

$$A^{\mathscr{G}} = \left\{ \sum_{l=1}^{\omega} x_l\,\mathbf{D}_l \,:\, x_l \in \mathbb{R} \right\}$$

with dimension dim $A^{\mathscr{G}} = \omega$.

Exploiting symmetry in semidefinite program is possible thank to the following crucial property of the matrices $(\mathbf{D}_l)$:

Theorem 5.22. *The mapping* $\mathbf{D}_l \to \mathbf{L}_l$ *is a* $\star$*-isomorphism called the regular* $\star$*-representation of* $A^{\mathscr{G}}$*, and in particular:*

$$\sum_{j=1}^{\omega} x_l \, \mathbf{D}_l \succeq 0 \quad \Longleftrightarrow \quad \sum_{j=1}^{\omega} x_l \, \mathbf{L}_l \succeq 0. \tag{5.40}$$

Application to semidefinite programming

Consider the semidefinite program

$$\sup_{\mathbf{X} \succeq 0} \{ \langle \mathbf{C}, \mathbf{X} \rangle \, : \, \langle \mathbf{A}_k, \mathbf{X} \rangle = b_k, \quad k = 1, \ldots, p \} \tag{5.41}$$

and assume it is invariant under the action of $\mathscr{G}$, that is, $\mathbf{C} \in A^{\mathscr{G}}$ and the feasible region is globally invariant, meaning that if $\mathbf{X}$ feasible in (5.41) then so is $\rho(g)(\mathbf{X})$ for every $g \in \mathscr{G}$.

By convexity, for every feasible $\mathbf{X}$ of (5.41), the matrix $\mathbf{X}_0 := \frac{1}{|\mathscr{G}|} \sum_{g \in \mathscr{G}} \rho(g)(\mathbf{X})$ is feasible, invariant under action of $\mathscr{G}$ and with same objective value as $\mathbf{X}$. Therefore, we can include in the semidefinite program (5.41) the additional linear constraint $\mathbf{X} \in A^{\mathscr{G}}$ without affecting the optimal value.

Therefore, writing $\mathbf{X} = \sum_{l=1}^{\omega} x_l \mathbf{D}_l$ and setting

$$c_l := \langle \mathbf{C}, \mathbf{D}_l \rangle, \quad a_{kl} := \langle \mathbf{A}_k, \mathbf{D}_l \rangle, \qquad \forall l = 1, \ldots, \omega; \; k = 1, \ldots, p,$$

the semidefinite program (5.41) has same optimal value as

$$\sup_{\mathbf{x} \in \mathbb{R}^{\omega}} \{ \mathbf{c}' \mathbf{x} \, : \, \mathbf{a}_k' \, \mathbf{x} = b_k, \quad k = 1, \ldots, p; \; \sum_{l=1}^{\omega} x_l \, \mathbf{L}_l \succeq 0 \}. \tag{5.42}$$

Observe that in (5.41) we have n variables and a $n \times n$ positive semidefinite matrix $\mathbf{X}$, whereas in (5.42) we only have ω variables and a $\omega \times \omega$ positive semidefinite matrix.

5.10 Summary

In this chapter we have considered the general global optimization problem with polynomial data. We have provided a hierarchy of semidefinite relaxations to approximate as closely as desired the optimal value of the original problem. Sometimes the optimal value is obtained exactly by solving (few) finitely many semidefinite programs of the hierarchy.

> Like *primal* approaches in nonlinear programming that search for a local minimizer $\tilde{\mathbf{x}} \in \mathbb{R}^n$, the moment approach should also be regarded as a primal approach where one now searches not only for a global minimizer $\mathbf{x}^* \in \mathbb{R}^n$ but also for the sequence of its moments $\mathbf{y}^* = ((\mathbf{x}^*)^{\boldsymbol{\alpha}})$, $\boldsymbol{\alpha} \in \mathbb{N}^n$, i.e., a search in a *lifted* space (of moments). Similarly, like *dual* approaches in nonlinear programming (e.g. Lagrangian and extended Lagrangian) that search for *scalar* Lagrange Karush-Kuhn-Tucker *multipliers* associated with the constraints, the s.o.s. approach should also be viewed as a dual approach as one also searches for multipliers of the constraints, but now s.o.s. (instead of scalar) multipliers, hence also in a lifted space (of polynomials).

A nice feature of the hierarchy (5.14) is finite convergence in the s.o.s.-convex case (as the first semidefinite program in the hierarchy is exact), and in the strict convex case as well. In other words, the moment-s.o.s. approach recognizes easier convex problems.

Next, we have provided a convergent hierarchy of linear relaxations which unfortunately suffers from several drawbacks; in particular, and in contrast with the hierarchy of semidefinite relaxations, convergence is only asymptotic in general (except for discrete problems). We have also defined a hierarchy of sparse semidefinite relaxations that exploits sparsity in the problem definition, that is when each of the polynomials involved in the description of the set $\mathbb{K}$ involves a few variables only, and when the polynomial f to minimize is a sum of polynomials that also involve a few variables only. Convergence still holds if the sparsity pattern satisfies the so-called "running intersection property" well-known in gaph theory. Finally, we have also extended the approach to find the global minimum of a rational function on a compact basic semi-algebraic set.

5.11 Exercises

Exercise 5.1. Prove Theorem 5.10 and Theorem 5.11.

Exercise 5.2. Let $f \in \mathbb{R}[x]$ be a univariate polynomial. Show that for solving the problem

$$f^*_{\mathbb{K}} = \min_x \{f(x) : \; x \geq 0\},$$

with $\mathbb{K} = [0, \infty)$, the relaxation (5.14) with $i := \lceil(\deg f)/2)\rceil$ is exact.

Exercise 5.3. Show that for solving the problem $\inf\{\, x \; : \; x^2 \; = \; 0\}$, the semidefinite relaxation (5.14) with $i = 1$ is exact. What about the dual? Is it solvable? Why?

Exercise 5.4. Consider the global minimization of the bivariate quartic polynomial $\mathbf{x} \mapsto f(\mathbf{x}) := x_1^2 + (1 - x_1 x_2)^2$. What can be said about the semidefinite relaxation (5.6)? Is there any optimal solution? Similarly, what can be said about the semidefinite relaxation (5.8)? Is there any optimal solution?

5.12 Notes and Sources

For a survey on semidefinite programming and its multiple applications, the interested reader is referred to Vandenberghe and Boyd (1996).

Most of the material in this Chapter is from Lasserre (2000, 2001, 2002d,b,c, 2004, 2006d). Shor (1987, 1998) was the first to prove that the global minimization of a univariate polynomial is a convex optimization problem. Later, Nesterov (2000) defined exact semidefinite formulations for the univariate case, while converging semidefinite relaxations for the general multivariate case were treated in Lasserre (2000, 2001, 2002b,c) and Parrilo (2000, 2003). de Klerk *et al.* (2006) provided a Polynomial Time Approximation Scheme for minimizing polynomials of fixed degree on the simplex.

Building on Shor's ideas, Sherali and Adams (1990, 1999) and Sherali *et al.* (1992), proposed their RLT (Reformulation-Linearization Technique), the earliest hierarchy of LP relaxations for polynomial optimization, and proved finite convergence for 0/1 problems. Notice that the LP-relaxations of the RLT formulation, and especially their dual, can be interpreted in the light of the (representation) Theorem 2.23 which provide a rationale for their convergence. Other linear relaxations for 0/1 programs have also been proposed by Balas *et al.* (1993), Lovász and Schrijver (1991), while a hierarchy of semidefinite relaxations for 0/1 problems were first proposed by Lovász and Schrijver (1991) who also proved finite convergence.

A comparison between semidefinite and linear relaxations (in the spirit of Sherali and Adams) for general polynomial optimization problems is made in Lasserre (2002d), in the light of the results of Chapter 2 on the representation of positive polynomials. For 0/1 problems, Laurent (2003) compares the linear relaxations of Sherali and Adams (1990) and Lovász and Schrijver (1991), and the semidefinite relaxations of Lovász and Schrijver (1991) and Lasserre (2002b) within the common framework of the moment matrix, and proved that the latter semidefinite relaxations are the strongest. This has motivated research on integrality gaps for difficult combinatorial optimization problems. (The integrality gap measures the ratio between the optimal value of the relaxation and that of the problem to solve.) In particular Chlamtac (2007) and Chlamtac and Singh (2008) showed that the hierarchy of semidefinite relaxations provides improved approximations algorithms for finding independent sets in graphs, and for colouring problems. See also the related work of Schoenebeck (2008).

Theorem 5.17 is from Parrilo (2003), an extension to the general setting (5.33) of the grid case studied in Lasserre (2002b,c). See also Laurent (2007a) for refinements. Recent approaches to unconstrained optimization via optimizing on the gradient ideal appears in Hanzon and Jibetean (2003) with matrix methods and in Jibetean and Laurent (2005) with semidefinite programming. In both approaches one slightly perturbates p (of say degree $2d$) by adding monomials $\{x_i^{2d+2}\}$, with a small coefficient ϵ, and obtain a sequence of polynomials f_ϵ, with the property that $V = \{\mathbf{x} : \nabla f_\epsilon(\mathbf{x}) = 0\}$, is finite and the minima f_ϵ^* converge to f^*. In particular, the polynomials $\{\partial f_\epsilon / \partial x_i\}$ form a Gröbner basis of the ideal they generate. On the other hand, Proposition 5.18 is Proposition 1 in Nie *et al.* (2006). To handle the case where no global minimizer exists, Schweighofer (2006) uses s.o.s. and the concept of gradient tentacles and Vui and Son (2009) use the truncated tangency variety.

Polynomials satisfying sparsity patterns were investigated in Kojima *et al.* (2005) and the sparse semidefinite relaxations (5.17) were first proposed in Waki *et al.* (2006) as a heuristic to solve global optimization problems with a large number of variables and which satisfy some sparsity pattern. Their convergence in Theorem 5.9 was proved in Lasserre (2006a) if the sparsity pattern satifies the running intersection property.

The sparse relaxations have been implemented in the SparsePOP software of Waki *et al.* (2009) and numerical experiments show that one may then solve global optimization problem with $n = 1000$ variables for which even the first non sparse semidefinite relaxation of the hierarchy (5.14) cannot be implemented. Kim *et al.* (2009, to appear) provides a nice application for sensor network localization. Section 5.6 is from Lasserre (2009b) while Section 5.8 is inspired from Jibetean and de Klerk (2006), the first to prove Theorem 5.21.

Section 5.9 is inspired from Laurent (2008) and Theorem 5.22 is from de Klerk *et al.* (2007). For exploiting symmetry in the context of sums of squares and semidefinite programming see also the work of Gaterman and Parrilo (2004) and Vallentin (2009). For instance, these ideas have been used sucessfully in coding theroy for large error correcting codes based on computing the stability number of some related graph; see e.g. Laurent (2007b) and Schrijver (2005).

Chapter 6

Systems of Polynomial Equations

This chapter is about solving systems of polynomial equations. Of course if the goal is to search for just one solution, one may minimize any polynomial criterion and see the problem as the global optimization of Chapter 5. But we also consider the case where one searches for *all* complex and/or real solutions and show that the moment approach is well-suited to solve this problem as it provides the first algorithm to compute all real solutions without computing all the complex solutions, in contrast with the usual algebraic approaches based on Gröbner bases or homotopy.

6.1 Introduction

Computing all complex and/or all real solutions of a system of polynomial equations is a fundamental problem in mathematics with many important practical applications. Namely, let $J \subseteq \mathbb{R}[\mathbf{x}]$ be an ideal generated by a set of polynomials $g_j \in \mathbb{R}[\mathbf{x}]$, $j = 1, \ldots, m$. Two basic important problems are:

(I) The computation of the algebraic variety

$$V_{\mathbb{C}}(J) := \{\mathbf{z} \in \mathbb{C}^n \; : \; g_j(\mathbf{z}) = 0 \quad \forall j = 1, \ldots, n\} \tag{6.1}$$

(II) The computation of the *real* variety

$$V_{\mathbb{R}}(J) := \{\mathbf{x} \in \mathbb{R}^n \; : \; g_j(\mathbf{x}) = 0 \quad \forall j = 1, \ldots, n\} = V_{\mathbb{C}}(J) \cap \mathbb{R}^n \tag{6.2}$$

both associated with J, and possibly, a set of generators for the radical ideal $I(V_{\mathbb{L}}(J))$ for $\mathbb{L} = \mathbb{R}$ or $\mathbb{C}$, assuming $V_{\mathbb{L}}(J)$ is finite.

One way to solve problem (II) is to first compute all complex solutions and to sort out $V_{\mathbb{R}}(J) = \mathbb{R}^n \cap V_{\mathbb{C}}(J)$ from $V_{\mathbb{C}}(J)$ afterwards. This is certainly possible when I is a zero-dimensional ideal, but even in this case one might perform many unnecessary computations, particularly if $|V_{\mathbb{R}}(J)| \ll |V_{\mathbb{C}}(J)|$, i.e. in case there are many more complex roots than real roots. In addition, there are cases where $V_{\mathbb{R}}(J)$ is finite whereas $V_{\mathbb{C}}(J)$ is not! These two reasons alone provide a rationale for designing a method tailored to problem (II), that is, a method that takes into account right from the beginning the *real* algebraic nature of the problem.

In this chapter we first consider obtaining (at least) *one* (real) solution instead of *all* solutions in $V_{\mathbb{C}}(J)$ as do some algebraic approches based on Gröbner bases when I is zero-dimensional (but at a high computational cost and with exact arithmetic). One may search for solutions $\mathbf{x} \in V_{\mathbb{R}}(J)$ that minimize some chosen polynomial criterion $f \in \mathbb{R}[\mathbf{x}]$, which reduces to solving a particular case of the polynomial optimization problem studied in Chapter 5. We show on a significant sample of benchmark problems that in most cases, the semidefinite relaxations (5.14) deliver a solution in a few iterations, thus providing a viable complement and/or alternative to the more ambitious but computationally expensive, algebraic approaches with exact arithmetic. In addition, it also works even when the ideal I is not zero-dimensional.

Then we provide a unified treatment of problems **(I)** and **(II)** to obtain the 0-dimensional variety $V_{\mathbb{L}}(J)$ with $\mathbb{L} = \mathbb{R}$ or $\mathbb{L} = \mathbb{C}$. Remarkably:

(a) All information needed is contained in the moment matrix $\mathbf{M}_i(\mathbf{y})$ provided the order i is sufficiently large.

(b) In the algorithm, what differentiates the search for all real solutions from the search of all real and complex solutions is the presence or absence of a semidefiniteness constraint on the moment matrix.

6.2 Finding a Real Solution to Systems of Polynomial Equations

If one is interested in finding only at least one point $\mathbf{x} \in \mathbb{K}$ with good (but not infinite) precision, then an alternative to algebraic methods is to specify an arbitrary polynomial criterion $f \in \mathbb{R}[\mathbf{x}]$ to minimize, and solve the associated optimization problem (5.2) of Chapter 5 with $\mathbb{K}$ as in (6.3), and with optimal value $f^*_{\mathbb{K}}$.

More specifically, the basic closed semi-algebraic feasible set $\mathbb{K} \subset \mathbb{R}^n$ is now of the form:

$$\mathbb{K} := \{\, \mathbf{x} \in \mathbb{R}^n \,:\, g_j(\mathbf{x}) = 0, \quad j = 1, \ldots, m; \\ h_k(\mathbf{x}) \geq 0,\; k = 1, \ldots, p\}, \tag{6.3}$$

for some polynomials $g_j, h_k \in \mathbb{R}[\mathbf{x}]$, for all $j = 1, \ldots, m$, and all $k = 1, \ldots, p$. That is, $\mathbb{K}$ is the intersection of a real variety with a basic closed semi-algebraic set. It is clearly a special case of the set $\mathbb{K}$ described in (5.3), as an inequality can be written as two reverse inequalities. Therefore the semidefinite relaxations (5.14) specialize to

$$\begin{aligned} \rho_i = \inf_{\mathbf{y}} \; & L_{\mathbf{y}}(f) \\ \text{s.t. } & y_0 = 1 \\ & \mathbf{M}_i(\mathbf{y}) \succeq \mathbf{0}, \\ & \mathbf{M}_{i-v_j}(g_j\mathbf{y}) = \mathbf{0}, \quad j = 1, \ldots, m \\ & \mathbf{M}_{i-w_k}(h_k\mathbf{y}) \succeq \mathbf{0}, \quad k = 1, \ldots, p \end{aligned} \tag{6.4}$$

with $2i \geq s := \max[\deg f, \max_j \deg g_j, \max_k \deg h_k]$.

Let $J := \langle g_1, \ldots, g_m \rangle \subset \mathbb{R}[\mathbf{x}]$ be the ideal generated by the g_j's. If $V_{\mathbb{R}}(J)$ is finite then the sequence of semidefinite relaxations (6.4) has *finite* convergence, that is, there is some $i_1 \in \mathbb{N}$, such that $\rho_i = f^*_{\mathbb{K}}$ for all $i \geq i_1$, and the inf is attained, that is, (6.4) has an optimal solution $\mathbf{y}$. Moreover, by running Algorithm 4.2, one extracts s ($:= \operatorname{rank} \mathbf{M}_i(\mathbf{y})$) points of $\mathbb{K}$, all global minimizers. In addition, if the ideal J is radical, then the dual SDPs of (6.4) also have finite convergence. More precisely, let $v := \lceil s/2 \rceil$.

Theorem 6.1. *Let $f \in \mathbb{R}[\mathbf{x}]$ and let $\mathbb{K}$ be compact and defined as in (6.3). Let Assumption (2.1) hold (considering an equality constraint $g_j(\mathbf{x}) = 0$ as two reverse inequality constraints), and consider the semidefinite relaxations defined in (6.4). Then:*
(a) *$\rho_i \uparrow f^*_{\mathbb{K}}$ and $\rho^*_i \uparrow f^*_{\mathbb{K}}$ as $i \to \infty$.*
(b) *If the ideal $J\,(= \langle g_1, \ldots, g_m \rangle)$ is such that $V_{\mathbb{R}}(J)$ is finite, then there exists i_1 such that (6.4) has an optimal solution and $\rho_i = f^*_{\mathbb{K}}$ for all $i \geq i_1$. Moreover, if $\mathbf{y}$ is an optimal solution of (6.4) for $i = i_1$, then*

$$\operatorname{rank} \mathbf{M}_{s-v}(\mathbf{y}) = \operatorname{rank} \mathbf{M}_s(\mathbf{y}) =: r, \tag{6.5}$$

for some $d \leq s \leq i_1$, and one may extract r global minimizers in $\mathbb{K}$ by Algorithm 4.2.
(c) *If $J = \langle g_1, \ldots, g_m \rangle$ is zero-dimensional and radical then there exists i_2 such that the dual SDP of (6.4) is also solvable with value $\rho^*_i = f^*_{\mathbb{K}}$ for all $i \geq i_2$.*

Theorem 6.1(a) is a particular case of Theorem 4.1 whereas to prove (b) and (c) one invokes Theorem 6.20 in Laurent (2008) and Theorem 2.26 respectively.

Notice that for (c) the radicality assumption of the ideal J cannot be removed as seen in the following trivial one-dimensional example.

Example 6.1.

$$f^*_{\mathbb{K}} = \inf_x \{x \in \mathbb{R} : x^2 = 0\},$$

for which we have

$$\rho_1 = \inf_{\mathbf{y}} \{ y_1 : y_0 = 1;\ y_2 = 0; \quad \begin{bmatrix} y_0 & y_1 \\ y_1 & y_2 \end{bmatrix} \succeq 0 \},$$

and so, evidently, $f^*_{\mathbb{K}} = 0 = \rho_1$ with optimal solution $y_0 = 1, y_1 = y_2 = 0$. On the other hand, the polynomial $x \mapsto x - f^*_{\mathbb{K}} = x$ *cannot* be written as $q_0 + x^2 q_1$ for some $q_1 \in \mathbb{R}[x]$ and some s.o.s. polynomial $q_0 \in \Sigma[x]$. Therefore, the dual of the semidefinite relaxation (6.4) has no optimal solution. However, there is no duality gap because for every $\epsilon > 0$,

$$x \mapsto x - (f^*_{\mathbb{K}} - \epsilon) = x + \epsilon = (\sqrt{\epsilon} + \frac{1}{2\sqrt{\epsilon}} x)^2 - \frac{1}{4\epsilon} x^2,$$

and so $\rho_1^* = f^*_{\mathbb{K}} = 0$. Of course, the ideal $J = \langle x^2 \rangle$ is zero-dimensional but not radical as $\sqrt{J} = \langle x \rangle$.

Numerical experiment.

The experiments were carried out by using the software GloptiPoly. In fact, instead of choosing an explicit criterion $f \in \mathbb{R}[\mathbf{x}]$ to minimize, we have decided to minimize the trace of the moment matrix in the semidefinite relaxations (6.4). Alternative criteria (such as e.g. minimum coordinate or minimum Euclidean-norm solution) are of course possible. When the rank test (6.5) is passed, then running Algorithm 4.2 returns r solutions in $\mathbb{K}$.

Tables 6.1 and 6.2 give a short description of the examples in the sample. Numerical results are reported in Tables 6.3 and 6.4. Column *sol* indicates the number of solutions successfully extracted by GloptiPoly. In the last

Table 6.1 Short descriptions of systems of polynomial equations. Part 1.

problem	short description
boon	neurophysiology problem
bifur	non-linear system bifurcation
brown	Brown's 5-dimensional almost linear system
butcher	Butcher's system from PoSSo test suite
cameral s	displacement of camera between two positions
caprasse	Caprasse's system from PoSSo test suite
cassou	Cassou-Nogues's system from PoSSo test suite
chemequ	chemical equilibrium of hydrocarbon combustion
cohn2	Cohn's modular equations for special algebraic number fields
cohn3	Cohn's modular equations for special algebraic number fields
comb3000	combustion chemistry example for a temperature of 3000 degrees
conform1	Emiris' conformal analysis of cyclic molecules ($b_{11} = -9$)
conform2	Emiris' conformal analysis of cyclic molecules ($b_{11} = -\sqrt{3}/2$)
conform3	Emiris' conformal analysis of cyclic molecules ($b_{11} = -310$)
conform4	Emiris' conformal analysis of cyclic molecules ($b_{11} = -13$)
cpdm5	5-dimensional system of Caprasse and Demaret
d1	sparse system by Hong and Stahl
des18_3	dessin d'enfant
des22_24	dessin d'enfant
discret3	from PoSSo test suite
eco5	5-dimensional economics problem
eco6	6-dimensional economics problem
eco7	7-dimensional economics problem
eco8	8-dimensional economics problem
fourbar	four-bar mechanical design problem
geneig	generalized eigenvalue problem
heart	heart dipole problem
i1	interval arithmetic benchmark
ipp	six-revolute-joint problem of mechanics
katsura5	problem of magnetism in physics
kinema	robot kinematics problem
kin1	inverse kinematics of an elbow manipulator
ku10	10-dimensional system of Ku
lorentz	equilibrium points of 4-dimensional Lorentz attractor
manocha	intersection of high-degree polynomial curves
noon3	neural network modeled by adaptive Lotka-Volterra system
noon4	neural network modeled by adaptive Lotka-Volterra system
noon5	neural network modeled by adaptive Lotka-Volterra system
proddeco	system with product-decomposition structure
puma	hand position and orientation of PUMA robot
quadfor2	Gaussian quadrature formula with 2 knots and 2 weights over [-1,+1]
quadgrid	interpolating quadrature formula for function defined on a grid

Table 6.2 Short descriptions of systems of polynomial equations. Part 2.

problem	short description
rabmo	optimal multi-dimensional quadrature formulas
rbpl	generic positions of parallel robot
redeco5	reduced 5-dimensional economics problem
redeco6	reduced 6-dimensional economics problem
redeco7	reduced 7-dimensional economics problem
redeco8	reduced 8-dimensional economics problem
rediff3	3-dimensional reaction-diffusion problem
reimer5	5-dimensional system of Reimer
rose	general economic equilibrium problem
s9_1	small system from constructive Galois theory
sendra	from PoSSo test suite
solotarev	from PoSSo test suite
stewart1	direct kinematic problem of parallel robot
stewart2	direct kinematic problem of parallel robot
trinks	from PoSSo test suite
virasoro	construction of Virasoro algebras
wood	system derived from optimizing the Wood function
wright	Wright's system

column, *mem* means that the error message "out of memory" was issued by SeDuMi. GloptiPoly successfully solved about 90% of the problems.

6.3 Finding All Complex and/or All Real Solutions: A Unified Treatment

We here consider the ideal $J := \langle g_1, \ldots, g_n \rangle \subset \mathbb{R}[\mathbf{x}]$ generated by some polynomials $g_j \in \mathbb{R}[\mathbf{x}]$, $j = 1, \ldots, n$, and its associated varieties $V_{\mathbb{L}}(J)$ defined in (6.1) and (6.2) with $\mathbb{L} = \mathbb{C}$ or $\mathbb{R}$.

One uses the vector space $\mathbb{R}[\mathbf{x}]_t \subset \mathbb{R}[\mathbf{x}]$ of polynomials of degree at most t, and certain subsets of its dual space $(\mathbb{R}[\mathbf{x}]_t)^*$, the space of linear functionals on $\mathbb{R}[\mathbf{x}]_t$. More precisely, for an integer $t \geq d := \max_j \deg g_j$, let

$$\mathscr{H}_t := \{\, \mathbf{x}^\alpha g_j \;:\; j = 1, \ldots, m, \text{ and } \boldsymbol{\alpha} \in \mathbb{N}^n \text{ with } |\boldsymbol{\alpha}| + \deg g_j \leq t \}. \quad (6.6)$$

In the algorithm for computing $V_{\mathbb{C}}(J)$ we consider the set

$$\mathscr{K}_t := \{\, \mathbf{y} \in (\mathbb{R}[\mathbf{x}]_t)^* \;:\quad L_{\mathbf{y}}(p) = 0 \quad \forall p \in \mathscr{H}_t \,\}, \quad (6.7)$$

Table 6.3 Systems of polynomial equations. CPU times and semidefinite relaxation orders required to reach global optimum. Part 1.

problem	var	cstr	deg	CPU	order	sol
boon	6	6	4	1220	4	8
bifur	3	3	9	8.20	6	2
brown	5	5	5	6.27	3	1
butcher	7	7	4	-	4	mem
camera1s	6	6	2	1.33	2	2
caprasse	4	4	4	0.58	3	2
cassou	4	4	8	-	8	mem
chemequ	5	5	3	9.48	3	1
chemequs	5	5	3	6.73	2	1
cohn2	4	4	6	0.48	3	1
cohn3	4	4	6	0.55	3	1
comb3000	10	10	3	24.6	2	1
conform1	3	3	4	0.22	3	2
conform2	3	3	4	0.19	3	2
conform3	3	3	4	3.89	5	4
conform4	3	3	4	12.2	6	2
cpdm5	5	5	3	0.24	2	1
d1	12	12	3	-	3	dim
des18_3	8	8	3	-	4	mem
des22_24	10	10	2	77.2	1	1
discret3	8	8	2	0.31	1	1

whereas for computing $V_{\mathbb{R}}(J)$, we rather consider the smaller set

$$\mathscr{K}_{t,\succeq} := \{\, \mathbf{y} \in \mathscr{K}_t \ : \quad \mathbf{M}_{\lfloor t/2 \rfloor}(\mathbf{y}) \succeq 0 \,\}. \tag{6.8}$$

So the only difference between the two cases is the presence or absence of the semidefiniteness constraint $\mathbf{M}_t(\mathbf{y}) \succeq 0$ on the moment matrix $\mathbf{M}_t(\mathbf{y})$. Notice that

$$\mathscr{K}_{2t} := \{\, \mathbf{y} \in (\mathbb{R}[\mathbf{x}]_{2t})^* \ : \ \mathbf{M}_t(g_j\, \mathbf{y}) = 0 \,, j = 1, \ldots, m\},$$

$$\mathscr{K}_{2t,\succeq} := \{\, \mathbf{y} \in (\mathbb{R}[\mathbf{x}]_{2t})^* \ : \ \mathbf{M}_t(\mathbf{y}) \succeq 0; \quad \mathbf{M}_t(g_j\, \mathbf{y}) = 0, \ j = 1, \ldots, m \,\},$$

and with $\mathbb{K} := V_{\mathbb{R}}(J)$ (i.e. no inequality constraints in (6.3)), the semidefinite relaxation (6.4) reads

$$\rho_i = \inf_{\mathbf{y}} \{\, L_{\mathbf{y}}(f) \ : \ \mathbf{y} \in \mathscr{K}_{2t,\succeq} \,;\ y_0 = 1 \,\}. \tag{6.9}$$

Table 6.4 Systems of polynomial equations. CPU times and semidefinite relaxation order required to reach global optimum. Part 2.

problem	var	cstr	deg	CPU	order	sol
eco5	5	5	3	5.98	3	1
eco6	6	6	3	57.4	3	1
eco7	7	7	3	256	3	1
eco8	8	8	3	1310	3	1
fourbar	4	4	4	0.16	2	1
geneig	6	6	3	33.2	3	1
heart	8	8	4	1532	3	2
i1	10	10	3	44.1	2	1
ipp	8	8	2	6.42	2	1
katsura5	6	6	2	0.74	2	1
kinema	9	9	2	26.4	2	1
kin1	12	12	3	-	3	dim
ku10	10	10	2	72.5	2	1
lorentz	4	4	2	0.64	2	2
manocha	2	2	8	1.27	6	1
noon3	3	3	3	0.22	3	1
noon4	4	4	3	0.65	3	1
noon5	5	5	3	4.48	3	1
proddeco	4	4	4	0.11	2	1
puma	8	8	2	1136	3	4
quadfor2	4	4	4	0.75	3	2
quadgrid	5	5	5	10.52	3	1
rabmo	9	9	5	-	3	mem
rbpl	6	6	3	36.9	3	1
redeco5	5	5	2	0.16	1	1
redeco6	6	6	2	0.13	1	1
redeco7	7	7	2	0.14	1	1
redeco8	8	8	2	0.13	1	1
rediff3	3	3	2	0.09	1	1
reimer5	5	5	6	-	6	mem
rose	3	3	9	79.5	7	2
s9_1	8	8	2	5.45	2	1
sendra	2	2	7	0.34	5	1
solotarev	4	4	3	0.24	2	1
stewart1	9	9	2	20.4	2	2
stewart2	12	10	2	372	2	1
trinks	6	6	3	0.78	2	1
virasoro	8	8	2	0.16	1	1
wood	4	3	2	0.20	2	1
wright	5	5	2	0.17	1	1

6.3.1 Basic underlying idea

Notice that if $\mathbb{K} = V_{\mathbb{R}}(J)$ (assumed to be finite) and $f = 1$, then every $\mathbf{x} \in V_{\mathbb{R}}(J)$ is an optimal solution. Denote $\delta_{\mathbf{x}}$ the Dirac measure at $\mathbf{x} \in \mathbb{R}^n$. The moment sequence $\mathbf{y} = (y_{\boldsymbol{\alpha}})$ of a probability measure

$$\mu := \sum_{\mathbf{x} \in V_{\mathbb{R}}(J)} c_{\mathbf{x}}\, \delta_{\mathbf{x}}, \quad c_{\mathbf{x}} \geq 0; \; \sum_{\mathbf{x} \in V_{\mathbb{R}}(J)} c_{\mathbf{x}} = 1$$

on $V_{\mathbb{R}}(J)$, is a feasible solution of the semidefinite relaxations (6.9). In addition, among all such probability measures, those with $c_{\mathbf{x}} > 0$ for all $\mathbf{x} \in V_{\mathbb{R}}(J)$, give maximum rank to the moment matrix $\mathbf{M}_i(\mathbf{y})$, and for all sufficiently large i, $\operatorname{rank} \mathbf{M}_i(\mathbf{y}) = |V_{\mathbb{R}}(J)|$. Moreover, as $V_{\mathbb{R}}(J)$ is finite, then by Theorem 6.1(b), there is an index i_1 such that the semidefinite relaxation (6.9) is exact and $\mathbf{y}$ is the moment sequence of some probability measure on $V_{\mathbb{R}}(J)$.

Now, a *generic* element of $\mathscr{K}_{2t,\succeq}$ (i.e., generated randomly according to a uniform probability distribution on $\mathscr{K}_{2t,\succeq}$) gives maximun rank to the moment matrix $\mathbf{M}_t(\mathbf{y})$. The idea is to impose in the semidefinite relaxation (6.9) the additional constraint that $\mathbf{y} \in \mathscr{K}_{t,\succeq}$ be generic (so as to give maximum rank to $\mathbf{M}_{\lfloor t/2 \rfloor}(\mathbf{y})$), and so, when (6.5) is satisfied, $\mathbf{y}$ is the moment sequence of a probability measure supported on *all* points of $V_{\mathbb{R}}(J)$. Finally, it turns out that a generic element of $\mathscr{K}_{t,\succeq}$ lies in the relative interior of $\mathscr{K}_{t,\succeq}$ and can be found with appropriate interior-point algorithm for semidefinite programming algorithms. Some of SDP solvers used in GloptiPoly (e.g. the SeDuMi solver) implement such interior point methods. So, any optimal solution $\mathbf{y}$ of (6.9) (obtained by running GloptiPoly with e.g. the SeDuMi solver of Sturm (1999)) is generic and provides the desired result when $\mathbf{y}$ satisfies (6.5).

At last but not least, for all t sufficiently large, every element $\mathbf{y}$ of $\mathscr{K}_t$ will also satisfy the rank condition (6.5), but $\mathbf{M}_t(\mathbf{y}) \succeq 0$ does not necessarily hold and so in general, $\mathbf{y} \in \mathscr{K}_t$ is *not* anymore the moment sequence of a probability measure. However, again, a generic element (i.e. in the relative interior of $\mathscr{K}_{2t}$) gives maximum rank to $\mathbf{M}_t(\mathbf{y})$ and along with the property (6.5), permits to extract *all* points of $V_{\mathbb{C}}(J)$ whenever $V_{\mathbb{C}}(J)$ is finite.

6.3.2 The moment-matrix algorithm

Before presenting the moment-matrix algorithm to compute the variety $V_{\mathbb{L}}(J)$ of an ideal $J \subset \mathbb{R}[\mathbf{x}]$ with $\mathbb{L} = \mathbb{C}$ or $\mathbb{R}$, we briefly explain how computing the points of $V_{\mathbb{C}}(J)$ reduces to an eigenvalue problem when J is 0-dimensional.

Recall that for an arbitrary variety $V \subset \mathbb{C}^n$, the vanishing ideal $I(V)$ is the set of polynomials $g \in \mathbb{R}[\mathbf{x}]$ that vanish on V, and by the *Hilbertnullstellenstaz*, for an arbitrary ideal $J \subseteq \mathbb{R}[\mathbf{x}]$, its associated radical ideal $\sqrt{J}$ is the vanishing ideal $I(V_{\mathbb{C}}(J))$. Similarly, by the *Realnullstellensatz*, the real radical $\sqrt[\mathbb{R}]{J}$ is the vanishing ideal $I(V_{\mathbb{R}}(J))$; see Section A.2

The quotient algebra $\mathbb{R}[\mathbf{x}]/J$.

Given an ideal $J \subseteq \mathbb{R}[\mathbf{x}]$, the elements of the quotient space $\mathbb{R}[\mathbf{x}]/J$ are the cosets $[f] = f + J := \{f + q : q \in J\}$ and if J is 0-dimensional, then $\mathbb{R}[\mathbf{x}]/J$ it is a finite-dimensional $\mathbb{R}$-vector space with addition $[f] + [g] = [f + g]$ and scalar multiplication $\lambda[f] = [\lambda f]$ for $\lambda \in \mathbb{R}$, and $f, g \in \mathbb{R}[\mathbf{x}]$. Its dimension is larger than $|V_{\mathbb{C}}(J)|$, with equality if J is radical (i.e. $J = \sqrt{J}$). It is also an algebra with multiplication $[f]\,[g] = [fg]$, and given an arbitrary $h \in \mathbb{R}[\mathbf{x}]$, one may define the linear multiplication operator $m_h : \mathbb{R}[\mathbf{x}]/J \to \mathbb{R}[\mathbf{x}]/J$ that gives $[f] \mapsto [fh]$ for all $f \in \mathbb{R}[\mathbf{x}]$. If $\mathscr{B}_J$ denotes a basis of $\mathbb{R}[\mathbf{x}]/J$ then the matrix $\mathbf{M}_h$ associated with m_h in that basis, plays a fundamental role in obtaining the points of $V_{\mathbb{C}}(J)$. Indeed:

Theorem 6.2. *Let $J \subset \mathbb{R}[\mathbf{x}]$ be a zero-dimensional ideal and let $\mathscr{B}_J := \{b_1, \ldots, b_N\}$ be a basis of $\mathbb{R}[\mathbf{x}]/J$. For every $\mathbf{v} \in V_{\mathbb{C}}(J)$ let $\mathbf{u}_{\mathbf{v}} := (b_i(\mathbf{v}))_{1 \le i \le N} \in \mathbb{C}^n$. Then the set $\{h(\mathbf{v}) : \mathbf{v} \in V_{\mathbb{C}}(J)\}$ is the set of eigenvalues of $\mathbf{M}_h$ and $\mathbf{M}_h\,\mathbf{u}_{\mathbf{v}} = h(\mathbf{v})\,\mathbf{u}_{\mathbf{v}}$ for all $\mathbf{v} \in V_{\mathbb{C}}(J)$.*

In particular, by applying Theorem 6.2 with the polynomial $\mathbf{x} \mapsto h(\mathbf{x}) := x_i$, one obtains the coordinates $v_i \in \mathbb{C}$ of all points $\mathbf{v} \in V_{\mathbb{C}}(J)$, as the eigenvalues of the multiplication matrix $\mathbf{M}_{x_i}$.

We next present a few results that are relevant to understand the moment-matrix algorithm to compute $V_{\mathbb{L}}(J)$ with $\mathbb{L} = \mathbb{C}$ or $\mathbb{L} = \mathbb{R}$.

Recall that $\mathbb{N}^n_t = \{\alpha \in \mathbb{N}^n : |\alpha| \le t\}$ and for a polynomial $g \in \mathbb{R}[\mathbf{x}]_t$, $\mathbf{g} \in \mathbb{R}^{\mathbb{N}^n_t}$ denotes its vector of coefficients in the usual canonical basis $(\mathbf{x}^\alpha)$.

For a sequence $\mathbf{y} = (y_\alpha) \in \mathbb{R}^{\mathbb{N}^n_{2t}}$, let $\langle \mathrm{Ker}\ \mathbf{M}_t(\mathbf{y}) \rangle \subset \mathbb{R}[\mathbf{x}]$ be the ideal generated by the polynomials $g \in \mathbb{R}[\mathbf{x}]_t$ such that $\mathbf{M}_t(\mathbf{y})\mathbf{g} = 0$. In other words, $\langle \mathrm{Ker}\ \mathbf{M}_t(\mathbf{y}) \rangle$ is the ideal generated by the polynomials g of degree at most t and whose vector of coefficients $\mathbf{g}$ lies in the kernel of the moment matrix $\mathbf{M}_t(\mathbf{y})$.

Let $d := \max_j \lceil \deg g_j/2 \rceil$, and let $\mathbb{L} = \mathbb{C}$ or $\mathbb{R}$. Let $K_t := \mathscr{K}_t$ when $\mathbb{L} = \mathbb{C}$ and $K_t := \mathscr{K}_{t\succeq}$ when $\mathbb{L} = \mathbb{R}$. The basic idea of the algorithm is to obtain an index s and a generic sequence $\mathbf{y} \in K_t$ such that the ideal $J_0 := \langle \text{Ker } \mathbf{M}_s(\mathbf{y}) \rangle$ satisfies

$$J \subseteq J_0 \subseteq I(V_{\mathbb{L}}(J)) \tag{6.10}$$

and the sequence $\mathbf{y}$ satisfies

$$\operatorname{rank} \mathbf{M}_{s-d}(\mathbf{y}) = \operatorname{rank} \mathbf{M}_s(\mathbf{y}). \tag{6.11}$$

The first condition (6.10) will ensure that $V_{\mathbb{L}}(J_0) = V_{\mathbb{L}}(J)$ while the second condition (6.11) will permit to build up multiplications matrices $\mathbf{M}_{x_i}$ for extracting all points of $V_{\mathbb{L}}(J)$.

It turns out that finding such a couple $(s, \mathbf{y}) \in \mathbb{N} \times K_t$ is possible thanks to very nice properties of the moment matrix's kernel Ker $\mathbf{M}_s(\mathbf{y})$ when $\mathbf{y} \in K_t$. More precisely, to get the first inclusion in (6.10) we use

Lemma 6.3. *Let $d \leq s \leq \lfloor t/2 \rfloor$ and $\mathbf{y} \in K_t$. If $\operatorname{rank} \mathbf{M}_{s-d}(\mathbf{y}) = \operatorname{rank} \mathbf{M}_s(\mathbf{y})$ then $J \subseteq \langle \text{Ker } \mathbf{M}_s(\mathbf{y}) \rangle$,*

while to get the second inclusion we use

Theorem 6.4. *Let $1 \leq s \leq \lfloor t/2 \rfloor$ and $\mathbf{y} \in K_t$ be such that*

$$\operatorname{rank} \mathbf{M}_s(\mathbf{y}) = \max \{ \operatorname{rank} \mathbf{M}_s(\mathbf{z}) \,:\, \mathbf{z} \in K_t \}. \tag{6.12}$$

Then $\langle \text{Ker } \mathbf{M}_s(\mathbf{y}) \rangle \subseteq I(V_{\mathbb{L}}(J))$.

Finally:

Theorem 6.5. *Let $\mathbf{y} \in K_t$ and $d \leq s \leq \lfloor t/2 \rfloor$. Assume that $\operatorname{rank} \mathbf{M}_s(\mathbf{y})$ is maximum (i.e. (6.12) holds), and let $J_0 := \langle \text{Ker } \mathbf{M}_s(\mathbf{y}) \rangle$.*

If $\operatorname{rank} \mathbf{M}_{s-d}(\mathbf{y}) = \operatorname{rank} \mathbf{M}_s(\mathbf{y})$ then

$$J \subseteq J_0 \subseteq I(V_{\mathbb{C}}(J)) \; (= \sqrt{J}), \quad \text{case } \mathbb{L} = \mathbb{C} \tag{6.13}$$

$$J_0 = I(V_{\mathbb{R}}(J)) \; (= \sqrt[\mathbb{R}]{J}), \quad \text{case } \mathbb{L} = \mathbb{R}, \tag{6.14}$$

and any basis of the column space of $\mathbf{M}_{s-1}(\mathbf{y})$ is a basis of $\mathbb{R}[\mathbf{x}]/J_0$.

Hence one can construct the multiplication matrices in $\mathbb{R}[\mathbf{x}]/J_0$ from the matrix $\mathbf{M}_s(\mathbf{y})$ and find the variety $V_{\mathbb{L}}(J_0) = V_{\mathbb{L}}(J)$ using the eigenvalue method.

In fact, (extraction) Algorithm 4.2 applied to the moment matrix $\mathbf{M}_s(\mathbf{y})$ of Theorem 6.5 is exactly the eigenvalue method applied to the ideal $\langle \mathrm{Ker}\, \mathbf{M}_s(\mathbf{y}) \rangle$.

The moment-matrix algorithm.

Assume that $\mathbb{L} = \mathbb{R}$ or $\mathbb{C}$ and $V_{\mathbb{L}}(J)| < \infty$.

Algorithm 6.1. The moment-matrix algorithm for $V_{\mathbb{L}}(J)$

Input: g_j, $j = 1, \ldots, m$; $t \geq d$.
If $\mathbb{L} = \mathbb{C}$ set $K_t := \mathscr{K}_t$ else if $\mathbb{L} = \mathbb{R}$ set $K_t := \mathscr{K}_{t,\succeq}$.

Output: A basis $\mathscr{B}_J \subseteq \mathbb{R}[\mathbf{x}]_{s-1}$ of $\mathbb{R}[\mathbf{x}]/\langle \mathrm{Ker}\, \mathbf{M}_s(\mathbf{y}) \rangle$ (which enables the computation of $V_{\mathbb{L}}(J)$).

1 Find a generic element $\mathbf{y} \in K_t$ (i.e., such that (6.12) holds).
2 Check if $\mathrm{rank}\, \mathbf{M}_{s-d}(\mathbf{y}) = \mathrm{rank}\, \mathbf{M}_s(\mathbf{y})$ for some $d \leq s \leq \lfloor t/2 \rfloor$.
3 If yes **then**
4   **return** a basis $\mathscr{B}_J \subseteq \mathbb{R}[\mathbf{x}]_{s-1}$ of the column space of $\mathbf{M}_{s-1}(\mathbf{y})$, and extract $V_{\mathbb{L}}(J)$ by applying Algorithm 4.2 to the moment matrix $\mathbf{M}_s(\mathbf{y})$.
5 else
6   **iterate** (go to 1) replacing t with $t+1$
7 end if

Algorithm 6.1 has finite convergence because:

Theorem 6.6. *Let $\mathbb{L} = \mathbb{C}$ or $\mathbb{R}$, and let $J := \langle g_1, \ldots, g_m \rangle \subset \mathbb{R}[\mathbf{x}]$ satisfy $1 \leq |V_{\mathbb{L}}(J)| < +\infty$. Then there exists $t_1 \geq d, t_2 \in \mathbb{N}$, such that for every $\lfloor t/2 \rfloor \geq t_1 + t_2$:*

$$\mathrm{rank}\, \mathbf{M}_{t_1}(\mathbf{y}) = \mathrm{rank}\, \mathbf{M}_{t_1-d}(\mathbf{y}) \qquad \forall \mathbf{y} \in K_t.$$

And so, Theorem 6.6 combined with Theorem 6.5 yields the finite convergence of Algorithm 6.1.

It is worth noticing that when $\mathbb{L} = \mathbb{R}$, then Algorithm 6.1 reduces to solving the optimization problem (5.2) with objective (polynomial) function $f = 1$, using the hierarchy of semidefinite relaxations (5.14), provided that one uses a primal-dual interior point algorithm, as proposed in the

GloptiPoly software. This is because with $f = 1$, all points of $V_{\mathbb{R}}(J)$ are optimal solutions, and an optimal solution $\mathbf{y}$ of (5.14) will lie in the relative interior of $\mathscr{K}_{t\succeq}$, the desired property needed in Algorithm 6.1 to get all real solutions.

Example 6.2. Consider the following example in $\mathbb{R}^3$

$$\begin{aligned} h_1 &= x_1^2 - 2x_1x_3 + 5 \\ h_2 &= x_1x_2^2 + x_2x_3 + 1 \\ h_3 &= 3x_2^2 - 8x_1x_3 \end{aligned}$$

which has 8 complex solutions among which only 2 are real.

Table 6.5 Rank of $\mathbf{M}_s(y)$ for generic $\mathbf{y} \in \mathscr{K}_t$ in Example 6.2.

	$t=2$	$t=4$	$t=6$	$t=8$
$s=0$	1	1	1	1
$s=1$	4	4	4	4
$s=2$		8	8	**8**
$s=3$			9	**8**
$s=4$				9

For the algorithm with $\mathbb{L} = \mathbb{C}$, Table 6.5 shows the ranks of the matrices $\mathbf{M}_s(\mathbf{y})$ for generic $\mathbf{y} \in K_t = \mathscr{K}_t$, as a function of s and t. The rank condition $\operatorname{rank} \mathbf{M}_{s-d}(\mathbf{y}) = \operatorname{rank} \mathbf{M}_s(\mathbf{y})$ for some $d \leq s \leq \lfloor t/2 \rfloor$ holds for $t = 8$ and $s = 3$ as we have: $\operatorname{rank} \mathbf{M}_3(\mathbf{y}) = \operatorname{rank} \mathbf{M}_2(\mathbf{y})$ with $\mathbf{y} \in K_8$. One obtains the following 8 complex solutions:

$$\begin{aligned} v_1 &= \left[-1.101, -2.878, -2.821\right], \\ v_2 &= \left[0.07665 + 2.243i, 0.461 + 0.497i, 0.0764 + 0.00834i\right], \\ v_3 &= \left[0.07665 - 2.243i, 0.461 - 0.497i, 0.0764 - 0.00834i\right], \\ v_4 &= \left[-0.081502 - 0.93107i, 2.350 + 0.0431i, -0.274 + 2.199i\right], \\ v_5 &= \left[-0.081502 + 0.93107i, 2.350 - 0.0431i, -0.274 - 2.199i\right], \\ v_6 &= \left[0.0725 + 2.237i, -0.466 - 0.464i, 0.0724 + 0.00210i\right], \\ v_7 &= \left[0.0725 - 2.237i, -0.466 + 0.464i, 0.0724 - 0.00210i\right], \\ v_8 &= \left[0.966, -2.813, 3.072\right] \end{aligned}$$

with maximum error of $\epsilon := \max_{i \leq 8, j \leq 3} |h_j(v_i)| \leq 3 \cdot 10^{-10}$.

For the algorithm with $\mathbb{L} = \mathbb{R}$ and the same example, Table 6.6 displays the ranks of the matrices $\mathbf{M}_s(\mathbf{y})$ for generic $\mathbf{y} \in \mathscr{K}_{t,\succeq}$; now the rank condition is satisfied at $s = 2$ and $t = 6$; that is, $\operatorname{rank} \mathbf{M}_2(\mathbf{y}) = \operatorname{rank} \mathbf{M}_1(\mathbf{y})$ with

Table 6.6 Rank of $M_s(y)$ for generic $\mathbf{y} \in \mathscr{K}_{t,\succeq}$ in Example 6.2.

	$t = 2$	$t = 4$	$t = 6$
$s = 0$	1	1	1
$s = 1$	4	4	**2**
$s = 2$		8	**2**
$s = 3$			3

$\mathbf{y} \in K_{6,\succeq}$. The real roots extracted are

$$v_1 = (-1.101, -2.878, -2.821); \quad v_2 = (0.966, -2.813, 3.072)$$

with a maximum error of $\epsilon \leq 9 \cdot 10^{-11}$.

6.4 Summary

In this chapter we have considered two applications of the method of moments for solving and analyzing systems of polynomial equations. In the first case we use the machinery developed in Chapter 5 and propose a hierarchy of semidefinite relaxations to obtain (at least) one solution. Experimental results show that in many cases a solution is obtained at a relaxation of low order.

In the second case we have provided an algorithm to compute all complex or all real solutions of a system of polynomial equations when the ideal generated by the polynomials defining the equations has finitely many complex or real roots. To do so we use ideal-like properties of the kernel of moment matrices and the resulting algorithm uses only numerical linear algebra in the complex case and numerical linear algebra plus semidefinite programming for extracting only the real roots. Remarkably:

- all information needed to compute the roots is contained in the moment matrix and its kernel.

- the only difference between the real and complex cases is the presence or absence of a semidefiniteness constraint on the moment matrix. In the former (real) case, Algorithm 6.1 is the same as solving the optimization problem (5.2) with objective criterion $f = 1$, using the hierarchy of semidefinite relaxations (5.14), provided that one uses a primal-dual interior point algorithm, as proposed in the Gloptipoly software.

The resulting algorithm for the real case is numerical in nature as it uses semidefinite programming and could provide an alternative to algebraic approaches based on Gröbner bases with exact arithmetic. The latter are more ambitious as they provide all solutions with infinite precision but at a high computational cost.

6.5 Exercises

Exercise 6.1. Consider the ideal $I := \langle x_i(x_i^2+1)\rangle \subset \mathbb{R}[\mathbf{x}]$ generated by the polynomials $\mathbf{x} \mapsto g_i(\mathbf{x}) = x_i(x_i^2+1)$, $i = 1, \ldots, n$.

- Compute $V_{\mathbb{C}}(I)$ and $V_{\mathbb{R}}(I)$; compare their cardinality.
- Compute $\mathscr{K}_{4,\succeq}$. What happens?
- When does Algorithm 6.1 with $\mathbb{L} = \mathbb{R}$ stop?

Exercise 6.2. With $n = 2$, consider the ideal $I := \langle x_1^2 + x_2^2\rangle \subset \mathbb{R}[\mathbf{x}]$ generated by the polynomial $\mathbf{x} \mapsto g(\mathbf{x}) = x_1^2 + x_2^2$.

- Compute $V_{\mathbb{C}}(I)$ and $V_{\mathbb{R}}(I)$; compare their cardinality.
- Compute $\mathscr{K}_{4,\succeq}$. What happens?
- When does Algorithm 6.1 with $\mathbb{L} = \mathbb{R}$ stop?

6.6 Notes and Sources

There exist various methods for solving problem **(I)** which range from numerical continuation methods see e.g. Sommese and Wampler (2005)), to exact symbolic methods (e.g. Rouillier (1999)), or more general symbolic/numeric methods (e.g. Mourrain and Trébuchet (2005) or Zhi and Reid (2004), see also the monograph Stetter (2004)). For instance, Verschelde (1999) proposes a numerical algorithm via homotopy continuation methods (cf. also Sommese and Wampler (2005)) whereas Rouillier (1999) solves a zero-dimensional system of polynomials symbolically by giving a rational univariate representation (RUR) for its solutions, of the form $f(t) = 0$, $x_1 = \frac{g_1(t)}{g(t)}$, …, $x_n = \frac{g_n(t)}{g(t)}$, where $f, g, g_1, \ldots, g_n \in \mathbb{R}[t]$ are univariate polynomials. The computation of the RUR relies in an essential way on the multiplication matrices in the quotient algebra $\mathbb{R}[\mathbf{x}]/I$ which thus requires knowledge of a corresponding linear basis of the quotient space.

The literature concerned with solving problem **(II)**, is by far not as broad as for problem **(I)**. With the exception of Lasserre *et al.* (2008b), most algorithms are based on real-root counting algorithms using e.g. Hermite's

Quadratic forms or variants of Sturm sequences; see e.g. Basu *et al.* (2003) or Rouillier and Zimmermann (2003) for a discussion.

The moment matrix algorithm was first derived for computing $V_{\mathbb{R}}(I)$ in Lasserre *et al.* (2008b) and then extended to cover both cases $V_{\mathbb{C}}(I)$ and $V_{\mathbb{R}}(I)$ in Lasserre *et al.* (2008c). Finally, one may also adapt to the real case some existing methods tailored to the complex case. One still uses positive semidefiniteness of moment matrices but now combined with a stopping criterion different from the rank test of Theorem 6.6. For instance, in Lasserre *et al.* (2009) one uses a stopping criterion borrowed from Zhi and Reid (2004) for the complex case.

Chapter 7

Applications in Probability

This chapter covers some applications in probability. We first consider the problem of computing an upper bound on $\mu(\mathbf{S})$ over all measures μ that satisfy certain moment conditions. We then consider the problem of computing (or at least approximating) the volume of a compact basic semi-algebraic set. We end up with the mass-tranfer (or Monge-Kantorovich) problem.

With $\mathbf{S} \subset \mathbb{R}^n$, the problem of computing bounds on $\mu(\mathbf{S})$ over all measures μ that satisfy certain moment conditions is motivated by providing bounds on the probability that a $\mathbb{R}^n$-valued random variable belongs to $\mathbf{S}$, given the only information that some of its moments are known. This latter problem is old as it dates back to famous mathematicians and probabilists like e.g. Markov, Chebyshev and Chernoff whose names are now associated with some celebrated bounds in probability.

In this chapter we apply results of earlier chapters to provide a general optimization based methodology to address this problem. We also consider the important problem of measuring a basic closed semi-algebraic set $\mathbb{K} \subset \mathbb{R}^n$, i.e. computing $\mu(\mathbb{K})$ for a given measure μ on $\mathbb{R}^n$ when all moments of μ are known. In particular if μ is the normalized Lebesgue measure on a box that contains $\mathbb{K}$, then $\mu(\mathbb{K})$ is just the (Lebesgue) volume of $\mathbb{K}$. Finally, we consider the problem of bounding the integral $\int_{\mathbb{K}} f\, d\mu$ over all measures μ on the cartesian product $\mathbb{K} := \prod_{j=1}^{p} \mathbb{K}_j$, with given marginals ν_j on $\mathbb{K}_j$, $j = 1, \ldots, p$. For $p = 2$ this is just the celebrated mass transfer (or Monge-Kantorovich) problem.

7.1 Upper Bounds on Measures with Moment Conditions

Let $\gamma_{\boldsymbol{\alpha}}$, $\boldsymbol{\alpha} \in \Gamma \subset \mathbb{N}^n$, be a sequence of moments and let

$$\mathbf{S} = \{\mathbf{x} \in \mathbb{R}^n \,:\, p_j(\mathbf{x}) \geq 0, \quad j = 1, \ldots, t\} \tag{7.1}$$

be a given basic semi-algebraic set. Then, given a $\mathbb{R}^n$-valued random variable $\mathbf{X}$, the problem of finding an optimal bound on $\mathrm{Prob}\,(\mathbf{X} \in \mathbf{S})$, given some of its moments $\gamma_{\boldsymbol{\alpha}}$, can be formulated as the problem: of computing:

$$\begin{aligned} \rho_{\mathrm{mom}} = \sup_{\mu \in \mathscr{M}(\mathbb{R}^n)_+} & \int_{\mathbb{R}^n} 1_{\mathbf{S}}\, d\mu \\ \text{s.t.} & \int_{\mathbb{R}^n} \mathbf{x}^{\boldsymbol{\alpha}}\, d\mu = \gamma_{\boldsymbol{\alpha}}, \quad \boldsymbol{\alpha} \in \Gamma, \end{aligned} \tag{7.2}$$

which is (4.2) with $\mathbb{K} = \mathbb{R}^n$, $f = 1_{\mathbf{S}}$ and $\mathbf{x} \mapsto h_{\boldsymbol{\alpha}}(\mathbf{x}) := \mathbf{x}^{\boldsymbol{\alpha}}$, $\boldsymbol{\alpha} \in \Gamma$. Note that $\gamma_0 = 1$.

As we do not want to deal with indicator functions, without loss of generality we decompose $\mu \in \mathscr{M}(\mathbb{R}^n)_+$ into a sum $\phi + \nu$, with $\phi, \nu \in \mathscr{M}(\mathbb{R}^n)_+$, ϕ supported on $\mathbf{S}$ and ν supported on $\mathbb{R}^n$. And so we now consider the generalized moment problem:

$$\begin{aligned} \rho_{\mathrm{mom}} = \sup_{\nu \in \mathscr{M}(\mathbb{R}^n)_+, \phi \in \mathscr{M}(\mathbf{S})_+} & \int_{\mathbf{S}} d\phi \\ \text{s.t.} & \int_{\mathbf{S}} \mathbf{x}^{\boldsymbol{\alpha}}\, d\phi + \int_{\mathbb{R}^n} \mathbf{x}^{\boldsymbol{\alpha}}\, d\nu = \gamma_{\boldsymbol{\alpha}}, \quad \boldsymbol{\alpha} \in \Gamma. \end{aligned} \tag{7.3}$$

The multi-measures moment problem (7.3) is equivalent to (7.2). Indeed, consider an arbitrary feasible solution ϕ, ν such that $\nu(\mathbf{S}) > 0$ and let ν_1, ν_2 be the restrictions of ν to $\mathbf{S}$ and $\mathbb{R}^n \setminus \mathbf{S}$ respectively. Then (ϕ', ν') with $\phi' := \phi + \nu_1$ and $\nu' := \nu_2$ is another feasible solution with value $\phi'(\mathbf{S}) = \phi(\mathbf{S}) + \nu_1(\mathbf{S}) \geq \phi(\mathbf{S})$. Therefore as we maximize $\mu(\mathbf{S})$ there is no need to impose that ν is supported on $\mathbb{R}^n \setminus \mathbf{S}$.

The multi-measures moment problem has been studied in Chapter 4. So introduce the moment variables $\mathbf{y} = (y_{\boldsymbol{\alpha}})$ and $\mathbf{z} = (z_{\boldsymbol{\alpha}})$,

$$y_{\boldsymbol{\alpha}} = \int_{\mathbf{S}} \mathbf{x}^{\boldsymbol{\alpha}} d\phi, \qquad z_{\boldsymbol{\alpha}} = \int_{\mathbb{R}^n} \mathbf{x}^{\boldsymbol{\alpha}} d\nu, \qquad \forall \boldsymbol{\alpha} \in \mathbb{N}^n.$$

Let $2v_j$ or $2v_j - 1$ be the degree of $p_j \in \mathbb{R}[\mathbf{x}]$, $j = 1, \ldots, t$. Using the decomposition of μ, the moment condition $\int \mathbf{x}^{\boldsymbol{\alpha}} d\mu = \gamma_{\boldsymbol{\alpha}}$ reads:

$$y_{\boldsymbol{\alpha}} + z_{\boldsymbol{\alpha}} = \gamma_{\boldsymbol{\alpha}}, \quad \forall \boldsymbol{\alpha} \in \Gamma.$$

Since the measure ϕ has support on the basic semi-algebraic set $\mathbf{S}$, then for

$i \geq v := \max_{j=1,\ldots,t} v_j$, one imposes the semidefinite constraints:

$$\mathbf{M}_i(\mathbf{y}) \succeq \mathbf{0}, \qquad \mathbf{M}_{i-v_j}(p_j\,\mathbf{y}) \succeq \mathbf{0},\ j = 1, \ldots, t.$$

On the other hand, since ν is supported on $\mathbb{R}^n$ we only require that $\mathbf{M}_i(\mathbf{z}) \succeq \mathbf{0}$. Therefore introduce the sequence of semidefinite optimization problems:

$$\begin{aligned} \rho_i = \sup_{\mathbf{y},\mathbf{z}}\ & y_0 \\ \text{s.t. } & y_{\boldsymbol{\alpha}} + z_{\boldsymbol{\alpha}} = \gamma_{\boldsymbol{\alpha}}, \qquad \boldsymbol{\alpha} \in \Gamma \\ & \mathbf{M}_i(\mathbf{y}) \succeq \mathbf{0}, \\ & \mathbf{M}_{i-v_j}(p_j\,\mathbf{y}) \succeq \mathbf{0},\ j = 1, \ldots, t \\ & \mathbf{M}_i(\mathbf{z}) \succeq \mathbf{0} \end{aligned} \tag{7.4}$$

which are the semidefinite relaxations analogues of (4.26) for problem (7.3). But we cannot apply Theorem 4.5 because ν is supported on $\mathbb{R}^n$, a non compact set. However, we obtain the following sequence of upper bounds. (Recall that $v := \max_{j=1,\ldots,t} v_j$.)

Theorem 7.1. *Let ρ_i be the optimal value of the semidefinite program (7.4). Then:*
(a) *For every $i \geq v$, $\rho_i \geq \rho_{\text{mom}}$ and moreover, $\rho_i \downarrow \rho^* \geq \rho_{\text{mom}}$ as $i \to \infty$.*
(b) *If ρ_i is attained at an optimal solution $(\mathbf{y}, \mathbf{z})$ which satisfies*

$$\begin{cases} \operatorname{rank} \mathbf{M}_i(\mathbf{y}) = \operatorname{rank} \mathbf{M}_{i-v}(\mathbf{y}) \\ \operatorname{rank} \mathbf{M}_i(\mathbf{z}) = \operatorname{rank} \mathbf{M}_{i-1}(\mathbf{z}) \end{cases} \tag{7.5}$$

then $\rho_i = \rho_{\text{mom}}$.

If Condition (7.5) is satisfied, we can apply Algorithm 4.2 to extract a finitely supported measure that achieves the bound ρ_{mom}.

If in problem (7.3) the sup is now over measures μ with support on a basic semi-algebraic set $\mathbb{K} \subset \mathbb{R}^n$ as defined in (4.1), we need to slightly modify the approach by requiring that the measure ν has support on $\mathbb{K}$, i.e., with $w_j = \lceil \deg g_j/2 \rceil$, one includes the additional semidefinite constraints $\mathbf{M}_{i-w_j}(g_j\mathbf{z}) \succeq 0$, $j = 1, \ldots, m$, in (7.4); see Exercise 7.2.

> If both $\mathbb{K}$ and $\mathbf{S}$ are compact and satisfy Assumption 2.1 then we can directly apply Theorem 4.5 to the generalized moment problem (7.3) and so in Theorem 7.1(a), the equality $\rho^* = \rho_{\text{mom}}$ now holds.

Let us illustrate the approach in this section with some examples.

Example 7.1. Let $n = 2$ and

$$(1, \gamma_{10}, \gamma_{01}, \gamma_{20}, \gamma_{11}, \gamma_{02}) = (1, 20, 20, 500, 390, 500) \tag{7.6}$$

be the vector of first and second order moments and let $\mathbf{S} = \{x^2 + y^2 \leq 1\}$. The first relaxation yields $\rho_1 \approx 0.1079$ and Algorithm 4.2 constructs an optimal measure μ to be a Dirac measure at the point $(\sqrt{2}/2, \sqrt{2}/2)$ on $\mathbf{S}$ with mass 0.1079. Thus, the first relaxation is exact. If in addition, to this set of moments we add the moments $\gamma_{40} = \gamma_{04} = 251000$, we have $v_2 = 2$, and thus $v = 2$. We find that $\rho_2 \approx 0.003264$ and a Dirac measure at the point $(\sqrt{2}/2, \sqrt{2}/2)$ on $\mathbf{S}$ with mass 0.003264. Note that the uper bound has decreased significantly with only two fourth order moment conditions.

To illustrate the dependence of the bound on the set $\mathbf{S}$, let $\mathbf{S} := \{x^2/2 + y^2 \leq 1\}$, and use the moment conditions (7.6) to obtain $\rho_1 \approx 0.10944$. Algorithm 4.2 constructs an optimal measure μ to be a Dirac measure at the point $(1.1454, 0.5865)$ on $\mathbf{S}$ with mass 0.10944.

Example 7.2. This example shows that the approach applies for nonconvex and disconnected sets $\mathbf{S}$. Let $n = 2$,

$$(1, \gamma_{10}, \gamma_{01}, \gamma_{20}, \gamma_{11}, \gamma_{02}, \gamma_{40}, \gamma_{04}) = (1, 0, 0, 20, 0, 20, 500, 500),$$

and let $\mathbf{S} := \{x^2/2 + y^2 \leq 1,\ x^2 + y^2/2 \geq 1\}$. The set $\mathbf{S}$ being an intersection of an ellipsoid with a complement of another ellipsoid, consists of two disconnected nonconvex sets. The two ellipsoids intersect in four points $(\mathbf{x}(i))_{i=1}^4 := (\pm\sqrt{2/3}, \pm\sqrt{2/3})$. The second relaxation yields $\rho_2 \approx 0.2111$ and Algorithm 4.2 yields an optimal measure

$$\mu = 0.2111 \left[\frac{1}{4} \sum_{i=1}^{4} \delta_{\mathbf{x}_i} \right],$$

where $(\mathbf{x}(i))_{i=1}^4 = (\pm\sqrt{2/3}, \pm\sqrt{2/3})$, are the intersection points of the two ellipsoids defining the set $\mathbf{S}$.

Upper bounds in the univariate case

We here consider the univariate case, that is, upper bounds on a probability measure μ on $\mathbb{R}$. Introduce the matrices

$$\mathbf{M}_i(\mathbf{y}) := \begin{bmatrix} y_0 & y_1 & y_2 & \cdots & y_i \\ y_1 & y_2 & \cdots & \cdots & y_{i+1} \\ \vdots & \vdots & \vdots & \vdots & \vdots \\ y_i & y_{i+1} & \cdots & y_{2i-1} & y_{2i} \end{bmatrix},$$

$$\mathbf{B}_i(\mathbf{y}) := \begin{bmatrix} y_1 & y_2 & \cdots & y_{i+1} \\ y_2 & y_3 & \cdots & y_{i+2} \\ \vdots & \vdots & \vdots & \vdots \\ y_{i+1} & y_{i+2} & \cdots & y_{2i+1} \end{bmatrix}, \quad \mathbf{C}_i(\mathbf{y}) := \begin{bmatrix} y_2 & y_3 & \cdots & y_{i+1} \\ y_3 & y_4 & \cdots & y_{i+2} \\ \vdots & \vdots & \vdots & \vdots \\ y_{i+1} & y_{i+2} & \cdots & y_{2i} \end{bmatrix}.$$

While the Hankel matrix $\mathbf{M}_i(\mathbf{y})$ is the moment matrix associated with the sequence $\mathbf{y}$, in $\mathbf{B}_i(\mathbf{y})$ (resp. in $\mathbf{C}_i(\mathbf{y})$) one recognizes the localizing matrix $\mathbf{M}_i(g\,\mathbf{y})$ (resp. $\mathbf{M}_{i-1}(h\,\mathbf{y})$) associated with $\mathbf{y}$ and the polynomial $x \mapsto g(x) := x$ (resp. $x \mapsto h(x) := x^2$).

With $\mathbf{S} = [a, b]$ and (γ_α), $\alpha \in \Gamma \subset \mathbb{N}$, being a finite collection of moments, consider the semidefinite optimization problem:

$$\begin{aligned} \rho_i = \sup_{\mathbf{y},\mathbf{z}} \; & y_0 \\ \text{s.t. } & y_\alpha + z_\alpha = \gamma_\alpha, \quad \alpha \in \Gamma \\ & \mathbf{M}_i(\mathbf{y}) \succeq \mathbf{0}, \\ & (a+b)\mathbf{B}_{i-1}(\mathbf{y}) \succeq ab\,\mathbf{M}_{i-1}(\mathbf{y}) + \mathbf{C}_i(\mathbf{y}) \\ & \mathbf{M}_i(\mathbf{z}) \succeq \mathbf{0}. \end{aligned} \tag{7.7}$$

Theorem 7.2. *Let $\gamma = (\gamma_\alpha)$ be such that $\alpha \leq 2i$ for all $\alpha \in \Gamma$.*

(a) *Let $\mathbf{S} = [a, b] \subset \mathbb{R}$. Then $\rho_i = \sup_{\mu \in \mathscr{M}(\mathbb{R})_+} \{\mu(\mathbf{S}) : \mu \sim \gamma\}$, where $\mu \sim \gamma$ means that $\int x^\alpha d\mu = \gamma_\alpha$ for all $\alpha \in \Gamma$, that is, ρ_i is a tight upper bound.*

(b) *Let $\mathbf{S} = [a, b] \subset \mathbb{R}^+$ and include the additional constraint $\mathbf{B}_{i-1}(\mathbf{z}) \succeq \mathbf{0}$ in the semidefinite program (7.7). Then $\rho_i = \sup_{\mu \in \mathscr{M}(\mathbb{R}^+)_+} \{\mu(\mathbf{S}) : \mu \sim \gamma\}$, that is, ρ_i is a tight upper bound.*

Note that the additional constraint $\mathbf{B}_{i-1}(\mathbf{z}) \succeq \mathbf{0}$ ensures that we restrict ourselves to measures μ with support on $\mathbb{R}^+$.

Example 7.3. Let $\mathbf{S} = [1,3]$ and $(1, \gamma_1, \gamma_2) = (1, 2, 10)$. In this case $\rho_1 = \sup_\mu \mu(\mathbf{S}) = 1$. With $\gamma_3 = 15$ and $\gamma_4 = 150$, one obtains $\rho_2 = \sup_\mu \approx 0.8815$ and $\mu = \phi + \nu$ with $\phi = 0.8815\,(\lambda\delta_{\{1\}} + (1-\lambda)\delta_{\{3\}})$ and $\lambda = 0.1102$.

Lower bounds

Suppose one now wishes to find a lower bound on $\mu(\mathbf{S})$. A natural idea is to replace "sup" by "inf" in the semidefinite optimization problem (7.4). But if we decompose $\mu = \phi + \nu$ with the support of ϕ contained in $\mathbf{S}$, we now must impose ν to have support in $\mathbb{R}^n \setminus \mathbf{S}$ (which is not closed if $\mathbf{S}$ is), otherwise we would obtain that $\rho_i = 0$, with $\phi = 0$ as long as $\mathbf{0} \in \mathbf{S}$. Alternatively, one may wish to find an upper bound on $1 - \mu(\mathbf{S}) = \mu(\mathbb{R}^n \setminus \mathbf{S})$ and apply the above methodology if $\mathbb{R}^n \setminus \mathbf{S}$ is a basic closed semi-algebraic set. Let us illustrate the approach by an example.

Example 7.4. Let $\mathbf{S} = \{\mathbf{x} \in \mathbb{R}^2 \,:\, x_1^2/100 + x_2^2 < 1\}$ and

$$\gamma = (1, 0, 0, 0.1, 0.0.1, 0.5, 0.5) = (1, \gamma_{10}, \gamma_{01}, \gamma_{20}, \gamma_{11}, \gamma_{02}, \gamma_{40}, \gamma_{04}).$$

In this case $\mathbf{S}^c (= \mathbb{R}^2 \setminus \mathbf{S}) = \{x_1^2/100 + x_2^2 \geq 1\}$. The optimal value $\rho_2 = 0.1010$ leads to the upper bound $\mu(\mathbf{S}^c) \leq 0.101$, which in turn leads to the lower bound $\mu(\mathbf{S}) \geq 0.899$.

7.2 Measuring Basic Semi-algebraic Sets

Given a measure μ on $\mathbb{R}^n$ with all its moments known (or which can be computed), and a compact basic semi-algebraic set $\mathbf{S} \subset \mathbb{R}^n$, computing (or approximating) $\mu(\mathbf{S})$ is an important problem with applications not only in probability but also in operations research and applied Mathematics.

For instance, with μ being the Lebesgue measure on a box $\mathbb{K} := [a,b]^n \subset \mathbb{R}^n$ with mass y_0 and with $\mathbf{S} \subset \mathbb{K}$, then $\mu(\mathbf{S})$ provides the volume of $\mathbf{S}$ (scaled by $y_0/(b-a)^n$). Computing the volume and/or integrating on a subset $\mathbb{K} \subset \mathbb{R}^n$ is a challenging problem with potentially many important applications. One possibility is to use a basic Monte Carlo technique that generates points uniformly in a box containing $\mathbb{K}$ and then compute the proportion of points falling into $\mathbb{K}$. To the best of our knowledge, most of all other approximate (deterministic or randomized) or exact techniques deal with polytopes or convex bodies only. In fact, it is well-known that even approximating the volume is hard (and even for convex polytopes). For numerical integration against a weight function on simple sets (like e.g. simplex, box) powerful cubature formulas are available, but not for abritrary basic semi-algebraic sets.

As we next see, the moment approach developed earlier is well-suited to address this problem. So with $\mathbb{K} \subset \mathbb{R}^n$ being a basic closed semi-algebraic set as described in (4.1) and $\mathbf{S}$ being the basic semi-algebraic set described in (7.1), consider the following optimization problem:

$$\begin{aligned} \rho = \sup_{\nu \in \mathscr{M}(\mathbb{K})_+} & \int_{\mathbb{K}} 1_{\mathbf{S}}\, d\nu \\ \text{s.t.} & \int_{\mathbb{K}} \mathbf{x}^{\boldsymbol{\alpha}} d\nu = \gamma_{\boldsymbol{\alpha}}, \qquad \forall \boldsymbol{\alpha} \in \mathbb{N}^n, \end{aligned} \tag{7.8}$$

where $\gamma = (\gamma_{\boldsymbol{\alpha}})$ is the (known) moment vector of μ. Hence (7.8) is the generalized moment problem (4.2) with $f = 1_{\mathbf{S}}$ and with countable set $\Gamma = \mathbb{N}^n$. Recall that measures on compact sets are determinate, that is, completely determined by their moments. Therefore, if $\mathbb{K}$ is compact, problem (7.8) is a fake optimization problem since because of the moment constraints, ν is uniquely determined. Therefore, necessarily $\nu = \mu$ and $\rho = \mu(\mathbf{S})$.

Again, we do not want to deal with indicator functions. So, as we did for problem (7.2), we consider the multi-measures generalized moment problem:

$$\begin{aligned} \rho_{\text{mom}} = \sup_{\nu \in \mathscr{M}(\mathbb{K})_+, \phi \in \mathscr{M}(\mathbf{S})_+} & \int_{\mathbf{S}} d\phi \\ \text{s.t.} & \int_{\mathbf{S}} \mathbf{x}^{\boldsymbol{\alpha}} d\phi + \int_{\mathbb{K}} \mathbf{x}^{\boldsymbol{\alpha}} d\nu = \gamma_{\boldsymbol{\alpha}}, \quad \forall \boldsymbol{\alpha} \in \mathbb{N}^n, \end{aligned} \tag{7.9}$$

which is exactly (7.3) with $\Gamma = \mathbb{N}^n$ and with $\mathbb{K}$ instead of $\mathbb{R}^n$.

Let (ϕ, ν) be an arbitrary feasible solution of (7.9). As both ϕ and ν have support contained in $\mathbb{K}$, the constraints of (7.9) imply $\phi + \nu = \mu$ because measures with compact support are determinate. Hence $\phi \leq \mu$ which in turn yields $\phi(\mathbf{S}) \leq \mu(\mathbf{S})$, and so $\rho_{\text{mom}} \leq \mu(\mathbf{S})$. Therefore, a unique optimal solution of (7.9) is the couple (ϕ^*, ν^*) where ϕ^* (resp. ν^*) is the restriction of μ to $\mathbf{S}$ (resp. to $\mathbb{K} \setminus \mathbf{S}$). Then indeed, $\rho_{\text{mom}} = \phi^*(\mathbf{S}) = \mu(\mathbf{S}) = \rho$.

Next, the analogues of semidefinite relaxations (4.26) for problem (7.9) read:

$$\begin{aligned} \rho_i = \sup_{\mathbf{y},\mathbf{z}} \; & y_{\mathbf{0}} \\ \text{s.t. } & y_{\boldsymbol{\alpha}} + z_{\boldsymbol{\alpha}} = \gamma_{\boldsymbol{\alpha}}, && \boldsymbol{\alpha} \in \mathbb{N}^n_{2i} \\ & \mathbf{M}_i(\mathbf{y}) \succeq \mathbf{0}, \mathbf{M}_i(\mathbf{z}) \succeq \mathbf{0}. \\ & \mathbf{M}_{i-v_j}(p_j\,\mathbf{y}) \succeq \mathbf{0}, && j = 1, \ldots, t \\ & \mathbf{M}_{i-w_j}(g_j\,\mathbf{z}) \succeq \mathbf{0}, && j = 1, \ldots, m \end{aligned} \tag{7.10}$$

where $w_j = \lceil (\deg g_j)/2 \rceil$, $j = 1, \ldots, m$ and $\mathbb{N}^n_{2i} := \{\boldsymbol{\alpha} \in \mathbb{N}^n : |\boldsymbol{\alpha}| \leq 2i\}$. Of course one has $\rho_i \geq \rho_{\text{mom}}$ because (7.10) is a relaxation of (7.9). So we have a multi-measures moment problem with countably many moment constraints. The following result is a slight extension of Theorem 4.5 to handle the countably many moment constraints of (7.9).

Theorem 7.3. *Let Assumption 2.1 hold for $\mathbb{K} \subset \mathbb{R}^n$ in (4.1) and for $\mathbf{S} \subset \mathbb{R}^n$ in (7.1). Let ρ_{mom} and ρ_i be as in (7.9) and (7.10) respectively. Then as $i \to \infty$, $\rho_i \downarrow \rho_{\text{mom}} = \mu(\mathbf{S})$.*

Proof. We briefly sketch the proof which is very similar to that of Theorem 4.5. Let $(\mathbf{y}^i, \mathbf{z}^i)$ be a nearly optimal solution of (7.10), e.g. any feasible solution with value $\rho_i - 1/i \leq y_{\mathbf{0}} \leq \rho_i$. Then $(\mathbf{y}^{i_k}, \mathbf{z}^{i_k})$ converges pointwise to some $(\mathbf{y}, \mathbf{z})$ for some subsequence (i_k). Fix $\boldsymbol{\alpha} \in \mathbb{N}^n$, arbitrary. The pointwise convergence yields $\mathbf{y}_{\boldsymbol{\alpha}} + \mathbf{z}_{\boldsymbol{\alpha}} = \gamma_{\alpha}$ and as $\boldsymbol{\alpha}$ is arbitrary, $(\mathbf{y}, \mathbf{z})$ satisfies the moment constraints. Again, because of the pointwise convergence, for every $r = 1, \ldots,$ we also have $\mathbf{M}_r(\mathbf{y}), \mathbf{M}_r(p_j\,\mathbf{y}) \succeq 0$, $j = 1, \ldots, t$, and $\mathbf{M}_r(\mathbf{z}), \mathbf{M}_r(g_j\,\mathbf{z}) \succeq 0$, $j = 1, \ldots, m$. Therefore, by Theorem 3.8, $\mathbf{y}$ (resp. $\mathbf{z}$) is the moment vector of some measure ϕ (res. ν) supported on $\mathbf{S}$ (resp. on $\mathbb{K}$), and so (ϕ, ν) is feasible for (7.9) with value $\phi(\mathbf{S}) = \lim_{k\to\infty} y_{\mathbf{0}}^{i_k} = \lim_{k\to\infty} \rho_{i_k} \geq \rho_{\text{mom}}$. And so (ϕ, ν) is an optimal solution of (7.9). As ρ_i is monotone we deduce the desired result $\lim_{i\to\infty} \rho_i = \rho_{\text{mom}}$. On the other hand we have already seen that $\rho_{\text{mom}} = \mu(\mathbf{S})$. □

Remark 7.1. One may always rescale the problem so as to make the set

$\mathbb{K}$ contained in the box $[-1,1]^n$. In this case, in (7.10) one may ignore the constraint $\mathbf{M}_{i-w_j}(g_j\mathbf{z}) \succeq 0$, to obtain the slightly simpler semidefinite realaxation:

$$\begin{aligned} \rho_i = \sup_{\mathbf{y},\mathbf{z}} \; & y_{\mathbf{0}} \\ \text{s.t.} \; & y_{\boldsymbol{\alpha}} + z_{\boldsymbol{\alpha}} = \gamma_{\boldsymbol{\alpha}}, \qquad & \boldsymbol{\alpha} \in \mathbb{N}^n_{2i} \\ & \mathbf{M}_i(\mathbf{y}) \succeq \mathbf{0}, \mathbf{M}_i(\mathbf{z}) \succeq \mathbf{0}. & \\ & \mathbf{M}_{i-v_j}(p_j\,\mathbf{y}) \succeq \mathbf{0}, & j = 1,\ldots,t, \end{aligned} \tag{7.11}$$

and Theorem 7.3 remains valid with (7.11) instead of (7.10).

Indeed, in the proof of Theorem 7.3, from $y^i_{\boldsymbol{\alpha}} + z^i_{\boldsymbol{\alpha}} = \gamma_{\boldsymbol{\alpha}}$, for every $\boldsymbol{\alpha} \in \mathbb{N}^n_{2i}$, one obtains $0 \preceq \mathbf{M}_i(\mathbf{z}^i) \preceq \mathbf{M}_i(\gamma)$ for every i. In particular, $z^i_{2\beta} \leq \gamma_{2\beta}$ for every $\beta \in \mathbb{N}^n_i$. Next, as μ is supported on $\mathbb{K} \subseteq [-1,1]^n$, one has $|\gamma_{\boldsymbol{\alpha}}| \leq 1$ for every $\boldsymbol{\alpha} \in \mathbb{N}^n$, and so by Proposition 3.6, $|z^i_{\boldsymbol{\alpha}}| \leq 1$ for all $\boldsymbol{\alpha} \in \mathbb{N}^n_{2i}$. This implies that the limiting sequence $\mathbf{z}$ in the proof of Theorem 7.3 satisfies $|z_{\boldsymbol{\alpha}}| \leq 1$ for all $\boldsymbol{\alpha} \in \mathbb{N}^n$, which in turn by Theorem 3.5(b), implies that $\mathbf{z}$ has a representing measure ν supported on $[-1,1]^n$. But by $\mathbf{y} + \mathbf{z} = \gamma$ and the fact that measures with compact support are determinate, we have $\nu \leq \mu$. Therefore since μ is supported on $\mathbb{K}$ then so is ν. The rest of the proof is the same.

The dual problem

As problem (7.8) is the same as $\rho = \sup\,\{\int_{\mathbf{S}} d\nu \;:\; \nu = \mu\,;\; \nu \in \mathscr{M}(\mathbb{K})_+\}$, consider its dual

$$\rho^* = \inf_f \,\{\int_{\mathbf{S}} f\,d\mu \;:\; f \geq 1_{\mathbf{S}}\,;\; f \in C(\mathbb{K})\,\}, \tag{7.12}$$

where $C(\mathbb{K})$ is the Banach space of continuous functions on $\mathbb{K}$.

Of course, instead of $C(\mathbb{K})$ one could have chosen the space $B(\mathbb{K})$ of bounded measurable functions on $\mathbb{K}$, in which case a trivial optimal solution is $f := 1_{\mathbf{S}}$. But one next shows that by using the much smaller class of functions $C(\mathbb{K})$, one also obtains the same optimal value. And as we will see, the dual of the semidefinite relaxation (7.11) is strongly related to (7.12).

Let $\mathbf{x} \mapsto d(\mathbf{x},\mathbf{S})$ be the euclidean distance to the set $\mathbf{S}$, and for $\epsilon > 0$ let $D_\epsilon := \{\mathbf{x} \in \mathbb{K} \;:\; d(\mathbf{x},\mathbf{S}) \geq \epsilon\}$.

Lemma 7.4. *Assume that* $\mathbf{S}$ *and* $\mathbb{K} \setminus \mathbf{S}$ *have non empty interior, and let* ρ *and* ρ^* *be as in (7.8) and (7.12), respectively.*

Then $\rho = \rho^* = \mu(\mathbf{S})$. *Moreover, there is a minimizing sequence in* $\mathbb{R}[\mathbf{x}]$, *i.e. a sequence of polynomials* $(p_\epsilon) \subset \mathbb{R}[\mathbf{x}]$ *that satisfy the constraints of (7.12) and such that* $\lim_{\epsilon\to 0}\int_{\mathbf{S}} p_\epsilon d\mu = \rho^*$.

Proof. As $\mathbf{S}$ and $\mathbb{K}\setminus\mathbf{S}$ have non empty interior, there is some $\epsilon_0 > 0$ such that for every $\epsilon < \epsilon_0$, the closed set D_ϵ is non empty and disjoint from $\mathbf{S}$. As $\mathbb{K}$ is a metric space (hence a normal space), by Urysohn's Lemma[1] there is a continuous function $f_\epsilon : \mathbb{K}\to[0,1]$ such that $f_\epsilon = 1$ on $\mathbf{S}$ and $f_\epsilon = 0$ on D_ϵ. Therefore,

$$\mu(\mathbf{S}) \le \int_{\mathbb{K}} f_\epsilon\, d\mu \le \mu(\{\mathbf{x}\in\mathbb{K} : d(\mathbf{x},\mathbf{S}) < \epsilon\}).$$

and so $\int_{\mathbb{K}} f_\epsilon d\mu \to \mu(\mathbf{S})$ as $\epsilon\to 0$.

Let $(\epsilon_k)_k \subset \mathbb{R}_+$ be a sequence with $\epsilon_k > 0$ for every k, $\epsilon_k\to 0$ as $k\to\infty$, and let f_{ϵ_k} be as above. As $\mathbb{K}$ is compact, by the Stone-Weierstrass theorem, for every $k\in\mathbb{N}$, there is a polynomial $p_k\in\mathbb{R}[\mathbf{x}]$ such that $\sup\{|f_{\epsilon_k}(\mathbf{x}) - p_k(\mathbf{x})| : \mathbf{x}\in\mathbb{K}\} \le \epsilon_k$. Thefore the sequence $(p_k+\epsilon_k)_k \subset \mathbb{R}[\mathbf{x}]$ is such that $p_k + \epsilon_k \ge 1_{\mathbf{S}}$ on $\mathbb{K}$ and $\lim_{k\to\infty}\int_{\mathbb{K}}(p_k+\epsilon_k)d\mu = \lim_{k\to\infty}\int_{\mathbb{K}} f_{\epsilon_k} d\mu = \rho^*$. □

Hence in solving (7.12), one wishes to minimize $\int_{\mathbb{K}} f d\mu$ over all f in $C(\mathbb{K})$ minorized by $1_{\mathbf{S}}$. And in fact, there is a minimizing sequence in $\mathbb{R}[\mathbf{x}]$ ($\subset C(\mathbb{K})$) with the desired limiting value ρ^*.

On the other hand, with $\mathbb{K}\subseteq[-1,1]^n$, the dual of the semidefinite relaxation (7.11) reads:

$$\begin{array}{rl} \rho_i^* = \displaystyle\inf_{\boldsymbol{\lambda},\sigma_j,\psi_0} & \displaystyle\sum_{\boldsymbol{\alpha}} \lambda_{\boldsymbol{\alpha}}\gamma_{\boldsymbol{\alpha}} \\ \text{s.t.} & \displaystyle\sum_{\boldsymbol{\alpha}} \lambda_{\boldsymbol{\alpha}}\mathbf{x}^{\boldsymbol{\alpha}} - 1 = \sigma_0 + \sum_{j=1}^t \sigma_j\, p_j, \quad (\sigma_j)_{j=0}^t \subset \Sigma[\mathbf{x}], \\ & \sum_{\boldsymbol{\alpha}} \lambda_{\boldsymbol{\alpha}}\mathbf{x}^{\boldsymbol{\alpha}} = \psi_0, \quad \psi_0\in\Sigma[\mathbf{x}], \\ & \deg\psi_0,\ \deg\sigma_0,\ \deg\sigma_j p_j \le 2i, \end{array}$$

or, equivalently:

[1] Urysohn's Lemma states that for any two disjoint closed sets A, B of a normal Hausdorff space X, there is a continuous function $X\to[0,1]$ such that $f = 1$ on A and $f = 0$ on B; see Ash (1972)[A4.2, p. 379]. A metric space is normal.

$$\begin{aligned} \rho_i^* = \inf_{f,\sigma_j} \, & \int_{\mathbb{K}} f \, d\mu \\ \text{s.t. } & f - 1 = \sigma_0 + \sum_{j=1}^{t} \sigma_j \, p_j, \qquad f, \, \sigma_0, \ldots, \sigma_t \in \Sigma[\mathbf{x}] \\ & \deg f, \, \deg \sigma_0, \, \deg \sigma_j p_j \leq 2i. \end{aligned} \tag{7.13}$$

That is, among the s.o.s. polynomials f such that $f \geq 1_{\mathbf{S}}$, one searches for the one that minimizes $\int_{\mathbb{K}} f d\mu$. Equivalently, with this criterion, one searches for the best *polynomial* approximation from above of the function $1_{\mathbf{S}}$. So obviously $\rho_i^* \geq \rho^*$.

Proposition 7.5. *Let $\mathbb{K} \subseteq [-1,1]^n$ and let both $\mathbf{S}$ and $\mathbb{K} \setminus \mathbf{S}$ have a non empty interior. Then for every $i \geq \max_{j=1,\ldots,t} v_j$, the semidefinite relaxations (7.11) and (7.13) have same optimal value, i.e., $\rho_i = \rho_i^*$. And so $\rho_i^* \to \rho_{\text{mom}} = \mu(\mathbf{S})$.*

Proof. Let $\mathbf{y}$ (resp. $\mathbf{z}$) be the moment sequence of the restriction of μ to $\mathbf{S}$ (resp. to $\mathbb{K} \setminus \mathbf{S}$). As both $\mathbf{S}$ and $\mathbb{K} \setminus \mathbf{S}$ have a non empty interior, then necessarily, for every i one has $\mathbf{M}_i(\mathbf{z}) \succ 0$, $\mathbf{M}_i(\mathbf{y}) \succ 0$, and $\mathbf{M}_i(p_j \, \mathbf{y}) \succ 0$ for every $j = 1, \ldots, t$. Indeed, for every $0 \neq h \in \mathbb{R}[\mathbf{x}]_i$ and $0 \neq g \in \mathbb{R}[\mathbf{x}]_{i-v_j}$, with respective vectors of coefficients $\mathbf{h}$ and $\mathbf{g}$,

$$\langle \mathbf{h}, \mathbf{M}_i(\mathbf{y})\mathbf{h} \rangle = \int_{\mathbf{S}} h^2 \, d\mu \neq 0, \quad \langle \mathbf{h}, \mathbf{M}_i(\mathbf{z})\mathbf{h} \rangle = \int_{\mathbb{K} \setminus \mathbf{S}} h^2 \, d\mu \neq 0,$$

and $\langle \mathbf{g}, \mathbf{M}_i(p_j \, \mathbf{y})\mathbf{g} \rangle = \int_{\mathbf{S}} g^2 \, p_j d\mu \neq 0$, because neither h nor g can vanish on an open set. Moreover, by definition, $\mathbf{y}_{\boldsymbol{\alpha}} + \mathbf{z}_{\boldsymbol{\alpha}} = \gamma_{\boldsymbol{\alpha}}$ for every $\boldsymbol{\alpha} \in \mathbb{N}^n$. Therefore $(\mathbf{y}, \mathbf{z})$ is a *strictly* feasible solution of (7.11), which by a standard result in semidefinite optimization, yields absence of a duality gap, i.e., the desired result $\rho_i = \rho_i^*$; see Section C.2. □

Example 7.5. Consider the two-dimensional example with $\mathbb{K} := [-1,1]^2$ and

$$\mathbf{S} := \{ \mathbf{x} \in \mathbb{R}^2 \; : \; -(x_1^2 + x_2^2)^3 + 4x_1^2 x_2^2 \geq 0 \}$$

which is the 2-dimensional *folium* displayed in figure 7.1. Notice that $\mathbf{S}$ is not convex and its interior is formed with several disjoint connected components. Figure 7.2 displays a polynomial approximation $g \in \mathbb{R}[\mathbf{x}]_{20}$

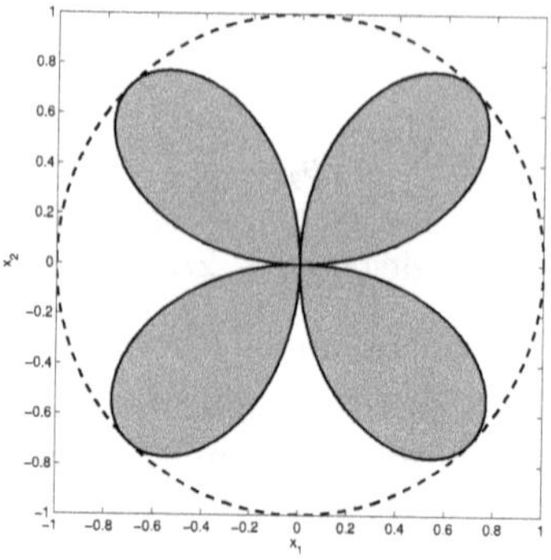

Fig. 7.1 The $2d$-Folium surface $\mathbf{S}$.

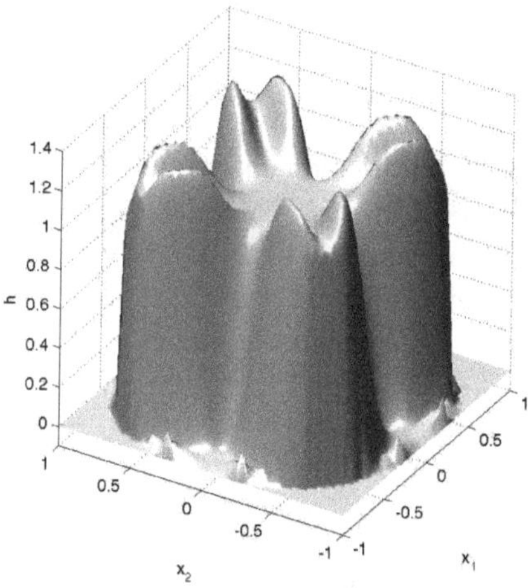

Fig. 7.2 Degree-20 polynomial approximation of $1_{\mathbf{S}}$.

of $1_{\mathbf{S}}$, from above, obtained by solving the semidefinite relaxation (7.11) (and its dual (7.13)), a semidefinite program with 231 variables $\mathbf{y}$, i.e., with moments variables (y_{α}) up to order 20. The obtained relative error $(\rho_{20} - \mu(\mathbf{S}))/\mu(\mathbf{S})$ is 1.2%. For visualization purposes, $\max(5/4, g)$ rather than g is displayed in Figure 7.2. Even if typical oscillations occur near the boundary regions, we can recognize the shape of Figure 7.1.

Integration against a weight function

In fact, from the proof of Theorem 7.3, it follows that for any given finite subset $\mathbb{N}_d^n = \{\boldsymbol{\alpha} \in \mathbb{N}^n : |\boldsymbol{\alpha}| \leq d\}$, one also has:

$$\lim_{i\to\infty} y^i_{\boldsymbol{\alpha}} = \int_{\mathbf{S}} \mathbf{x}^{\boldsymbol{\alpha}} d\phi = \int_{\mathbf{S}} \mathbf{x}^{\boldsymbol{\alpha}} d\mu, \qquad \forall \boldsymbol{\alpha} \in \mathbb{N}_d^n, \tag{7.14}$$

where $\mathbf{y}^i$ is any optimal (or nearly optimal) solution of (7.10). Hence let $\mathbb{K}$ be a box or a simplex of $\mathbb{R}^n$ that contains the basic semi-algebraic set $\mathbf{S}$, and suppose that one wishes to approximate the integral

$$J := \int_{\mathbf{S}} f(\mathbf{x})\, W(\mathbf{x})\, d\mathbf{x}$$

where $W : \mathbb{R}^n \to \mathbb{R}$ is a nonnegative measurable function, and $f \in \mathbb{R}[\mathbf{x}]$ has degree at most d. One may then proceed as follows:

(1) Fix some integer i with $d \ll i$, and compute or approximate the collection of moments $\boldsymbol{\gamma} = (\gamma_{\boldsymbol{\alpha}})$, $\boldsymbol{\alpha} \in \mathbb{N}_{2i}^n$, of the measure $d\mu = W dx$ on $\mathbb{K}$, e.g. via cubature formulas. As $\mathbb{K}$ is a box or a simplex, powerful techniques are available for such a task; see e.g. Gautschi (1981, 1997).
(2) Compute ρ_i in (7.10) and obtain an optimal (or nearly optimal solution) $\mathbf{y}^i$.

Then from (7.14), one obtains $J \approx \sum_{\boldsymbol{\alpha}\in\mathbb{N}_d^n} f_{\boldsymbol{\alpha}}\, y^i_{\boldsymbol{\alpha}}$ provided that i is sufficiently large.

7.3 Measures with Given Marginals

Let $\mathbb{K} := \mathbb{K}_1 \times \mathbb{K}_2$ with $\mathbb{K}_1, \mathbb{K}_2$ being compact Borel subsets of $\mathbb{R}^{n_1}$ and $\mathbb{R}^{n_2}$ respectively, and let $\pi_i : \mathscr{M}(\mathbb{K})_+ \to \mathscr{M}(\mathbb{K}_i)_+$, $i = 1, 2$, be the projection mappings. That is, for every measure μ in $\mathscr{M}(\mathbb{K})_+$,

$$(\pi_1\mu)(B) := \mu(B \times \mathbb{K}_2) \quad \text{and} \quad (\pi_2\mu)(B) := \mu(\mathbb{K}_1 \times B)$$

for every Borel subset B of $\mathbb{K}_1$ and $\mathbb{K}_2$, respectively.

With $f : \mathbb{K} \to \mathbb{R}$ being a measurable function, and $\nu_1 \in \mathscr{M}(\mathbb{K}_1)_+, \nu_2 \in \mathscr{M}(\mathbb{K}_2)_+$ being two given probability measures on $\mathbb{K}_1$ and $\mathbb{K}_2$, respectively, consider the linear program:

$$\begin{aligned} \mathrm{MT}: \rho_{\mathrm{mom}} = & \sup_{\mu \in \mathscr{M}(\mathbb{K})_+} \int_{\mathbb{K}} f \, d\mu \\ & \text{s.t. } \pi_i \, \mu = \nu_i \text{ for } i = 1, 2. \end{aligned} \tag{7.15}$$

That is, one searches for the supremum of $\int_{\mathbb{K}} f d\mu$ over all probability measures μ on $\mathbb{K}$ whose marginal on $\mathbb{K}_1$ (resp. on $\mathbb{K}_2$) matches ν_1 (resp. ν_2). Of course one may generalize to more than two sets and consider a finite family $(\mathbb{K}_i)$, $i = 1, \ldots, p$.

MT (i.e. when $p = 2$) is also the celebrated transportation (or *mass-transfer*) problem, sometimes also called the Monge-Kantorovich problem. To see that (7.15) is a particular instance of the generalized moment problem (1.1), it suffices to notice that if $\mathbb{K}_i$ is compact for $i = 1, 2$, every finite measure on $\mathbb{K}_i$ is *moment determinate*, i.e. is completely determined by its moments; see Definition 3.2. One may thus replace the constraint $\pi_i \mu = \nu_i$ with the countably many linear equality constraints

$$\int_{\mathbb{K}} \mathbf{x}^{\boldsymbol{\alpha}} \, d\mu = \int_{\mathbb{K}_i} \mathbf{x}^{\boldsymbol{\alpha}} \, d\nu_i, \quad \forall \alpha \in \mathbb{N}^{n_i}; \quad i = 1, 2.$$

Therefore (7.15) is also the GMP (1.1) with $\Gamma := \mathbb{N}^{n_1} \cup \mathbb{N}^{n_2}$, $h_{\boldsymbol{\alpha}} = \mathbf{x}^{\boldsymbol{\alpha}}$ and $\gamma_{\boldsymbol{\alpha}} = \int \mathbf{x}^{\boldsymbol{\alpha}} \, d\nu_i$ whenever $\boldsymbol{\alpha} \in \mathbb{N}^{n_i}$, $i = 1, 2$.

Proposition 7.6. *Let f be bounded and upper-semicontinuous. Then (7.15) has an optimal solution and there is no duality gap.*

Proof. Observe that the product probability measure $\mu := \nu_1 \otimes \nu_2$ on $\mathbb{K}$ is a feasible solution of (7.15). Also the assumptions of Theorem 1.3 are satisfied ($h_0 = 1$) and so Corollary 1.4 applies. □

Next, let $f \in \mathbb{R}[\mathbf{x}]$ and let

$$\mathbb{K}_t := \{\mathbf{x} \in \mathbb{R}^{n_t} \, : \, g_{tj}(\mathbf{x}) \geq 0, \quad j = 1, \ldots, m_t\}, \qquad t = 1, 2,$$

with $v_{tj} := \lceil (\deg g_{tj})/2 \rceil$, $j = 1, \ldots, m_t$, $t = 1, 2$. For $2i \geq$

max[deg f, $\max_{t,j}$ deg g_{tj}], the semidefinite programs

$$
\begin{aligned}
\rho_i = \sup_{\mathbf{y}} \; & L_{\mathbf{y}}(f) \\
\text{s.t.}\; & \mathbf{M}_i(\mathbf{y}) \succeq \mathbf{0}, \\
& \mathbf{M}_{i-v_{tj}}(g_{tj}\,\mathbf{y}) \succeq \mathbf{0},\; j = 1,\ldots,m_t; \quad t = 1,2 \\
& y_{\boldsymbol{\alpha}} = \gamma_{\boldsymbol{\alpha}} \qquad\qquad \boldsymbol{\alpha} \in \Gamma; \quad |\boldsymbol{\alpha}| \leq 2i,
\end{aligned}
\tag{7.16}
$$

are the semidefinite relaxations (4.22) adapted to the generalized moment problem (7.15). And so the following convergence result is a direct consequence of Theorem 4.3.

Proposition 7.7. *Let $f \in \mathbb{R}[\mathbf{x}]$ and let ρ_{mom} (resp. ρ_i) be as in (7.15) (resp. as in (7.16)). Then $\rho_i \downarrow \rho_{\mathrm{mom}}$ as $i \to \infty$.*

7.4 Summary

In this chapter, we have defined a sequence of semidefinite relaxations that provide increasingly stronger bounds on $\mu(\mathbf{S})$ whenever $\mathbf{S}$ is a basic closed semi-algebraic set, and μ satisfies some moment constraints. When μ is restricted to be supported on a compact basic semi-algebraic set $\mathbb{K} \supset \mathbf{S}$ then the associated sequence of upper bounds converge to a tight upper bound on $\mu(\mathbf{S})$. As a consequence, we can approximate the volume of a basic semi-algebraic set to any desired accuracy by solving an appropriate hierarchy of semidefinite relaxations. Similarly, one may also approximate as closely as desired the integral $\int_{\mathbf{S}} W dx$ for any nonnegative measurable weight function W provided one is able to approximate the moments of the measure $d\mu = W dx$ on a box (or simplex) $\mathbb{K} \supset \mathbf{S}$, e.g. by cubature formulas. Finally, we has also proposed a hierarchy of semidefinite relaxations to solve (or at least approximate) the mass-transfer problem.

7.5 Exercises

Exercise 7.1. With $\mathbf{S}$ as in (7.1) what happens if in (7.2) one replaces "sup" with "inf". If we now assume that $\mathbf{S} \subset \mathbb{K}$ is open (with $\mathbb{K}$ still compact), what can be said? (Hint: $\mathbb{K} \setminus \mathbf{S}$ is closed hence compact and $\inf_\mu \mu(\mathbf{S}) = 1 - \sup_\mu \mu(\mathbb{K} \setminus \mathbf{S})$. What about the semidefinite relaxations? Can we always describe the set $\mathbb{K} \setminus \mathbf{S}$ easily? What is the conclusion?

Exercise 7.2. Let $\mathbb{K} = \{ \mathbf{x} \in \mathbb{R}^n : g_j(\mathbf{x}) \geq 0,\ j = 1, \ldots, m \}$ and $\mathbf{S} = \{ \mathbf{x} \in \mathbb{R}^n : p_k(\mathbf{x}) \geq 0,\ k = 1, \ldots, t \} \subseteq \mathbb{K}$, with $g_j, p_k \in \mathbb{R}[\mathbf{x}]$. Suppose that both sets $\mathbb{K}$ and $\mathbf{S}$ satisfy Assumption 2.1. Let $\gamma_{\boldsymbol{\alpha}}$ be a finite collection of moments. We are interested in solving

$$\begin{aligned} \sup_{\mu \in \mathscr{M}(\mathbb{K})_+} \quad & \mu(\mathbf{S}) \\ \text{s.t.} \quad & \int_{\mathbb{K}} \mathbf{x}^{\boldsymbol{\alpha}} d\mu = \gamma_{\boldsymbol{\alpha}}, \quad \boldsymbol{\alpha} \in \Gamma. \end{aligned} \tag{7.17}$$

(a) By modifying Problem (7.4), propose a sequence of relaxations that provide increasingly better upper bounds on Problem (7.17). Does convergence $\rho_i \downarrow \rho_{\text{mom}}$ hold?
(b) Let $n = 2$, $\mathbb{K} = \{\mathbf{x} : 0 \leq x_1, x_2 \leq 1\}$, $\mathbf{S} = \{\mathbf{x} : x_1^2 + x_2^2 \leq 1\}$ and

$$(1, \gamma_{10}, \gamma_{01}, \gamma_{20}, \gamma_{11}, \gamma_{02}) = (1, 20, 20, 500, 390, 500).$$

Solve Problem (7.17).

Exercise 7.3. Prove Theorem 7.2.

7.6 Notes and Sources

7.1. Part of this chapter is inspired from Lasserre (2002a). A detailed discussion on the complexity and results on finding bounds under moment constraints can be found in Bertsimas and Popescu (2005) and in particular, the following historical notes. Research on bound problems (also called *generalized Chebyshev inequalities*) dates back to the work of Gauss, de la Vallée Poussin, Cauchy, Chebyshev, Markov, Stieltjes. The univariate case was first proposed and formulated in Chebyshev (1874) and later resolved by his student Markov (1884) in his PhD thesis, using continued fractions techniques. For surveys on Chebyshev systems see e.g. Shohat and Tamarkin (1943) Godwin (1955, 1964), and Karlin and Studden (1966).

Citing (Shohat and Tamarkin, 1943, p. 10), "the problem of moments lay dormant for more than 20 years." It surfaced with the book of Tong (1980) and later in Landau (1987b) with the volume *Moments in Mathematics* which includes the survey Landau (1987a), as well as several relevant papers among which Kemperman (1987) and Diaconis (1987). More recent contributions are from e.g. Lasserre (2002a) and Bertsimas and Popescu (2005).

The use of optimization and duality to address moment inequalities in probability first appeared in 1960, independently and simultaneously in Isii (1960) and in Karlin and Studden (1966). Isii (1963) extended these duality results for random vectors on complete regular spaces, whereas in Smith (1995) one finds new interesting applications in decision analysis, dynamic programming, statistics and finance. On the other hand, Shapiro (2001) provided a rigorous and detailed discussion on conditions for strong duality to hold. A detailed analysis of strong duality and sensitivity analysis for semi-infinite programming problems can be found in Bonnans and Shapiro (2000).

7.2. Computing or even approximating the volume of a convex body is hard theoretically and in pracice as well. Even for a convex polytope $\Omega \subset \mathbb{R}^n$, exact computation of its volume vol(Ω) or integrating over Ω is difficult. Its computational complexity is discussed in e.g. Bollobás (1997) and Dyer and Frieze (1988). Any deterministic algorithm with polynomial time complexity that would compute upper and lower bounds $\overline{v}$ and $\underline{v}$ on vol(Ω), cannot yield an upper bound $g(n)$ on $\overline{v}/\underline{v}$ better than polynomial in the dimension n. Methods for exact volume computation use either triangulations or simplicial decompositions depending on whether Ω has a half-space description or a vertex description. See e.g. Cohen and Hickey (1979), Lasserre (1983), Lawrence (1991) and see Büeler *et al.* (2000) for a comparison. Another set of methods which use generating functions are described in e.g. Barvinok (1993) and Lasserre and Zeron (2001). Concerning integration on simple sets (e.g. simplex, box) via cubature formulas, the interested reader is referred to e.g. Gautschi (1981, 1997).

In contrast with these negative results, and if one accepts randomized algorithms that fail with small probability, then the situation is much better. Indeed, the celebrated probabilistic approximation algorithm of Dyer *et al.* (1991) computes the volume to fixed arbitrary relative precision ϵ, in time polynomial in ϵ^{-1}. The latter algorithm uses approximations schemes based on rapidly mixing Markov chains and isoperimetric inequalities. See also hit-and-run algorithms for sampling points according to a given distribution, described in e.g. Belisle *et al.* (1993). Theorem 7.3 which permits to approximate as closely as desired the volume of a compact basic semi-algebraic set, is from Henrion *et al.* (2009b). In this latter reference, numerical issues associated with solving the semidefinite programs are discussed, as well as the use of other bases for $\mathbb{R}[\mathbf{x}]_i$ than the usual canonical basis of monomials $(\mathbf{x}^\alpha)$. In particular, in some examples, much better results (in terms of numerical precision and stability) are obtained with the basis of Chebyshev polynomials.

7.3. The mass transfer problem is a very hold engineering problem stated by the french geometer Monge in the 18th century for military applications, and is a special case of moment problems involving measures with *given marginals.* For instance, with $\mathbb{K}_1 = \mathbb{K}_2$ and a specific *distance* function f, its optimal value also measures the distance between two probability measures ν_1 and ν_2 on $\mathbb{K}_1$, and so induces a metric on the space of probability measures on $\mathbb{K}_1$. For the interested reader, a nice discussion on moment problems involving measures with given marginals can be found in Kemperman (1987) and the many references therein.

Chapter 8

Markov Chains Applications

This chapter is about Markov chains and invariant probabilities. We first address the problem of computing an upper bound on $\mu(\mathbf{S})$ over all invariant probability measures μ of a given Markov chain on $\mathbb{R}^n$. We then consider the problem of approximating the value of an ergodic criterion, as an alternative to simulation which only provides a random estimate.

Consider a discrete-time Markov chain (MC) $\Phi = (\Phi_0, \Phi_1, \ldots)$ on a measurable space $(X, \mathscr{B})$ with transition probability function (or stochastic kernel) $P : X \times \mathscr{B} \to \mathbb{R}$, that is:

- For every $\mathbf{x} \in X$, $P(\mathbf{x}, \cdot)$ is a probability measure on X.
- For every $B \in \mathscr{B}$, $\mathbf{x} \mapsto P(\mathbf{x}, B)$ is a measurable function.

When X is finite then P is just a stochastic matrix, the matrix of transition probabilities with all rows summing up to one.

One may consider P from two dual viewpoints. As a linear operator acting on $B(X)$ (the Banach space of bounded measurable functions on X, equipped with the sup-norm) by:

$$f \longmapsto Pf(\mathbf{x}) := \int_X P(\mathbf{x}, d\mathbf{y}) f(\mathbf{y}), \qquad \forall \mathbf{x} \in X,$$

and as a linear operator acting on $\mathscr{M}(X)$ (the Banach space of finite signed measures on X) by:

$$\mu \longmapsto \mu P(B) := \int_X P(x, B)\, \mu(dx), \qquad \forall B \in \mathscr{B}.$$

With $X \subseteq \mathbb{R}^n$, let $C(X)$ denote the Banach space of bounded continuous functions on X, equipped with the sup-norm.

A Markov chain Φ on $X \subseteq \mathbb{R}^n$, with associated stochastic kernel P, is said to be *weak-Feller* if P maps $C(X)$ into $C(X)$.

A probability measure $\mu \in \mathscr{M}(X)_+$ is called *invariant* if it satisfies

$$\mu P = \mu; \qquad \mu(X) = 1. \tag{8.1}$$

The spaces $B(X)$ and $\mathscr{M}(X)$ form a dual pair[1] of vector spaces with duality bracket

$$\langle f, \mu \rangle = \int_X f \, d\mu, \qquad \forall f \in B(X), \, \mu \in \mathscr{M}(X). \tag{8.2}$$

Therefore (8.1) is equivalent to:

$$\mu(X) = 1; \quad \langle f, \mu \rangle = \langle f, \mu P \rangle \, (= \langle Pf, \mu \rangle), \quad \forall f \in B(X).$$

When P is the operator acting on $\mathscr{M}(X)$, one sometimes write $P^*\mu$ instead of μP, as it is the adjoint operator of $P : B(X) \to B(X)$.

Similarly, the spaces $C(X)$ and $\mathscr{M}(X)$ also form a dual pair with same duality bracket as in (8.2). So if P is weak-Feller, (8.1) is equivalent to

$$\mu(X) = 1; \quad \langle f, \mu \rangle = \langle Pf, \mu \rangle, \quad \forall f \in C(X). \tag{8.3}$$

If $X := \mathbb{K} \subset \mathbb{R}^n$ is compact, then $C(\mathbb{K})$ is separable and the space of polynomials (restricted to $\mathbb{K}$) is a dense subset of $C(\mathbb{K})$. Therefore, the canonical basis of monomials $(\mathbf{x}^{\boldsymbol{\alpha}})_{\boldsymbol{\alpha} \in \mathbb{N}^n}$ is a countable dense subset of $C(\mathbb{K})$, and (8.3) is equivalent to

$$\langle P\mathbf{x}^{\boldsymbol{\alpha}}, \mu \rangle = \langle \mathbf{x}^{\boldsymbol{\alpha}}, \mu \rangle; \quad \forall \boldsymbol{\alpha} \in \mathbb{N}^n; \quad \mu(\mathbb{K}) = 1. \tag{8.4}$$

The following is a basic result for Markov chains on a compact space.

Proposition 8.1. *Let $X \subset \mathbb{R}^n$ be compact and let P be weak-Feller. Then P has an invariant probability measure.*

As seen in the following example, the weak-Feller assumption is crucial and cannot be removed.

Example 8.1. Let $X \subset \mathbb{R}$ be the compact space $[0, 1]$, and let Φ be the Markov chain with transitions defined by:

$$\Phi_{t+1} := \begin{cases} \Phi_t/2 \text{ if } \Phi_t \neq 0 \\ 1 \text{ if } \Phi_t = 0. \end{cases}$$

[1] See Section C.3.

The weak-Feller property fails because Pf is not continuous a $x = 0$ whenever f is continuous. And P has no invariant probability.

8.1 Bounds on Invariant Measures

8.1.1 *The compact case*

Let $X := \mathbb{K} \subseteq \mathbb{R}^n$ and $\mathbf{S} \subset \mathbb{K}$ be given compacts sets, and consider the problem of computing an upper bound on $\mu(\mathbf{S})$ over all invariant probability measures μ of P. That is, evaluate or approximate

$$j^* := \sup_{\mu \in \mathscr{M}(\mathbb{K})_+} \{\, \mu(\mathbf{S}) \,:\, \mu P = \mu\,; \qquad \mu(\mathbb{K}) = 1\}. \tag{8.5}$$

Let $\Gamma := \mathbb{N}^n$ and consider the problem:

$$\begin{aligned} \rho_{\mathrm{mom}} := \sup_{\mu \in \mathscr{M}(\mathbb{K})_+} & \langle 1_{\mathbf{S}}, \mu\rangle \\ \text{s.t.}\ & \langle P\mathbf{x}^{\boldsymbol{\alpha}} - \mathbf{x}^{\boldsymbol{\alpha}}, \mu\rangle = 0,\ \forall \boldsymbol{\alpha} \in \Gamma \\ & \mu(\mathbb{K}) = 1, \end{aligned} \tag{8.6}$$

where $1_{\mathbf{S}}$ is the indicator function of $\mathbf{S}$. In (8.6) one recognizes an instance of the generalized moment problem (1.1) with countably many moment constraints.

Lemma 8.2. *Assume that $\mathbb{K} \subset \mathbb{R}^n$ and $\mathbf{S} \subset \mathbb{K}$ are compact. If the transition probability function P is weak-Feller then in (8.6) the sup is attained and there is no duality gap. Moreover, $j^* = \rho_{\mathrm{mom}}$.*

Proof. By Proposition 8.1, as $\mathbb{K}$ is compact and P is weak-Feller, P admits an invariant probability measure μ; hence (8.6) has a feasible solution. As $\mathbf{S}$ is compact, its indicator function $1_{\mathbf{S}}$ is upper-semicontinuous. By Corollary 1.4, the sup is attained and there is no duality gap. Finally, that $j^* = \rho_{\mathrm{mom}}$ follows from the equivalence of (8.4), (8.3), and (8.1) when P is weak-Feller. □

Observe that if in addition to be weak-Feller, P maps $\mathbb{R}[\mathbf{x}]$ into $\mathbb{R}[\mathbf{x}]$, then the constraints of (8.6) state linear constraints on the moments of μ. However, Problem (8.6) is still untractable numerically because $\mathbb{K}$ and

$\mathbf{S}$ are still too general and in addition, the indicator function $1_{\mathbf{S}}$ is not a polynomial. However, observe that (8.6) is equivalent to

$$\begin{aligned} \rho_{\text{mom}} := \sup_{\mu,\nu}\ & \langle 1, \mu \rangle \\ \text{s.t. } & \langle P\mathbf{x}^{\boldsymbol{\alpha}} - \mathbf{x}^{\boldsymbol{\alpha}}, \mu + \nu \rangle = 0, \quad \forall \boldsymbol{\alpha} \in \Gamma \\ & \langle 1, \mu + \nu \rangle = 1 \\ & \mu \in \mathscr{M}(\mathbf{S})_+,\ \nu \in \mathscr{M}(\mathbb{K})_+. \end{aligned} \tag{8.7}$$

Indeed, as we have already seen before, if (μ, ν) is a feasible solution of (8.6) with $\nu(\mathbf{S} > 0)$ then one may obtain another feasible solution (μ', ν') with $\mu'(\mathbf{S}) > \mu(\mathbf{S})$. It suffices to take $\nu'(B) := \nu(B \cap (\mathbb{K} \setminus \mathbf{S}))$ and $\mu'(B) := \mu(B) + \nu(B \cap \mathbb{K})$ for every $B \in \mathscr{B}$.

Now, if $\mathbb{K}$ and $\mathbf{S}$ are compact basic semi-algebraic sets and P maps $\mathbb{R}[\mathbf{x}]$ into $\mathbb{R}[\mathbf{x}]$, then (8.7) is an instance of the generalized moment problem with polynomial data and countably many moment constraints. And so, one may invoke results of Chapter 4 and build up convergent semidefinite relaxations for (8.7).

So let $\mathbb{K} \subset \mathbb{R}^n$ be as in (4.1) for some polynomials $(g_j)_{j=1}^m \subset \mathbb{R}[\mathbf{x}]$, and let

$$\mathbf{S} := \{\mathbf{x} \in \mathbb{R}^n : h_k(\mathbf{x}) \geq 0, \quad k = 1, \ldots, p\}, \tag{8.8}$$

with $h_k \in \mathbb{R}[\mathbf{x}]$ for all $k = 1, \ldots, p$. Let $v_j := \lceil \deg g_j/2 \rceil$ and $w_k := \lceil \deg h_k/2 \rceil$ for all $j = 1, \ldots, m$ and $k = 1, \ldots, p$. Assume that P maps $\mathbb{R}[\mathbf{x}]$ into $\mathbb{R}[\mathbf{x}]$ and consider the following semidefinite program:

$$\begin{aligned} \rho_i := \sup_{\mathbf{y},\mathbf{z}}\ & y_{\mathbf{0}} \\ \text{s.t. } & L_{\mathbf{y}+\mathbf{z}}(P\mathbf{x}^{\boldsymbol{\alpha}} - \mathbf{x}^{\boldsymbol{\alpha}}) = 0, \quad \forall \boldsymbol{\alpha}; \quad \deg P\mathbf{x}^{\boldsymbol{\alpha}} \leq 2i \\ & y_{\mathbf{0}} + z_{\mathbf{0}} = 1 \\ & \mathbf{M}_i(\mathbf{y}), \mathbf{M}_i(\mathbf{z}) \succeq 0 \\ & \mathbf{M}_{i-v_j}(\mathbf{z}\, g_j) \succeq 0, \quad j = 1, \ldots, m \\ & \mathbf{M}_{i-w_k}(\mathbf{y}\, h_k) \succeq 0, \quad k = 1, \ldots, p. \end{aligned} \tag{8.9}$$

Theorem 8.3. *Let $\mathbb{K}$ be as in (4.1), $\mathbf{S} \subset \mathbb{K}$ as in (8.8), and let Assumption 2.1 hold for both $\mathbb{K}$ and $\mathbf{S}$. Assume that P is weak-Feller and maps $\mathbb{R}[\mathbf{x}]$ into $\mathbb{R}[\mathbf{x}]$. Let $\rho_{\text{mom}}, \rho_i$ be the respective optimal values of (8.6) and (8.9). Then $\rho_i \downarrow \rho_{\text{mom}} = j^*$ as $i \to \infty$.*

Sketch of the proof. Let $(\mathbf{y}^i, \mathbf{z}^i)$ be a feasible solution of (8.9) such that $y_0^i \geq \rho_i - 1/i$. First, one shows that there exist $\mathbf{y}, \mathbf{z}$ and a subsequence (i_j) such that pointwise convergence holds, i.e., as $j \to \infty$,

$$y_{\boldsymbol{\alpha}}^{i_j} \to y_{\boldsymbol{\alpha}}\,; \qquad z_{\boldsymbol{\alpha}}^{i_j} \to z_{\boldsymbol{\alpha}}, \qquad \forall\, \boldsymbol{\alpha} \in \mathbb{N}^n.$$

Next, this pointwise convergence implies that for every $i \in \mathbb{N}$,

$$\mathbf{M}_i(\mathbf{y}) \succeq 0, \quad \mathbf{M}_i(\mathbf{z}) \succeq 0, \quad \mathbf{M}_i(g_j\,\mathbf{y}) \succeq 0, \quad \mathbf{M}_i(h_k\,\mathbf{z}) \succeq 0,$$

for all $j = 1, \dots, m$ and all $k = 1, \dots, p$. Invoking Theorem 3.8, $\mathbf{y}$ and $\mathbf{z}$ are the respective moment sequences of two measures, μ on $\mathbf{S}$ and ν on $\mathbb{K}$. For fixed $\boldsymbol{\alpha} \in \mathbb{N}^n$, pointwise convergence implies $L_{\mathbf{y}+\mathbf{z}}(P\mathbf{x}^{\boldsymbol{\alpha}} - \mathbf{x}^{\boldsymbol{\alpha}}) = 0$, and so (μ, ν) is feasible for (8.7). But we also have $\rho_{\mathrm{mom}} \leq y_0^{i_j} \to y_0 = \mu(\mathbf{S})$, which imples that (μ, ν) is an optimal solution of (8.7), and so $\mu(\mathbf{S}) = \rho_{\mathrm{mom}}$. Finally, as the converging subsequence (i_j) was arbitrary and the sequence (ρ_i) is monotone, we obtain the desired result $y_0^i \downarrow \rho_{\mathrm{mom}}$. □

8.1.2 *The non compact case*

We now consider the non compact case where $X = \mathbb{R}^n$. We still assume that P is weak-Feller and maps $\mathbb{R}[\mathbf{x}]$ into $\mathbb{R}[\mathbf{x}]$. Let $\delta, M > 0$ be fixed. We are now less ambitious and consider the problem

$$\rho_{\mathrm{mom}} := \sup_{\mu \in \mathscr{M}(X)_+^{M\delta}} \{\, \mu(\mathbf{S}) \,:\, \mu P = \mu; \quad \mu(X) = 1\} \tag{8.10}$$

where $\mathscr{M}(X)_+^{M\delta} \subset \mathscr{M}(X)_+$ is the set defined by:

$$\mathscr{M}(X)_+^{M\delta} := \{\, \mu \in \mathscr{M}(X)_+ \,:\, \int_X (\mathrm{e}^{\delta\, x_i} + \mathrm{e}^{-\delta\, x_i})\, d\mu < M, \quad \forall i = 1, \dots, n\}. \tag{8.11}$$

That is, we only consider invariant measures μ whose all maginals μ_i have a tail with exponential decay. Let $\mathbf{S} \subset \mathbb{R}^n$ be as in (8.8) and consider the following semidefinite program:

$$\begin{array}{rll} \rho_i := & \sup\limits_{\mathbf{y},\mathbf{z}} \; y_0 & \\ & \text{s.t. } L_{\mathbf{y}+\mathbf{z}}(P\,\mathbf{x}^{\boldsymbol{\alpha}} - \mathbf{x}^{\boldsymbol{\alpha}}) = 0, \quad \forall\, \boldsymbol{\alpha}; \quad \deg P\,\mathbf{x}^{\boldsymbol{\alpha}} \leq 2i & \\ & \quad y_0 + z_0 \;=\; 1 & \\ & \quad \mathbf{M}_i(\mathbf{y}),\, \mathbf{M}_i(\mathbf{z}) \succeq 0 & \\ & \quad \mathbf{M}_{i-w_k}(\mathbf{y}\, h_k) \succeq 0, \quad k = 1, \dots, p, & \\ & \quad \sum_{t=1}^{i} L_{\mathbf{z}}(x_k^{2t})/(2t!) \leq M, \quad k = 1, \dots, n. & \end{array} \tag{8.12}$$

Theorem 8.4. *Let $\mathbf{S} \subset \mathbb{R}^n$ as in (8.8), and let Assumption 2.1 hold for $\mathbf{S}$. Assume that P maps $\mathbb{R}[\mathbf{x}]$ into $\mathbb{R}[\mathbf{x}]$ and P has an invariant measure in $\mathscr{M}(X)_+^{M\delta}$. Let ρ_{mom} and ρ_i be the respective optimal values of (8.10) and (8.12). Then $\rho_i \downarrow \rho_{\text{mom}}$ as $i \to \infty$.*

Sketch of the proof. The proof mimics that of Theorem 8.3 except that to prove the pointwise convergence $z_{\boldsymbol{\alpha}}^{i_j} \to z_{\boldsymbol{\alpha}}$, one now invokes a property of the moment matrix $\mathbf{M}_{i_j}(\mathbf{z})$ coupled with the constraint $\sum_{t=1}^{i_j} L_{\mathbf{z}}(x_k^{2t})/(2t!) \leq M$, $k = 1, \ldots, n$, which implies that $\sup_j |z_{\boldsymbol{\alpha}}^{i_j}| \leq \tau_s$ whenever $|\boldsymbol{\alpha}| \leq s$, for some scalars τ_s, $s = 1, \ldots$

Next, the limit sequence $\mathbf{z}$ satisfies $\sum_{t=1}^{\infty} L_{\mathbf{z}}(x_k^{2t})/(2t!) \leq M$, $k = 1, \ldots, n$, from which we deduce that the generalized Carleman condition (3.12) holds. Therefore, by Proposition 3.5, $\mathbf{z}$ is the moment sequence of a determinate measure ν on $\mathbb{R}^n$, and $\mathbf{y}$ is the moment sequence of a measure μ on $\mathbf{S}$ (hence also determinate). For fixed $\boldsymbol{\alpha} \in \mathbb{N}^n$, pointwise convergence implies $L_{\mathbf{y}+\mathbf{z}}(P\mathbf{x}^{\boldsymbol{\alpha}} - \mathbf{x}^{\boldsymbol{\alpha}}) = 0$; this and the fact that both μ and ν are determinate imply that $(\mu+\nu)P = \mu+\nu$, and so $\mu+\nu$ is feasible for (8.10). The rest of the proof is similar. □

Example 8.2. On the real line $\mathbb{R}$, and with $r \leq 4$, consider the deterministic dynamical system:

$$\Phi_{t+1} = r\,\Phi_t\,(1 - \Phi_t), \qquad t = 0, 1, \ldots$$

which is called the *logistic map*. Depending on r, the above system has interesting properties. In particular, when $r > r_0$ (for some r_0) it provides an example of a very simple dynamical system with *chaotic* behavior. With $r \leq 4$ observe that if $\Phi_0 \in [0,1]$ then all iterates Φ_t also remain in $[0,1]$. Furthermore, the transition probability function P of the associated Markov chain is obsviously weak-Feller and the moments $\mathbf{y} = (y_j)$ of an invariant probability measure μ of P on $[0,1]$ satisfy the countably many linear constraints

$$L_{\mathbf{y}}(x^j) = y_j = L_{\mathbf{y}}((rx(1-x))^j) = r^j \sum_{k=0}^{j} (-1)^k \binom{j}{k} y_{j+k}, \qquad j = 0, 1, \ldots$$

So let $x \mapsto g(x) = x(1-x)$ and $x \mapsto h(x) = (x-a)(b-x)$, with $0 \leq a \leq b \leq 1$. With $\mathbb{K} := [0,1]$ and $\mathbf{S}$ the closed interval $[a,b] \subset [0,1]$, the semidefinite relaxations (8.9), which read

$$\begin{array}{rl} \rho_i := & \sup_{\mathbf{y},\mathbf{z}} y_0 \\ & \text{s.t. } y_j + z_j - r^j \sum_{k=0}^{j} (-1)^k \binom{j}{k} (y_{j+k} + z_{j+k}) = 0, \; j = 0, \ldots i \\ & y_0 + z_0 = 1 \\ & \mathbf{M}_i(\mathbf{y}), \, \mathbf{M}_i(\mathbf{z}) \succeq 0 \\ & \mathbf{M}_{i-1}(\mathbf{y}\, g) \succeq 0, \quad j = 1, \ldots, m \\ & \mathbf{M}_{i-1}(\mathbf{z}\, h) \succeq 0, \quad k = 1, \ldots, p, \end{array}$$

provide a monotone sequence of upper bounds that converges to $\max_\mu \mu(\mathbf{S})$ where the max is over all invariant probability measures of P.

Example 8.3. Iterated Functions Systems (IFS). For every $k = 1, \ldots, N$, let $f_k \in \mathbb{R}[\mathbf{x}]$ and consider the Markov chain on $\mathbb{R}^n$ defined by

$$\Phi_{t+1} = f_{\xi_t}(\Phi_t), \qquad t = 0, 1, \ldots$$

where (ξ_t) are i.i.d. random variables[2] with values in $\Delta = \{1, \ldots, N\}$ and associated probabilities $\mathbf{p} = (p_j) \in \mathbb{R}^N$. The transition probability function P of the Markov chain Φ is weak-Feller, because if $h : \mathbb{R}^n \to \mathbb{R}$ is bounded and continuous then so is the function

$$\mathbf{x} \longmapsto Ph(\mathbf{x}) = \sum_{j \in \Delta} p_j \, h(f_j(\mathbf{x})), \quad \mathbf{x} \in \mathbb{R}^n.$$

If for some compact basic semi-algebraic set $\mathbb{K} \subset \mathbb{R}^n$, f_k maps $\mathbb{K}$ into $\mathbb{K}$ for every k, then the semidefinite relaxations (8.9) converge to ρ_{mom}. Otherwise one may implement the semidefinite relaxations (8.12).

8.2 Evaluation of Ergodic Criteria

Evaluation of ergodic criteria is an important particular (and simpler) case of the generalized moment problem considered in the previous section. Given a Markov chain $\Phi = (\Phi_0, \Phi_1, \ldots)$ on $X \subset \mathbb{R}^n$, with associated stochastic kernel P, one is often interested in its expected long-run average behavior. More precisely, with $f : \mathbb{R}^n \to \mathbb{R}$ being a given measurable function, one is interested in computing or approximating the long-run expected average cost

[2] i.i.d. stands for independent and identically distributed.

$$J(\mathbf{x}) := \limsup_{N\to\infty} \mathrm{E}_{\mathbf{x}} \left[\frac{1}{N} \sum_{t=0}^{N-1} f(\Phi_t) \right] \tag{8.13}$$

where $\mathrm{E}_{\mathbf{x}}$ is the expectation operator[3] associated with $\mathbf{P}_{\mathbf{x}}$. A Borel set $B \in \mathscr{B}$ is called *invariant* if $P1_B \geq 1_B$, and a probability measure μ is called *ergodic* if $\mu(B) \in \{0,1\}$ for every invariant set $B \in \mathscr{B}$.

Let μ be an invariant probability measure of the transition probability function P associated with Φ, and let $f \in L_1(\mu)$. If μ is ergodic or μ is the unique invariant probability measure of P, then as a consequence of Birkhoff's Individal Ergodic Theorem, for μ-almost all $\mathbf{x}$:

$$J(\mathbf{x}) := \limsup_{N\to\infty} \mathrm{E}_{\mathbf{x}} \left[\frac{1}{N} \sum_{t=0}^{N-1} f(\Phi_t) \right] = \int_X f\,d\mu. \tag{8.14}$$

$$J'(\mathbf{x}) := \limsup_{N\to\infty} \left[\frac{1}{N} \sum_{t=0}^{N-1} f(\Phi_t) \right] = \int_X f\,d\mu, \quad \mathbf{P}_{\mathbf{x}} - \text{a.s.}$$

See e.g. Theorem 2.3.4 and Proposition 2.4.2 in Hernández-Lerma and Lasserre (2003). Therefore, with $\Gamma = \mathbb{N}^n$ and $\mathbb{K} \subset \mathbb{R}^n$, consider the following infinite-dimensional linear programs:

$$\begin{aligned} \rho^1_{\text{mom}}\ (\rho^2_{\text{mom}}) := \sup_{\mu}\ (\inf_{\mu})\ & \langle f, \mu \rangle \\ \text{s.t. } & \langle P\,\mathbf{x}^{\boldsymbol{\alpha}} - \mathbf{x}^{\boldsymbol{\alpha}}, \mu \rangle = 0, \quad \forall \boldsymbol{\alpha} \in \Gamma \\ & \langle 1, \mu \rangle = 1 \\ & \mu \in \mathscr{M}(\mathbb{K})_+, \end{aligned} \tag{8.15}$$

each of which is an instance of the generalized moment problem (1.1) with countably many moment constraints. Solving (8.15) provides an ergodic probability measure μ_1 (resp. μ_2) that maximizes (resp. minimizes) (8.14) over all invariant probability measures with support contained in $\mathbb{K}$. If P admits a unique invariant probability measure on $\mathbb{K}$ then $\rho^1_{\text{mom}} = \rho^2_{\text{mom}}$.

If $\mathbb{K}$ as in (4.1) is compact, Assumption 2.1 holds, and P maps $\mathbb{R}[\mathbf{x}]$ into $\mathbb{R}[\mathbf{x}]$, then the semidefinite relaxations

[3] Let $(\Omega, \mathscr{F})$ be the measurable space consisting of the canonical sample path $\Omega = X^\infty := X \times X \times \cdots$ and the corresponding product σ-algebra $\mathscr{F}$. Then every initial probability distribution ϕ determines a probability measure on Ω, denoted $\mathbf{P}_\phi$, and the notation $\mathbf{P}_{\mathbf{x}}$ corresponds to the initial distribution $\phi := \delta_{\mathbf{x}}$ with all mass concentrated on $\mathbf{x} \in X$.

$$\begin{aligned}\rho_i^1\,(\rho_i^2) \;:=\; \sup_{\mathbf{y}}\,(\inf_{\mathbf{y}})\; & L_{\mathbf{y}}(f)\\ \text{s.t. } & L_{\mathbf{y}}(P\,\mathbf{x}^{\boldsymbol\alpha}-\mathbf{x}^{\boldsymbol\alpha})=0, \quad \forall\boldsymbol\alpha;\quad \deg P\,\mathbf{x}^{\boldsymbol\alpha}\le 2i\\ & \mathbf{M}_i(\mathbf{y}),\,\mathbf{M}_{i-v_j}(\mathbf{y}\,g_j)\succeq 0,\quad j=1,\ldots,m\\ & y_0 \;=\; 1,\end{aligned} \tag{8.16}$$

provide upper bounds $\rho_i^1 \geq \rho_{\text{mom}}^1$ (resp. lower bounds $\rho_i^2 \leq \rho_{\text{mom}}^2$) and in addition, for $j=1,2$, $\rho_i^j \to \rho_{\text{mom}}^j$ as $i\to\infty$.

So if P admits a unique invariant probability measure on $\mathbb{K}$, the exact value $J^* = \rho_{\text{mom}}^1 = \rho_{\text{mom}}^2$ satisfies

$$\rho_i^2 \;\leq\; J^* \;\leq\; \rho_i^1, \quad \forall i \qquad \text{and} \quad \rho_i^2 \uparrow J^*, \quad \rho_i^1 \downarrow J^* \quad \text{as } i\to\infty. \tag{8.17}$$

In the non compact case $X = \mathbb{R}^n$, and like in Section 8.1, one may consider the subspace of invariant probability measures μ of P, that have a tail with exponential decay, that is, such that $\mu \in \mathscr{M}^{M\delta}(X)_+$ with $\mathscr{M}^{M\delta}(X)_+$ as in (8.11). Then one consider the moment problem (8.15) with $\mu \in \mathscr{M}^{M\delta}(X)_+$ instead of $\mu \in \mathscr{M}(\mathbb{K})_+$.

If P admits an invariant probability measure $\mu \in \mathscr{M}^{M\delta}(X)_+$ then the semidefinite relaxations

$$\begin{aligned}\rho_i^1\,(\rho_i^2) \;:=\; \sup_{\mathbf{y}}\,(\inf_{\mathbf{y}})\; & L_{\mathbf{y}}(f)\\ \text{s.t. } & L_{\mathbf{y}}(P\mathbf{x}^{\boldsymbol\alpha}-\mathbf{x}^{\boldsymbol\alpha})=0, \quad \forall\boldsymbol\alpha;\quad \deg P\,\mathbf{x}^{\boldsymbol\alpha}\le 2i\\ & \mathbf{M}_i(\mathbf{y})\succeq 0,\quad j=1,\ldots,m\\ & y_0 \;=\; 1\\ & \textstyle\sum_{t=1}^{i} L_{\mathbf{y}}(x_k^{2t})/(2t!) \;\leq M, \qquad k=1,\ldots,n,\end{aligned}$$

provide upper bounds $\rho_i^1 \geq \rho_{\text{mom}}^1$ (resp. lower bounds $\rho_i^2 \leq \rho_{\text{mom}}^2$) and in addition, $\rho_i^j \to \rho_{\text{mom}}^j$ as $i\to\infty$, for all $j=1,2$.

So, again, if P admits a unique invariant probability measure $\mu \in \mathscr{M}^{M\delta}(X)_+$ then (8.17) also holds for every i.

8.3 Summary

In this chapter we have considered the moment approach to provide tight bounds on $j^* = \sup_\mu \mu(\mathbf{S})$ where the "sup" is over all invariant measures of a given Markov chain Φ whose transition probability function P maps polynomials into polynomials. The upper bounds provided by appropriate semidefinite relaxations converge to the exact value j^*. This approach also permits to approximate the exact value of the ergodic functional $J(\mathbf{x}) = \limsup_{N\to\infty} E_{\mathbf{x}}[\,(1/N)\sum_{k=0}^{N-1} f(\Phi_k)]$ for some given polynomial $f \in \mathbb{R}[\mathbf{x}]$. This deterministic approach which provides an arbitrarily close approximation of j^* (and of $J(\mathbf{x}) = J^*$ when the invariant probability measure is unique) is to be contrasted with simulation techniques that provide random *estimates* of j^* or J^*. In particular, if μ is the unique invariant probability measure of P, then (8.17) holds and so one provides smaller and smaller intervals in which J^* lies, something that cannot be done with random estimators.

8.4 Exercises

Exercise 8.1. In (8.5) replace "sup" with "inf". Do we have an analogue of Lemma 8.2? What happens? If we now assume that $\mathbf{S} \subset \mathbb{K}$ is open (with $\mathbb{K}$ still compact), what can be said? (Hint: $\mathbb{K} \setminus \mathbf{S}$ is closed hence compact and $\inf_\mu \mu(\mathbf{S}) = 1 - \sup_\mu \mu(\mathbb{K} \setminus \mathbf{S})$. What about the semidefinite relaxations? Can we always describe the set $\mathbb{K} \setminus \mathbf{S}$ easily? What is the conclusion?)

Exercise 8.2. Consider again the IFS like in Example 8.3 but now suppose that the (ξ_t) are not i.i.d. any longer but $\xi = (\xi_0, \ldots)$ is itself a Markov chain on Δ with associated transition probability function $Q = (Q_{ij})$, i.e., $\text{Prob}\,[\xi_{t+1} = j \,|\, \xi_t = i] = Q_{ij}$, for every $i, j \in \Delta$, and every $t = 0, 1, \ldots$.

Is (Φ_t, ξ_t) a Markov chain on $\mathbb{R}^n \times \Delta$? Describe its transition probability function P and write explicitly $Ph(\Phi, k)$ for an arbitrary polynomial $h \in \mathbb{R}[\mathbf{x}]$. Is P weak-Feller? Write the semidefinite relaxations (8.9) for approximating $\sup_\mu \mu(\mathbf{S})$ over all invariant probability measures μ of P.

Exercise 8.3. Suppose that for some i, the semidefinite relaxation (8.12) has no feasible solution. What can be concluded?

Exercise 8.4. Consider Example 8.2 with $r := 4$, and let $f \in \mathbb{R}[x]$ be given. Write the corresponding semidefinite relaxations (8.15). What can be said? is there a unique invariant probability measure for P (Hint: try to compute fixed points (i.e. cycles of length 1) of the mapping $x \mapsto 4x(1-x)$, then cycles of length 2). What can be concluded?

8.5 Notes and Sources

Proposition 8.1 is from Krylov and Bogolioubov (1937). For a numerical approximation of j^* and J^*, the most popular methods are Monte Carlo schemes that simulate the Markov chain Φ and provide random estimates of j^* and J^*. For basics on the Monte Carlo approach, the interested reader is referred to Niederreiter (1992). On the other hand there exist general numerical schemes to approximate j^* directly by (a) discretization of the space X into a grid and (b) looking for a measure that is a convex combinations of Dirac measures on that grid, and such that $|\langle f, \mu P - \mu \rangle| \leq \epsilon$ for all $f \in F$, where $\epsilon > 0$ is fixed and F is a finite set of test functions. This results in a linear program that computes the optimal weights of the convex combination of Dirac measures. One then decreases ϵ and enlarge F to provide a sequence of bounds that converges to j^*. See e.g. Hernández-Lerma and Lasserre (1998b,a).

Chapter 9

Application in Mathematical Finance

This chapter considers an application in mathematical finance, namely the pricing of some exotic options. First under a no-arbitrage assumption and knowledge of some moments of the distribution of the underlying asset price. Then when one assumes that the asset price obeys some Ito's stochastic differential equation.

An important problem in mathematical finance is to evaluate the price of a derivative security given information of the underlying asset. For instance, an European Call Option with *strike* K gives its holder the option (but no obligation) to buy the underlying security at time T at price K. If the price x_T at time T is above K, then the holder will exercise the option to make a profit of $x_T - K$, whereas if $x_T \leq K$ he will not exercise the option. Therefore, the payoff $\max(x_T - K, 0)$ of this option being nonnegative, it has some value. Under no arbitrage,[1] this value is given by $\mathrm{E}[(x-K)^+]$ where E is the expectation operator with respect to the distribution of the price of the underlying asset x, and the notation x^+ stands for $\max[0, x]$. In this chapter we investigate this issue in the case where one has some moment information of the distribution and in the case where the price of the underlying asset obeys a stochastic differential equation.

9.1 Option Pricing with Moment Information

Finding an optimal upper bound on the price of an European Call Option with strike K, given the first $p+1$ moments $\{\gamma_j\}_{j=0}^p$ of the price of the underlying asset, reduces to solving:

[1]For the concept of arbitrage, see the discussion in Section 1.1.

$$\rho_{\text{mom}} := \sup_{\mu \in \mathscr{M}(\mathbb{R}_+)_+} \left\{ \int_{\mathbb{R}_+} (x-K)^+ d\mu \,:\, \int_{\mathbb{R}_+} x^j \, d\mu = \gamma_j, \quad j \in \Gamma \right\} \tag{9.1}$$

with $\Gamma := \{0, \ldots, p\}$. Obviously, (9.1) is an instance of the generalized moment problem (1.1) with $\mathbb{K} = \mathbb{R}_+$, $x \mapsto h_j(x) = x^j$ for all $j \in \Gamma$, and $x \mapsto f(x) = (x-K)^+$.

Problem (9.1) is similar to problem (7.2) in the univariate case. Observe that f is *not* a polynomial, and as in Chapter 7 we replace μ with the sum of two measures ϕ and ν with support on $\mathbf{S} := [K, +\infty)$ and $\mathbb{K} = \mathbb{R}_+$ respectively. That is, we again want to solve a variant of the generalized moment problem with two unknown measures ϕ and ν; namely :

$$\begin{aligned} \rho_{\text{mom}} = \sup_{\phi,\nu} & \int_{\mathbf{S}} (x-K)\, d\phi \\ \text{s.t.} & \int_{\mathbf{S}} x^j \, d\phi + \int_{\mathbb{R}_+} x^j \, d\nu = \gamma_j \quad j \in \Gamma \\ & \phi \in \mathscr{M}(\mathbf{S})_+, \ \nu \in \mathscr{M}(\mathbb{R}_+)_+. \end{aligned} \tag{9.2}$$

Again we should have imposed that $\nu(\mathbf{S}) = 0$. However, since we maximize, if a feasible solution (ϕ, ν) of (9.2) is such that $\nu(\mathbf{S}) > 0$ then one easily constructs a feasible solution (ϕ', ν') with $\nu'(\mathbf{S}) = 0$ and with value at least as good. Write ν as the sum $\nu_1 + \nu_2$ of two mutually singular measures ν_1, ν_2, with $\nu_1(B) = \nu(B \cap \mathbf{S})$ and $\nu_2(B) = \nu(B \cap (\mathbb{R}_+ \setminus \mathbf{S}))$ for all Borel sets B of $\mathbb{R}_+$. Then $(\phi', \nu') := (\phi + \nu_1, \nu_2)$ is feasible for (9.2) with value $\int_{\mathbf{S}}(x-K)d\phi' = \int_{\mathbf{S}}(x-K)d\phi + \int_{\mathbf{S}}(x-K)d\nu_1 \geq \int_{\mathbf{S}}(x-K)d\phi$ because $(x-K) \geq 0$ on $\mathbf{S}$.

The dual of (9.1) reads

$$\begin{aligned} \inf_{\lambda \in \mathbb{R}^{p+1}} \ & \sum_{j \in \Gamma} \lambda_j \gamma_j \\ \text{s.t.} \ & \sum_{j \in \Gamma} \lambda_j x^j - (x-K) \geq 0 \text{ on } \mathbf{S} \\ & \sum_{j \in \Gamma} \lambda_j x^j \geq 0 \qquad\qquad \text{on } \mathbb{R}_+ \end{aligned} . \tag{9.3}$$

Then (9.3) has a complete description as a semidefinite program because one has an appropriate description of polynomials nonnegative on an interval, in terms of a weighted sum of squares. Indeed by Theorem 2.7, a nonnegative univariate polynomial $f \in \mathbb{R}[x]$ nonnegative on $[a,+\infty)$ can be written as $q_0 + (x-a)q_1$ for two s.o.s. polynomials $q_0, q_1 \in \Sigma[x]$ such that $\deg q_0$, $\deg (x-a)q_1 \leq \deg f$. Therefore solving (9.3) reduces to solving the single semidefinite program:

$$\begin{aligned}
\rho^* = \inf_{\lambda\in\mathbb{R}^{p+1}} & \sum_{j\in\Gamma} \lambda_j \gamma_j \\
\text{s.t. } & \sum_{j\in\Gamma} \lambda_j x^j - (x-K) = q_0 + (x-K)\, q_1 \\
& \sum_{j\in\Gamma} \lambda_j x^j = q_2 + x\, q_3 \\
& q_j \in \Sigma[x],\ j=0,\ldots,3 \\
& \deg\,(q_0, (x-K)q_1, q_2, xq_3) \leq p.
\end{aligned} \tag{9.4}$$

Let $\tilde{p} := \lceil p/2 \rceil$. The semidefinite program (9.4) is the dual of

$$\begin{aligned}
\rho = \sup_{\mathbf{y},\mathbf{z}}\ & L_{\mathbf{y}}(x-K)\ (= y_1 - Ky_0) \\
\text{s.t. } & \mathbf{M}_{\tilde{p}}(\mathbf{y}), \mathbf{M}_{\tilde{p}}(\mathbf{z}) \succeq 0 \\
& \mathbf{M}_{\tilde{p}-1}((x-K)\,\mathbf{y}) \succeq 0 \\
& \mathbf{M}_{\tilde{p}-1}(x\,\mathbf{z}) \succeq 0 \\
& y_j + z_j = \gamma_j, \quad j \in \Gamma,
\end{aligned} \tag{9.5}$$

where $\mathbf{M}_{\tilde{p}-1}((x-K)\,\mathbf{y})$ (resp. $\mathbf{M}_{\tilde{p}-1}((x-K)\,\mathbf{y})$) is the localizing matrix associated with $\mathbf{y}$ and the polynomial $x \mapsto x-K$ (resp. $\mathbf{z}$ and the polynomial $x \mapsto x$). Of course one has $\rho_{\text{mom}} \leq \rho \leq \rho^*$.

Theorem 9.1. *Asume that there exists a probability measure μ on $\mathbb{R}_+$ which has a density $f > 0$ with respect to the Lebesgue measure on $\mathbb{R}_+$, and such that*

$$\int_{\mathbb{R}_+} x^j\, d\mu \left(= \int_0^\infty x^j f(x)\, dx \right) = \gamma_j, \qquad j \in \Gamma.$$

Then:
(a) $\rho_{\text{mom}} = \rho = \rho^*$.
(b) *Let $\mathbf{y}$ be an optimal solution of (9.5). If*

$$\begin{aligned}
\operatorname{rank} \mathbf{M}_{\tilde{p}}(\mathbf{y}) &= \operatorname{rank} \mathbf{M}_{\tilde{p}-1}(\mathbf{y}) \\
\operatorname{rank} \mathbf{M}_{\tilde{p}}(\mathbf{z}) &= \operatorname{rank} \mathbf{M}_{\tilde{p}-1}(\mathbf{z})
\end{aligned}$$

there is a $\operatorname{rank} \mathbf{M}_{\tilde{p}}(\mathbf{y})$*-atomic measure ϕ on $\mathbf{S}$ and a* $\operatorname{rank} \mathbf{M}_{\tilde{p}}(\mathbf{z})$*-atomic measure ν on $\mathbb{R}_+ \setminus \mathbf{S}$ such that $\mu := \phi + \nu$ is an optimal solution of (9.1).*

Proof. (a) As there exists $d\mu = f dx$ which satisfies the moment constraints, it follows that the moment vector γ lies in interior of the moment space. Therefore, there is no duality gap between (9.1) and its dual (9.2), i.e. $\rho_{\text{mom}} = \rho^*$. But this also implies the desired result $\rho_{\text{mom}} = \rho = \rho^*$.

(b) As in addition $M_{\tilde{p}-1}((x-K)\,\mathbf{y}) \succeq 0$, then by Theorem 3.11, $\mathbf{y}$ is the (truncated) moment vector of a measure ϕ on $\mathbf{S}$ and for same reasons $\mathbf{z}$ is the (truncated) moment vector of a measure ν on $\mathbb{R}_+$. Therefore

$$\gamma_j = y_j + z_j = \int_{\mathbf{S}} x^j \, d\phi + \int_{\mathbb{R}_+} x^j \, d\nu = \int_{\mathbb{R}_+} x^j \, d\mu$$

for every $j \in \Gamma$. Moreover ν has its support in $\mathbb{R}_+ \setminus \mathbf{S}$ otherwise if some atom a of ν (with weight say $\lambda_a > 0$) is in $\mathbf{S}$ then taking $\phi' := \phi + \lambda_a \delta_a$ and $\nu' := \nu - \lambda_a \delta_a$ would yield a strictly better feasible solution for (9.5) (obtained from the moment vector $(\mathbf{y}', \mathbf{z}')$ of (ϕ', ν')), in contradiction with the optimality of $\mathbf{y}$. Hence $\nu(\mathbf{S}) = 0$ which in turn yields

$$\rho_{\text{mom}} = L_{\mathbf{y}}(x-K) = \int_{\mathbf{S}} (x-K) d\phi = \int_{\mathbb{R}_+} (x-K)^+ d\mu$$

since $\nu(\mathbf{S}) = 0$. □

9.2 Option Pricing with a Dynamic Model

In the previous section, there was no model for the dynamics of the price of the underlying asset. The only available information was knowledge of some moments of its distribution at the time to exercise the option. In this section we assume that the price of the underlying asset obeys some Ito's stochastic differential equation.

In this context, the basic idea of the moment approach is to associate the asset price of a European style option with suitably defined expected *occupation* measures. One then exploits the martingale property of certain associated stochastic integrals to derive an infinite system of linear equations satisfied by the moments of the measures considered. Then one relates the value of an option with the solution of an infinite-dimensional linear programming problem, an instance of the generalized moment problem (1.1) with countably many moment constraints.

9.2.1 *Notation and definitions*

Fix a filtered probability space $(\Omega, \mathscr{F}, \mathscr{F}_t, \mathbf{P})$ satisfying the usual conditions and supporting a standard, one-dimensional $(\mathscr{F}_t)$-Brownian motion W. We consider a number of derivatives, the underlying asset price process of which satisfies a stochastic differential equation (SDE) of the form

$$dX_t = b(X_t)\,dt + \sigma(X_t)\,dW_t, \quad X_0 = x_0 \in \mathscr{I}, \tag{9.6}$$

Here, $\mathscr{I}$ is either $(0, +\infty)$ or $\mathbb{R}$, and $b, \sigma : \mathscr{I} \to \mathscr{I}$ are given functions such that (9.6) has a unique strong solution with values in $\mathscr{I}$, for all $t \geq 0$, $\mathbf{P}$-a.s.. In particular, we assume that the underlying asset price process X is given by one of the following three models:

Model 1 $b(x) := bx$, $\sigma(x) := \sigma x$ and $\mathscr{I} = (0, +\infty)$, for some constants $b, \sigma \in \mathbb{R}$.

This is the familiar *geometric Brownian motion* underlying the Black and Scholes model.

Model 2 $b(x) := \gamma(\theta - x)$, $\sigma(x) := \sigma$ and $\mathscr{I} = \mathbb{R}$, for some constants $\gamma, \theta, \sigma \in \mathbb{R}$.

This mean-reverting diffusion is an *Ornstein-Uhlenbeck process*, which for instance, appears in Vasicek's interest rate model.

Model 3 $b(x) := \gamma(\theta - x)$, $\sigma(x) := \sigma\sqrt{x}$ and $\mathscr{I} = (0, +\infty)$, for some constants $\gamma, \theta, \sigma \in \mathbb{R}$ such that $\gamma\theta > \sigma^2/2$.

This diffusion models the short rate dynamics assumed in the Cox-Ingersoll-Ross interest rate model. Note that the inequality $\gamma\theta > \sigma^2/2$ is necessary and sufficient for the solution of (9.6) to be non-explosive, in particular, for the hitting time of 0 to be equal to $+\infty$ with probability one.

Payoff. Several types of derivative payoff structures may be considered. But here for illustration we consider *barrier* call options. The price of a typical *down-and-out* barrier call option written on the underlying process X, is given by:

$$J(x_0) := e^{-\rho T} E\left[(X_T - K)^+ 1_{\{T\}}(\tau)\right], \tag{9.7}$$

where $T > 0$ is the option's *maturity time*, K is the option's *strike price*, ρ is a constant discounting factor, x_0 is the initial underlying asset price, and τ is the $(\mathscr{F}_t)$-stopping time defined by

$$\tau := \inf\{t \geq 0 \,:\, X_t \leq H\} \wedge T, \tag{9.8}$$

and $H < x_0$ is the *knockout barrier*. So, starting with initial state $x_0 > H$, the process stops either when asset price hits the barrier H for the first time at time $\tau \leq T$ or stops at $\tau = T$ if the asset price sample path was always above H.

9.2.2 *The martingale approach*

The extended infinitesimal generator $\mathscr{A}$ of the price-time process (X_t, t) associated with the diffusion (9.6) is defined by

$$f \mapsto (\mathscr{A} f)(x,t) := \frac{1}{2}\sigma(x)^2 \nabla_x^2 f(x,t) + b(x)\frac{\partial f}{\partial x}(x,t) + \frac{\partial f}{\partial t}(x,t)$$

for all f in the domain $\mathscr{D}(\mathscr{A})$ (which contains the set $C_c^2(\mathbb{R}\times\mathbb{R}_+)$ of all twice-continuously differentiable functions $f : \mathbb{R}\times\mathbb{R}_+ \mapsto \mathbb{R}$ with compact support).

Assumption 9.1. $b, \sigma^2 \in \mathbb{R}[x]$ so that $\mathscr{A}$ maps polynomials into polynomials. Moreover,

$$\sup_{t\in[0,T]} \sum_{j=1}^{n} E\left[|X_t^j|^k\right] < \infty, \quad \text{for all } T > 0, \quad \text{for all } k \in \mathbb{N}.$$

Assumption 9.1 implies in particular that every polynomial $f \in \mathbb{R}[\mathbf{x}, t]$ belongs to $\mathscr{D}(\mathscr{A})$, and the process M^f defined by

$$M_t^f := f(X_t, t) - f(x_0, 0) - \int_0^t (\mathscr{A} f)(X_s, s)\, ds$$

is a square-integrable martingale. Moreover, fix any $(\mathscr{F}_t)$-stopping time τ that is bounded by a constant $T > 0$, $\mathbf{P}$-a.s.. With regard to Doob's optional sampling theorem, under Assumption 9.1, if $f \in \mathbb{R}[\mathbf{x}, t]$ then

$$E[f(X_\tau, \tau)] - f(x_0, 0) - E\left[\int_0^\tau (\mathscr{A} f)(X_s, s)\, ds\right] = 0. \tag{9.9}$$

Expected occupation measures

Consider:

• the *expected occupation measure* $\mu(\cdot) = \mu(\cdot\,; x_0)$ of the price-time process (X_t, t) up to time τ, supported on $\mathbf{Q} := [H, +\infty) \times [0, T]$ and defined by:

$$\mu(B \times C) := E\left[\int_{[0,\tau]\cap C} 1_B(X_s)\, ds\right], \quad B \in \mathscr{B}(\mathbb{R}^n),\, C \in \mathscr{B}([0,T]) \tag{9.10}$$

where $\mathscr{B}(\mathbb{R})$ (resp. $\mathscr{B}([0,T])$) is the Borel σ-algebra on $\mathbb{R}^n$ (resp. on $[0,T]$), and

• the *exit location probability measure* $\nu(\cdot) = \nu(\cdot\,; x_0)$, supported on $\mathbf{S} := [0,T] \times \{H\} \cup \{T\} \times [H, +\infty)$, and defined by:

$$\nu(B \times C) := P\left((X_\tau, \tau) \in B \times C\right), \quad B \in \mathscr{B}(\mathbb{R}^n),\, C \in \mathscr{B}([0,T]), \tag{9.11}$$

i.e., ν is the joint probability distribution of (X_τ, τ).

The generalized moment problem

From the definitions of μ and ν, one may rewrite the martingale property (9.9) as

$$\int_{\mathbf{S}} f\, d\nu - f(x_0, 0) - \int_{\mathbf{Q}} (\mathscr{A} f)\, d\mu = 0, \tag{9.12}$$

which is called the *basic adjoint equation* and which characterizes the measures μ and ν associated with the generator $\mathscr{A}$.

Wih no loss of generality and to simplify the exposition, we set the discount factor to zero so that the payoff (9.7) reads

$$J(x_0) := E\left[(X_T - K)^+ 1_{\{T\}}(\tau)\right] = \int_{\mathbf{S}} (x - K)^+\, d\nu, \tag{9.13}$$

and we consider the moment problems:

$$\sup_{\mu,\nu}\,(\inf_{\mu,\nu}) \int_{\mathbf{S}} (x-K)^+ d\nu$$
$$\text{s.t.} \int_{\mathbf{S}} f\,d\nu - \int_{\mathbf{Q}} (\mathscr{A} f)\,d\mu = f(x_0,0), \quad \forall f \in \mathbb{R}[x,t]$$
$$\mu \in \mathscr{M}(\mathbf{Q})_+, \nu \in \mathscr{M}(\mathbf{S})_+.$$

Equivalently, as ν is supported on $[0,T] \times \{H\} \cup \{T\} \times [H,+\infty)$, decompose ν into the sum of three mutally singular measures, namely

$$\nu = \nu_1 + \nu_2 + \nu_3$$

with ν_1 supported on $\mathbf{S}_1 := [0,T] \times \{H\}$, ν_2 supported on $\mathbf{S}_2 := \{T\} \times (H,K)$ and ν_3 supported on $\mathbf{S}_3 := \{T\} \times [K,+\infty)$. Therefore

$$J(x_0) = \int_{\mathbf{S}} (x-K)^+\,d\nu = \int_{\mathbf{S}_3} (x-K)\,d\nu_3$$

and so we consider the two moment problems

$$\begin{aligned} \rho^1_{\text{mom}} = \sup_{\mu,\nu_i} & \int_{\mathbf{S}_3} (x-K)\,d\nu_3 \\ \text{s.t.} & \sum_{i=1}^{3} \int_{\mathbf{S}_i} f\,d\nu_i - \int_{\mathbf{Q}} (\mathscr{A} f)\,d\mu = f(x_0,0), \quad \forall f \in \mathbb{R}[x,t] \\ & \mu \in \mathscr{M}(\mathbf{Q})_+, \nu_i \in \mathscr{M}(\mathbf{S}_i)_+,\ i=1,2,3, \end{aligned} \tag{9.14}$$

$$\begin{aligned} \rho^2_{\text{mom}} = \inf_{\mu,\nu_i} & \int_{\mathbf{S}_3} (x-K)\,d\nu_3 \\ \text{s.t.} & \sum_{i=1}^{3} \int_{\mathbf{S}_i} f\,d\nu_i - \int_{\mathbf{Q}} (\mathscr{A} f)\,d\mu = f(x_0,0), \quad \forall f \in \mathbb{R}[x,t] \\ & \mu \in \mathscr{M}(\mathbf{Q})_+, \nu_i \in \mathscr{M}(\mathbf{S}_i)_+,\ i=1,2,3. \end{aligned} \tag{9.15}$$

9.2.3 *Semidefinite relaxations*

Let $\mathbf{y}_1 = (y_{1k})$, $\mathbf{y}_2 = (y_{2k})$ and $\mathbf{y}_3 = (y_{3k})$ be the moment sequences of the measures ν_1, ν_2 and ν_3 respectively, whereas $\mathbf{z} = (z_{ij})$ denotes that of μ.

Observe that with $f(x,t) := x^k t^j$ the integral $\int_{\mathbf{Q}} (\mathscr{A} f)\, d\mu$ reads:

$$\begin{array}{r} j\, z_{k(j-1)} + k\, b\, z_{kj} + \frac{k(k-1)\sigma^2}{2} z_{(k-2)j} \text{ in Model 1} \\ j\, z_{k(j-1)} + k\gamma\, \theta\, z_{(k-1)j} - k\, \gamma z_{kj} + \frac{k(k-1)\sigma^2}{2} z_{(k-2)j} \text{ in Model 2} \\ j\, z_{k(j-1)} + k\, \gamma\, \theta\, z_{(k-1)j} - k\, \gamma z_{kj} + \frac{k(k-1)\sigma^2}{2} z_{(k-1)j} \text{ in Model 3} \end{array} \tag{9.16}$$

Let $t \mapsto g_1(t) = t(T-t)$, $x \mapsto g_2(x) = (H-x)(K-x)$, $x \mapsto g_3(x) := (x-K)$, $x \mapsto g_4(x) := (x-H)$, and consider the semidefinite programs:

$$\begin{array}{rl} \rho_i^1 = \sup\limits_{\mathbf{y}_i, \mathbf{z}} & y_{31} - K y_{30} \\ \text{s.t.} & \mathbf{M}_i(\mathbf{z}), \mathbf{M}_i(\mathbf{y}_k) \succeq 0, \quad k = 1,2,3 \\ & \mathbf{M}_{i-1}(g_1\, \mathbf{y}_1), \mathbf{M}_{i-1}(g_2\, \mathbf{y}_2), \mathbf{M}_{i-1}(g_3\, \mathbf{y}_3) \succeq 0 \\ & \mathbf{M}_{i-1}(g_1\, \mathbf{z}), \mathbf{M}_{i-1}(g_4\, \mathbf{z}) \succeq 0 \\ & H^k y_{1j} + T^j y_{2k} + T^j y_{3k} - (9.16) = 0^j x_0^k, \quad k+j \leq i \end{array} \tag{9.17}$$

and

$$\begin{array}{rl} \rho_i^2 = \inf\limits_{\mathbf{y}_i, \mathbf{z}} & y_{31} - K y_{30} \\ \text{s.t.} & \mathbf{M}_i(\mathbf{z}), \mathbf{M}_i(\mathbf{y}_k) \succeq 0, \quad k = 1,2,3 \\ & \mathbf{M}_{i-1}(g_1\, \mathbf{y}_1), \mathbf{M}_{i-1}(g_2\, \mathbf{y}_2), \mathbf{M}_{i-1}(g_3\, \mathbf{y}_3) \succeq 0 \\ & \mathbf{M}_{i-1}(g_1\, \mathbf{z}), \mathbf{M}_{i-1}(g_4\, \mathbf{z}) \succeq 0 \\ & H^k y_{1j} + T^j y_{2k} + T^j y_{3k} - (9.16) = 0^j x_0^k, \quad k+j \leq i. \end{array} \tag{9.18}$$

Theorem 9.2. *Suppose that the following conditions hold:*
(i) *The measure ν_3 is moment-determinate.*
(ii) *The infinite system*

$$\begin{cases} \mathbf{M}_i(\mathbf{z}), \mathbf{M}_i(\mathbf{y}_k) \succeq 0, \quad k = 1,2,3 \\ \mathbf{M}_i(g_1\, \mathbf{y}_1), \mathbf{M}_i(g_2\, \mathbf{y}_2), \mathbf{M}_i(g_3\, \mathbf{y}_3) \succeq 0 \\ \mathbf{M}_i(g_1\, \mathbf{z}), \mathbf{M}_i(g_4\, \mathbf{z}) \succeq 0 \\ H^k y_{1j} + T^j y_{2k} + T^j y_{3k} - (9.16) = 0^j x_0^k, \quad \forall\, k+j \leq i, \end{cases} \tag{9.19}$$

with $i = 0, 1, \ldots$, uniquely determines the moment sequence $\mathbf{y}_3$. Then as $i \to \infty$,

$$\rho_i^1 \downarrow J(x_0) \quad \text{and} \quad \rho_i^2 \uparrow J(x_0).$$

In view of the structure of the problem considered, Assumption (i) in the statement of Theorem 9.2 is satisfied if $\mathrm{E}(\exp\{c\,|X_T|\}) < +\infty$ for some $c > 0$, which is true for Models 2 and 3, but not for Model 1.

Numerical experiments

For barrier options, relative errors of the moment approach are displayed in Table 9.1 for several values of the volatility σ. The other parameters read:

Model 1: $x_0 = 1,\ K = 1,\ H = 0.8,\ T = 2,\ b = 0$
Models 2 and 3: $x_0 = 1,\ K = 1,\ H = 0.8,\ T = 2,\ \gamma = 1,\ \theta = 0.95.$

With regard to Model 1, the SDP solver ran into numerical problems for $i = 8$ and the value $\sigma = 0.25$. The relative error in Table 9.1 gives the accuracy of the $i = 8, 9$ approximation. These results indicate that the bounds provided by the moment approach are very tight. However, our experimentation was limited in size because of the numerical problems encountered with the semidefinite solver SeDuMi. As a result, we could not go beyond $i = 8$ or $i = 9$ for barrier payoff structures.

Table 9.1 Barrier options.

Model 1

	$\sigma = 0.10$	$\sigma = 0.15$	$\sigma = 0.20$	$\sigma = 0.25$
$i = 8$	2.63%	3.91%	0.52%	1.07%

Model 2

	$\sigma = 0.10$	$\sigma = 0.15$	$\sigma = 0.20$	$\sigma = 0.25$
$i = 9$	1.97%	2.19%	1.36%	2.8%

Model 3

	$\sigma = 0.10$	$\sigma = 0.15$	$\sigma = 0.20$	$\sigma = 0.25$
$i = 9$	6.3%	2.85%	1.47%	0.83%

9.3 Summary

In Section 9.2 we have described the moment approach for approximating the prices of several exotic options of European type. For concreteness, we have focused on fixed-strike, down-and-out barrier *call* option payoff structures. However, it can easily be modified to account for other payoff structures including double-barrier knockout and Parisian call options, or their *put* counterparts. An distinguishing feature of the approach is its ability to handle various dynamic models of the asset price, other than the geometric Brownian motion model (GBM). For instance, it can handle underlying price dynamics modelled by processes such as an Ornstein-Uhlenbeck process as in Vasicek's model, a standard square-root process with mean-reversion as in the Cox-Ingersoll-Ross (CIR) model, and others (provided that the parameters of the infinitesimal generator are polynomials) which are of particular interest in the fixed-income and the commodity markets.

9.4 Notes and Sources

9.1 The material in this section is from Lasserre (2008b). Bounding the price of an option without assuming any model for the price dynamics and only under the no-arbitrage assumption has a long history in the economic literature. For a detailed account as well as complexity results and exact formulas in specific cases, the interested reader is referred to Bertsimas and Popescu (2002).

9.2 The material of this section is from Lasserre *et al.* (2006). The study of exotic options has been a major research area in mathematical finance, and the literature is abundant in results such as exact formulas or numerical approximation techniques when the underlying asset price dynamics are modelled by a geometric Brownian motion (GBM). If we assume that the underlying asset follows a GBM, then there exists a closed form analytic expression for the price of a down-and-out barrier call option (e.g., in (Musiela and Rutkowski, 1997, Section 9.6)). On the other hand, this is not the case for arithmetic-average Asian options. Approaches to the approximate pricing of Asian options when the underlying is modelled by a GBM include quasi-analytic techniques based on Edgeworth and Taylor expansions, and the like (e.g., Turnbull and Wakeman (1991)), methods derived by means of probabilistic techniques as in Curran (1992), Rogers and Shi (1995)), the numerical solution of appropriate PDE's in Rogers and Shi (1995)) and Monte Carlo simulations in e.g., Glasserman *et al.* (1999). Most of these approximation techniques can be adapted to account for the pricing of Asian options when the underlying follows other diffusion processes such as a mean-reverting process. However, despite their practical importance, such extensions are still at their early stages.

The moment approach developed in this chapter derives from the methodology of moments introduced by Dawson (1980) for the analysis of geostochastic systems modelled as solutions, called *stochastic measure diffusions*, of measure-valued martingale problems with a view to a range of applications in areas such as statistical physics, population, ecology and epidemic modelling, and others. Modeling some problems associated with diffusions as an infinite-dimensional linear program on a space of appropriate measures, and approximating the optimal value using only finitely moments was already proposed in Helmes *et al.* (2001); for instance to evaluate the moments of certain exit time distribution and also for *change-point detection* problems. Also, Schwerer (2001) considered the evaluation of moments of the steady-state distribution of a reflected Brownian motion. In both the above works, the authors use Hausdorff type moment conditions of Theorem 3.10 which translate into linear constraints (and so LP relaxations), whereas we use semidefinite relaxations based on the moment conditions described in Theorem 3.8. For a comparison between LP and SDP relaxations in the general case, see Chapter 5 and in the context of some diffusions see Lasserre and Priéto-Rumeau (2004). For moment problems associated with SDE see also Stoyanov (2001). Finally, the infinite-dimensional LP approach has also been used for some continuous-time stochastic control problems (e.g., modeled by diffusions) and optimal stopping problems in Jung and Stockbridge (2002), Helmes and Stockbridge (2000), Kurtz and Stockbridge (1998), and also Fleming and Vermes (1989). In some of these works, the moment approach is also used to approximate the resulting infinite-dimensional LP.

Chapter 10

Application in Control

This chapter considers an application in control. We apply the moment approach to the so-called weak formulation of optimal control problems in which the initial problem is viewed as an infinite linear-programming model over suitable *occupation measures*, an instance of the generalized moment problem.

10.1 Introduction

Consider the following problem:

$$
\begin{aligned}
J(0,\mathbf{x}_0) := \inf_{\mathbf{u}\in\mathscr{U}} & \int_0^T h(s,\mathbf{x}(s),\mathbf{u}(s))\,ds + H(\mathbf{x}(T)) \\
\text{s.t. } \tfrac{d}{dt}\mathbf{x}(s) &= \mathbf{f}(s,\mathbf{x}(s),\mathbf{u}(s)), \qquad \text{a.e. on } [0,T] \\
(\mathbf{x}(s),\mathbf{u}(s)) &\in \mathbf{X}\times\mathbf{U}, \qquad \text{a.e. on } [0,T] \\
\mathbf{x}(T) &\in \mathbb{K},
\end{aligned}
\tag{10.1}
$$

with initial condition $\mathbf{x}(0)=\mathbf{x}_0\in\mathbf{X}$, which models a deterministic system where $\mathbf{x}(t)$ (resp. $\mathbf{u}(t)$) is the *state* (resp. the *control*) of the system at time t, and $\mathbf{x}(t)$ obeys the ordinary differential equation in (10.1) and where the goal is to find a control trajectory $\mathbf{u}:\ [0,T]\to\mathbf{U}$ such that the (control-state) trajectory lies in $\mathbf{X}\times\mathbf{U}$, the final state $\mathbf{x}(T)$ lies in $\mathbb{K}$ and such that the cost functional $\int_0^T h ds + H(\mathbf{x}(T))$ is minimized.

The above problem (10.1) is a so-called deterministic *optimal control problem* (OCP in short). In principle, solving an OCP is a difficult challenge, despite powerful theoretical tools are available, e.g. the celebrated *Maximum Principle* and *Hamilton-Jacobi-Bellman* (HJB) optimality equation. The problem is even more difficult in the presence of state and/or control constraints as is the case in (10.1) and in most practical applications. State constraints are particularly difficult to handle.

On the other hand, there is a so-called *weak* formulation of the OCP which consists of an infinite-dimensional linear program (LP) over two spaces of measures. The two unknown measures are the state-action *occupation measure up to* the final time T, and the state occupation measure *at* time T. The optimal value of the resulting LP always provides a lower bound on the optimal value of the OCP, and under some convexity assumptions, both values coincide.

The dual of this infinite dimensional LP has an interpretation in terms of *subsolutions* of related HJB-like optimality conditions, as for the unconstrained case. Another interesting feature of this LP approach with occupation measures is that state constraints, as well as state and/or action constraints, are all easy to handle; indeed they simply translate into constraints on the supports of the unknown occupation measures.

Although this LP approach is valid for any OCP, in general solving the corresponding (infinite-dimensional) LP is difficult. However, it turns out that this LP can be formulated as a particular instance of the generalized moment problem (1.1) and so, as we shall see, the general methodology of Chapter 4 can be easily adapted when the data of the OCP (10.1) are polynomials and basic semi-algebraic sets.

10.2 Weak Formulation of Optimal Control Problems

Consider the optimal control problem OCP (10.1) with initial condition $\mathbf{x}(0) = \mathbf{x}_0 \in \mathbf{X}$ and where:

- $\mathbf{X}, \mathbb{K} \subset \mathbb{R}^n$ and $\mathbf{U} \subset \mathbb{R}^m$ are compact basic semi-algebraic sets.
- $\mathscr{U}$ is the space of measurable functions $\mathbf{u} : [0, T] \to \mathbf{U}$.
- $h \in \mathbb{R}[t, \mathbf{x}, \mathbf{u}]$, $H \in \mathbb{R}[\mathbf{x}]$
- $\mathbf{f} : \mathbb{R} \times \mathbb{R}^n \times \mathbb{R}^m \to \mathbb{R}^n$ is a polynomial map, i.e. $f_k \in \mathbf{R}[t, \mathbf{x}, \mathbf{u}]$ for all $k = 1, \ldots, n$.

Before rewriting the OCP as a particular instance of the generalized moment problem, we need introduce some additional notation and definitions. For a compact Hausdorff space $\mathscr{X} \subset \mathbb{R}^n$ let $C(\mathscr{X})$ denote the

Banach space of continuous functions on $\mathscr{X}$, equipped with the sup-norm, so that its topological dual $C(\mathscr{X})^*$ is isometrically isomorphic to $\mathscr{M}(\mathscr{X})$, the Banach space of finite signed measures on $\mathscr{X}$, equipped with the total variation norm (denoted $C(\mathscr{X})^* \simeq \mathscr{M}(\mathscr{X})$). The notation $\mathscr{M}(\mathscr{X})_+$ denotes the positive cone of $\mathscr{M}(\mathscr{X})$, i.e. the space of bounded measures on $\mathscr{X}$.

Let $\Sigma := [0,T] \times \mathbf{X}$, $\mathbf{S} := \Sigma \times \mathbf{U}$, and let $C_1(\Sigma)$ be the Banach space of continuously differentiable functions $w \in C(\Sigma)$, endowed with the sup-norm $\|w\| + \|\nabla w\|$. Next, with $\mathbf{u} \in \mathbf{U}$, let $\mathscr{A} : C_1(\Sigma) \to C(\mathbf{S})$ be the mapping

$$w \mapsto \mathscr{A}\, w(t, \mathbf{x}, \mathbf{u}) := \frac{\partial w}{\partial t}(t, \mathbf{x}) + \langle \mathbf{f}(t, \mathbf{x}, \mathbf{u}), \nabla_{\mathbf{x}} w(t, \mathbf{x}) \rangle, \tag{10.2}$$

and let $\mathscr{L} : C_1(\Sigma) \to C(\mathbf{S}) \times C(\mathbb{K})$ be the mapping

$$w \mapsto \mathscr{L} w := (-\mathscr{A}\, w, w_T),$$

where $w_T(\mathbf{x}) := w(T, \mathbf{x})$, for all $\mathbf{x} \in \mathbb{K}$.

Occupation measures

Notice that for a feasible trajectory $(s, \mathbf{x}(s), \mathbf{u}(s))$, and $w \in C_1(\Sigma)$, one has

$$w_T(\mathbf{x}(T)) = w(0, \mathbf{x}_0) + \int_0^T \mathscr{A} w\,(s, \mathbf{x}(s), \mathbf{u}(s))\, ds. \tag{10.3}$$

Let $(\mu, \nu) \in \mathscr{M}(\mathbf{S})_+ \times \mathscr{M}(\mathbb{K})_+$ be defined as

$$\nu(D) = 1_D(\mathbf{x}(T)); \quad \mu(A \times B \times C) := \int_{[0,T] \cap A} 1_{B \times C}((\mathbf{x}(s), \mathbf{u}(s)))\, ds$$

for all Borel sets $D \subset \mathbb{K}$, $A \subset [0,T]$, $B \subset \mathbf{X}$, and $C \subset \mathbf{U}$ (and where $\mathbf{x} \mapsto 1_\bullet(\mathbf{x})$ stands for the indicator function of the set $\bullet$).

The measure μ is called the *state-action occupation measure* up to time T, whereas ν is the *state occupation measure* at time T, for the trajectory $(s, \mathbf{x}(s), \mathbf{u}(s))$.

Then the time integration (10.3) is the same as the *spatial* integration

$$\int_{\mathbb{K}} w_T\, d\nu = w(0, \mathbf{x}_0) + \int_{\mathbf{S}} \mathscr{A} w\, d\mu. \tag{10.4}$$

Similarly, the cost of this trajectory can be expressed via μ and ν by

$$J(0,\mathbf{x}_0,\mathbf{u}) = \int_0^T h(s,\mathbf{x}(s),\mathbf{u}(s))\,ds + H(\mathbf{x}(T)) = \int_{\mathbf{S}} h\,d\mu + \int_{\mathbb{K}} H\,d\nu. \tag{10.5}$$

And so, for an arbitrary feasible trajectory $(s,x(s),u(s))$, one has

$$\langle \mathscr{L}\,w,(\mu,\nu)\rangle = \langle w,\delta_{(0,\mathbf{x}_0)}\rangle \qquad \forall\, w \in C_1(\Sigma). \tag{10.6}$$

Notice that we have translated the dynamics of the system, i.e., the ordinary differential equation in (10.1), in terms of the system of linear equations (10.4) on the occupation measures μ and ν, induced by a family of test functions $w \in C_1$. Similarly, the cost functional is now the linear criterion (10.5) in μ,ν.

This formulation in terms of occupation measures is the deterministic analogue (with control) of the martingale approach that we have described in Section 9.2.2 for stochastic differential equations. Indeed, (9.12) and (9.13) are the stochastic analogues of (10.4) and (10.5), respectively. The only difference is that for the former, the infinitesimal generator $\mathscr{A}$ contains second-order partial derivatives because of the presence of the Brownian. Hence (10.4)-(10.5) will be the basis of the semidefinite relaxations for the OCP (10.1).

The generalized moment problem

Let $\mathscr{L}^* : \mathscr{M}(\mathbf{S}) \times \mathscr{M}(\mathbb{K}) \to C_1(\Sigma)^*$ be the adjoint mapping of $\mathscr{L}$, defined by

$$\langle(\mu,\nu),\mathscr{L}\,w\rangle = \langle\mathscr{L}^*(\mu,\nu),w\rangle,$$

for all $((\mu,\nu),w) \in \mathscr{M}(\mathbf{S}) \times \mathscr{M}(\mathbb{K}) \times C_1(\Sigma)$.

A function $g : [0,T] \times \mathbb{R}^n \to \mathbb{R}$, is a solution of the Hamilton-Jacobi-Bellman optimality equation if

$$\inf_{u\in\mathbf{U}} \{\mathscr{A}g\,(s,\mathbf{x},\mathbf{u}) + h(s,\mathbf{x},\mathbf{u})\} = 0, \quad \forall\,(s,\mathbf{x}) \in \operatorname{int}\Sigma \tag{10.7}$$

with boundary condition $g_T(\mathbf{x})\,(= g(T,\mathbf{x})) = H(\mathbf{x})$, for all $\mathbf{x} \in \mathbb{K}$.

On the other hand, a function $w \in C_1(\Sigma)$ is said to be a smooth *subsolution* of the Hamilton-Jacobi-Bellman equation (10.7) if

$$\mathscr{A}w + h \geq 0 \quad \text{on } \mathbf{S}, \quad \text{and} \quad w(T,\mathbf{x}) \leq H(\mathbf{x}), \quad \forall\,\mathbf{x} \in \mathbb{K}.$$

Then, consider the infinite-dimensional linear program:

$$\rho_{\text{mom}} = \inf_{(\mu,\nu)\in\Delta} \{\langle(\mu,\nu),(h,H)\rangle \ : \quad \mathscr{L}^*(\mu,\nu) = \delta_{(0,\mathbf{x}_0)}\} \tag{10.8}$$

(where $\Delta := \mathscr{M}(\mathbf{S})_+ \times \mathscr{M}(\mathbb{K})_+$). Its dual reads:

$$\rho_{\text{pop}} = \sup_{w\in C_1(\Sigma)} \{\langle\delta_{(0,\mathbf{x}_0)}, w\rangle \ : \quad \mathscr{L}w \leq (h,H)\}. \tag{10.9}$$

Note that the feasible solutions w of (10.9) are precisely smooth subsolutions of (10.7).

Under compactness-continuity conditions, there is no duality gap between the infinite dimensional linear programs (10.8) and (10.9).

Theorem 10.1. *Let $\mathbf{X}, \mathbf{U}, \mathbb{K}$ be all compact, and let h, H be continuous. Moreover, assume that for every $(t,\mathbf{x}) \in \Sigma$,*

$$\begin{cases} \text{the set } \mathbf{f}(t,\mathbf{x},\mathbf{U}) \subset \mathbb{R}^n \text{ is convex, and the function} \\ \mathbf{v} \mapsto g_{t,\mathbf{x}}(\mathbf{v}) := \inf_{\mathbf{u}\in\mathbf{U}} \{\, h(t,\mathbf{x},\mathbf{u}) \ : \ \mathbf{v} = \mathbf{f}(t,\mathbf{x},\mathbf{u})\} \text{ is convex,} \end{cases} \tag{10.10}$$

then the OCP (10.1) has an optimal solution and

$$\rho_{\text{pop}} \ = \ \rho_{\text{mom}} \ = \ J(0,\mathbf{x}_0). \tag{10.11}$$

To see that (10.8) is an instance of the generalized moment problem with polynomial data, it suffices to notice that since Σ is compact, the constraint $\mathscr{L}^*(\mu,\nu) = \delta_{(0,\mathbf{x}_0)}$ is equivalent to (10.6) for all w in $\mathbb{R}[t,\mathbf{x}]$, and so is equivalent to the countably many equality constraints

$$\langle \mathscr{L}(\mathbf{x}^\alpha t^k), (\mu,\nu)\rangle \ = \ \langle \mathbf{x}^\alpha\, t^k, \delta_{(0,\mathbf{x}_0)}\rangle \qquad \forall\, (\alpha,k) \in \mathbb{N}^n \times \mathbb{N}, \tag{10.12}$$

which in turn reads

$$T^k \int_{\mathbb{K}} \mathbf{x}^\alpha \, d\nu - \int_{\mathbf{S}} \left[kt^{k-1}\mathbf{x}^\alpha + t^k\langle\nabla_{\mathbf{x}}\mathbf{x}^\alpha, \mathbf{f}\rangle\right] d\mu \ = \begin{cases} \mathbf{x}_0^\alpha \text{ if } k=0 \\ 0 \ \ \text{ otherwise} \end{cases}, \tag{10.13}$$

for all $\alpha \in \mathbb{N}^n$ and $k \in \mathbb{N}$. Equivalently (10.13) reads:

$$\int_{\mathbb{K}} f^1_{\alpha k}\, d\nu - \int_{\mathbf{S}} f^2_{\alpha k}\, d\mu \ = \ b_{\alpha k}, \qquad (\alpha,k) \in \mathbb{N}^n \times \mathbb{N},$$

with $f^1_{\alpha k} \ = T^k\mathbf{x}^\alpha \in \mathbb{R}[\mathbf{x}]$, $f^2_{\alpha k} \ = \ kt^{k-1}\mathbf{x}^\alpha + t^k\langle\nabla_{\mathbf{x}}\mathbf{x}^\alpha, \mathbf{f}\rangle \in \mathbb{R}[t,\mathbf{x},\mathbf{u}]$, and $b_{\alpha k} = \mathbf{x}_0^\alpha$ if $k=0$ and 0 otherwise.

Hence, under (10.10) the OCP (10.1) has same optimal value as the infinite dimensional linear program

$$\rho_{\text{mom}} = \min_{\mu \in \mathscr{M}(\mathbf{S})_+, \nu \in \mathscr{M}(\mathbb{K})_+} \left\{ \int_{\mathbf{S}} h \, d\mu + \int_{\mathbb{K}} H \, d\nu \; : \quad (10.12) \text{ holds} \right\}, \tag{10.14}$$

a variant of the generalized moment problem (1.1) with two unknown measures μ, ν and countably many constraints.

10.3 Semidefinite Relaxations for the OCP

Let $\mathbf{y} = \{y_{\boldsymbol{\alpha}}\}$, $\boldsymbol{\alpha} \in \mathbb{N}^n$, and $\mathbf{z} = \{z_{\boldsymbol{\beta}}\}$, $\boldsymbol{\beta} \in \mathbb{N} \times \mathbb{N}^n \times \mathbb{N}^m$, be the moment variables associated with ν and μ respectively. For every $(\boldsymbol{\alpha}, k) \in \mathbb{N}^n \times \mathbb{N}$, let $d(\boldsymbol{\alpha}, k) := \max[|\boldsymbol{\alpha}|, |\boldsymbol{\beta}|]$ of $y_{\boldsymbol{\alpha}}$ and $z_{\boldsymbol{\beta}}$ in the constraint

$$T^k y_\alpha - L_{\mathbf{z}} \left(k t^{k-1} \mathbf{x}^\alpha + t^k \langle \nabla_{\mathbf{x}} \mathbf{x}^\alpha, \mathbf{f} \rangle \right) = \begin{cases} \mathbf{x}_0^\alpha \text{ if } k = 0 \\ 0 \quad \text{otherwise} \end{cases}. \tag{10.15}$$

The sets $\mathbb{K}$ and $\mathbf{S} = \Sigma \times \mathbf{U}$ are compact basic semi-algebraic sets, with $\mathbb{K}$ defined as in (2.10) and $\mathbf{S}$ defined by:

$$\mathbf{S} := \{(t, \mathbf{x}, \mathbf{u}) \in \mathbb{R} \times \mathbb{R}^n \times \mathbb{R}^m \quad : \quad h_k(t, \mathbf{x}, \mathbf{u}) \geq 0 \quad \forall k = 1, \ldots, p\} \tag{10.16}$$

for some polynomials $\{h_k\}_{k=1}^p \subset \mathbb{R}[t, \mathbf{x}, \mathbf{u}]$.

Depending on parity, let $\deg g_j = 2r_j$ or $2r_j - 1$ and similarly, $\deg h_k = 2l_k$ or $2l_k - 1$. For every $2i \geq i_0 := \max[\deg h, \deg H, \max_{j,k}[r_j, l_k]]$, the semidefinite relaxation of the GPM (10.14) reads:

$$\begin{aligned} \rho_i = \inf_{\mathbf{y}, \mathbf{z}} \; & L_{\mathbf{y}}(H) + L_{\mathbf{z}}(h) \\ \text{s.t } & \mathbf{M}_r(\mathbf{y}), \, \mathbf{M}_r(\mathbf{z}) \succeq 0 \\ & \mathbf{M}_{i-r_j}(g_j \, \mathbf{y}) \succeq 0, \quad j = 1, \ldots, m \\ & \mathbf{M}_{i-l_k}(h_k \, \mathbf{z}) \succeq 0, \quad k = 1, \ldots, p \\ & (10.15), \quad (\boldsymbol{\alpha}, k) \in \mathbb{N}^n \times \mathbb{N}, \, d(\boldsymbol{\alpha}, k) \leq 2i. \end{aligned} \tag{10.17}$$

Notice that with $(\boldsymbol{\alpha}, k) = (0, 0)$ (resp. $(\boldsymbol{\alpha}, k) = (0, 1)$) in (10.13), one obtains $\nu(\mathbb{K}) = 1$ (resp. $\mu(\mathbf{S}) = T$), so that $y_0 = 1$ and $z_0 = T$.

Theorem 10.2. *Let $\mathbb{K} \subset \mathbb{R}^n, \mathbf{S} \subset \mathbb{R} \times \mathbb{R}^n \times \mathbb{R}^m$ be the compact basic semi-algebraic sets defined in (2.10) and (10.16) respectively. Let Assumption 2.1 holds for $\mathbb{K}$ and for $\mathbf{S}$ as well. Let ρ_{mom} be as in (10.14) and let ρ_i be as in (10.17).*

Then $\rho_i \uparrow \rho_{\text{mom}}$ as $i \to \infty$, and if (10.10) holds, $\rho_i \uparrow \rho_{\text{mom}} = J(0, \mathbf{x}_0)$.

The dual relaxation

The dual of the semidefinite program (10.17), which reads:

$$\begin{aligned}
\rho_i^* = \sup_{\lambda_{k\boldsymbol{\alpha}}, \sigma_j, \psi_k} \quad & w(0, \mathbf{x}_0) \\
\text{s.t} \quad w(t, \mathbf{x}) &= \{ \sum_{k,\boldsymbol{\alpha}} \lambda_{k\boldsymbol{\alpha}}\, t^k \mathbf{x}^{\boldsymbol{\alpha}} \; : \; d(\boldsymbol{\alpha}, k) \leq 2i \} \\
\mathscr{A} w + h &= \psi_0 + \sum_{k=1}^{p} \psi_k\, h_k \qquad\qquad (10.18) \\
H - w_T &= \sigma_0 + \sum_{j=1}^{m} \sigma_j\, g_j \\
\deg \sigma_0, \psi_0, \sigma_j g_j, \psi_k h_k &\leq 2i, \quad \forall j \leq m, \, \forall k \leq p
\end{aligned}$$

is a strengthening of (10.9) because $w \in C_1(\Sigma)$ is now replaced with a polynomial of $\mathbb{R}[t, \mathbf{x}]$. Moreover, the constraint $h + \mathscr{A} w \geq 0$ on $\mathbf{S}$ (resp. $H - w_T \leq 0$ on $\mathbb{K}$) is replaced with the stronger requirement that $h + \mathscr{A} w$ (resp. $H - w_T$) belong to the quadratic module generated by the polynomials (h_k) (resp. (g_j)), with a degree bound on the weights $(\psi_k) \subset \mathbb{R}[t, \mathbf{x}, \mathbf{u}]$ (resp. $(\sigma_j) \subset \mathbb{R}[\mathbf{x}]$).

That is, in (10.18) one searches for a polynomial $w \in \mathbb{R}[t, \mathbf{x}]$ (with a degree bound that increases with i), which is a subsolution of the Hamilton-Jacobi equation (10.7), and which maximizes $w(0, \mathbf{x}_0)$.

Optimal solutions $w^* \in \mathbb{R}[t, \mathbf{x}]$ of the dual relaxation (10.18) can provide some useful information to compute an approximate control feedback for the OCP. Recall that $h + \mathscr{A} w^* \geq 0$ on $\mathbf{S}$. Therefore, with $(t, \mathbf{x}) \in [0, T] \times \mathbf{X}$ fixed, consider a control $\mathbf{u}^*$ that minimizes over the compact set $\mathbf{U}$, the polynomial $\mathbf{u} \mapsto h(t, \mathbf{x}, \mathbf{u}) + \mathscr{A} w^*(t, \mathbf{x}, \mathbf{u})$, which is nonnegative on $\mathbf{U}$. If the optimal value is close to zero, i.e., if $h(t, \mathbf{x}, \mathbf{u}^*) + \mathscr{A} w^*(t, \mathbf{x}, \mathbf{u}^*) \approx 0$, then such a global minimizer $\mathbf{u}^* \in \mathbf{U}$ is likely a good candidate for feedback control a time t and in state $\mathbf{x} \in \mathbf{X}$. Indeed, when using $\mathbf{u}^*$, the Hamilton-Jacobi-Bellman equation (10.7) is almost satisfied at the point $(t, \mathbf{x}) \in \Sigma$.

OCP with Free terminal time.

The above moment approach is also valid for *free terminal time* OCPs, that is, OCP where the final time T is itself unknown. For instance, think about computing the optimal trajectory for sending a satellite to some given orbit in minimum time.

For the OCP with *free* terminal time $T \leq T_0$ (for some T_0 fixed), we need adapt the notation because now T is also a variable. And so in this context, $\Sigma := [0, T_0] \times \mathbf{X}$ and $\mathbf{S} = \Sigma \times \mathbf{U}$. The measure ν in the infinite dimensional linear program (10.8) is now supported in $[0, T_0] \times \mathbb{K}$ instead of $\mathbb{K}$ previously. Hence, the sequence $\mathbf{y}$ associated with ν is now indexed in the canonical basis $\{t^k \mathbf{x}^\alpha\}$ of $\mathbb{R}[t, \mathbf{x}]$ instead of $\{\mathbf{x}^\alpha\}$ previously. The corresponding semidefinite relaxation is exactly the same as in (10.17) with appropriate definitions of the polynomials $g_j \in \mathbb{R}[t, \mathbf{x}]$ (resp. $h_k \in \mathbb{R}[t, \mathbf{x}, \mathbf{u}]$) that define the set $[0, T_0) \times \mathbb{K}$ (resp. $\mathbf{S}$). The particular case of minimal time problem is obtained with $h = 1$, $H = 0$.

10.3.1 *Examples*

We have tested the above methodology on two minimum time OCPs, namely the double and Brockett *integrators*, because the associated optimal value $T(\mathbf{x}_0)$ to steer an initial state $\mathbf{x}_0$ to the origin 0, can be calculated exactly.

The double integrator.

Consider the double integrator system in $\mathbb{R}^2$

$$\begin{aligned} \dot{x}_1(t) &= x_2(t), \\ \dot{x}_2(t) &= u(t), \end{aligned} \tag{10.19}$$

where $\mathbf{x} = (x_1, x_2)$ is the state and the control $u(t) \in \mathscr{U}$, satisfies the constraint $|u(t)| \leq 1$, for all $t \geq 0$. In addition, the state $\mathbf{x}(t)$ is constrained to satisfy $x_2(t) \geq -1$, for all t. Therefore $\mathbf{X} = \{\mathbf{x} \in \mathbb{R}^2 \; : \; x_2 \geq -1\}$, $\mathbb{K} = \{(0, 0)\}$, and $\mathbf{U} = [-1, 1]$.

For this very simple system (10.19), one is able to compute exactly the optimal minimum time $T(\mathbf{x})$ to steer an initial state $\mathbf{x}$ to the origin.

If $x_1 \geq -(x_2^2 - 2)/2$ then $T(x) = t_1 + t_2 + t_3$ with:

$$t_1 = 1 + x_2; \quad t_2 = -(t_1 + 1)/2 + t_1 x_2 + x_1; \quad t_3 = 1,$$

else if $x_1 \geq (-x_2^2/2)\,\mathrm{sign}\, x_2$ then

$$T(x) = 2\sqrt{x_1 + x_2^2/2} + x_2, \quad \text{else} \quad T(x) = 2\sqrt{x_1 + x_2^2/2} - x_2.$$

Observe that $\mathbf{X}$ is not compact and so the convergence result of Theorem 10.2 may not hold. In fact, one may impose the additional constraint $\|\mathbf{x}(t)\|_\infty \leq M$ for some large M (and modify $\mathbf{X}$ accordingly), because for initial states $\mathbf{x}_0$ with $\|\mathbf{x}_0\|_\infty$ relatively small when compared to M, the optimal trajectory remains in $\mathbf{X}$. However, in the numerical experiments of Table 10.19, one has maintained the original constraint $x_2 \geq -1$.

With $\rho_i(\mathbf{x}_0)$ the optimal value of the semidefinite relaxation (10.17) with initial state $\mathbf{x}_0 \in \mathbf{X}$, Table 10.1 displays the values of $\mathbf{x}_0 \in \mathbf{X}$ while Table 10.2 displays the values of the ratii $\rho_5(\mathbf{x}_0)/T(\mathbf{x}_0)$. One may see that with relatively few moments, the approximation is quite close (i.e. the ratio is close to 1), especially for values of the initial state corresponding to the right upper-triangle of Table 10.2.

Table 10.1 Double integrator: data initial state $\mathbf{x}_0 = (x_{01}, x_{02})$.

x_{01}	0.0	0.2	0.4	0.6	0.8	1.0	1.2	1.4	1.6	1.8	2.0
x_{02}	-1.0	-0.8	-0.6	-0.4	-0.2	0.0	0.2	0.4	0.6	0.8	1.0

Table 10.2 Double integrator: ratio $\rho_5(\mathbf{x}_0)/T(\mathbf{x}_0)$.

fifth semidefinite relaxation										
0.7550	0.5539	0.3928	0.9995	0.9995	0.9995	0.9994	0.9992	0.9988	0.9985	0.9984
0.6799	0.4354	0.9828	0.9794	0.9896	0.9923	0.9917	0.9919	0.9923	0.9923	0.9938
0.6062	0.9805	0.9314	0.9462	0.9706	0.9836	0.9853	0.9847	0.9848	0.9862	0.9871
0.5368	0.8422	0.8550	0.8911	0.9394	0.9599	0.9684	0.9741	0.9727	0.9793	0.9776
0.4713	0.6417	0.7334	0.8186	0.8622	0.9154	0.9448	0.9501	0.9505	0.9665	0.9637
0.0000	0.4184	0.5962	0.7144	0.8053	0.8825	0.9044	0.9210	0.9320	0.9544	0.9534
0.4742	0.5068	0.6224	0.7239	0.7988	0.8726	0.8860	0.9097	0.9263	0.9475	0.9580
0.5410	0.6003	0.6988	0.7585	0.8236	0.8860	0.9128	0.9257	0.9358	0.9452	0.9528
0.6106	0.6826	0.7416	0.8125	0.8725	0.9241	0.9305	0.9375	0.9507	0.9567	0.9604
0.6864	0.7330	0.7979	0.8588	0.9183	0.9473	0.9481	0.9480	0.9559	0.9634	0.9733
0.7462	0.8032	0.8564	0.9138	0.9394	0.9610	0.9678	0.9678	0.9696	0.9755	0.9764

The Brockett integrator

Consider the so-called *Brockett system* in $\mathbb{R}^3$

$$\begin{aligned}\dot{x}_1(t) &= u_1(t),\\ \dot{x}_2(t) &= u_2(t),\\ \dot{x}_3(t) &= u_1(t)x_2(t) - u_2(t)x_1(t),\end{aligned} \tag{10.20}$$

where $\mathbf{x} = (x_1, x_2, x_3)$, and the control $\mathbf{u}(t) = (u_1(t), u_2(t)) \in \mathscr{U}$ satisfies the constraint $u_1(t)^2 + u_2(t)^2 \leq 1$ for all $t \geq 0$. The goal is again to steer an initial state $\mathbf{x}_0 \in \mathbf{X}$ to the origin and therefore $\mathbf{X} = \mathbb{R}^3$, $\mathbb{K} = \{(0,0,0)\}$ and $\mathbf{U}$ is the closed unit ball of $\mathbb{R}^2$, centered at the origin.

Let $T(\mathbf{x}_0)$ be the minimum time needed to steer an initial state $\mathbf{x}_0 \in \mathbf{X}$ to the origin which is in fact equal to the minimum time to reach $\mathbf{x}_0$ from 0.

Proposition 10.3. *The minimum time $T(\mathbf{x})$ needed to steer the origin to a point $\mathbf{x} = (x_1, x_2, x_3) \in \mathbb{R}^3$ is given by*

$$T(\mathbf{x}) = \frac{\theta\sqrt{x_1^2 + x_2^2 + 2|x_3|}}{\sqrt{\theta + \sin^2\theta - \sin\theta\cos\theta}}, \tag{10.21}$$

where $\theta = \theta(\mathbf{x})$ is the unique solution in $[0, \pi)$ of

$$\frac{\theta - \sin\theta\cos\theta}{\sin^2\theta}(x_1^2 + x_2^2) = 2|x_3|. \tag{10.22}$$

Moreover, the function T is continuous on $\mathbb{R}^3$, and is analytic outside the line $x_1 = x_2 = 0$.

Again recall that the convergence result of Theorem 10.2 is guaranteed for $\mathbf{X}$ compact only whereas in the present case, $\mathbf{X} = \mathbb{R}^3$ is not compact. One possibility is to take for $\mathbf{X}$ a large ball of $\mathbb{R}^3$ centered at the origin because for initial states $\mathbf{x}_0$ with norm $\|\mathbf{x}_0\|$ relatively small, the optimal trajectory remains in $\mathbf{X}$. However, in the numerical experiments presented below, we have chosen to maintain $\mathbf{X} = \mathbb{R}^3$.

In Table 10.3 we have displayed the optimal values $\rho_i(\mathbf{x}_0)$ for 16 different values of the initial state $\mathbf{x}_0$, in fact, all 16 combinations of $x_1 = 0$, $x_2 = 0, 1, 2, 3$, and $x_3 = 0, 1, 2, 3$. So, the entry $(2, 3)$ of Table 10.3 for the second semidefinite relaxation is ρ_2 for the initial condition $\mathbf{x}_0 = (0, 1, 2)$. At some (few) places in the table, the * indicates that the SDP-solver encountered some numerical problems, which explains why one finds a lower bound

Table 10.3 Brockett integrator: semidefinite relaxations: $\rho_i(\mathbf{x}_0)$.

First semidefinite relaxation $i = 1$			
0.0000	0.9999	1.9999	2.9999
0.0140	1.0017	2.0010	3.0006
0.0243	1.0032	2.0017	3.0024
0.0295	1.0101	2.0034	3.0040
Second semidefinite relaxation $i = 2$			
0.0000	0.9998	1.9997*	2.9994*
0.2012	1.1199	2.0762	3.0453
0.3738	1.2003	2.1631	3.1304
0.4946	1.3467	2.2417	3.1943
Third semidefinite relaxation $i = 3$			
0.0000	0.9995	1.9987*	2.9984*
0.7665	1.3350	2.1563	3.0530
1.0826	1.7574	2.4172	3.2036
1.3804	2.0398	2.6797	3.4077
Fourth semidefinite relaxation $i = 4$			
0.0000	0.9992	1.9977	2.9952
1.2554	1.5925	2.1699	3.0478
1.9962	2.1871	2.5601	3.1977
2.7006	2.7390	2.9894	3.4254
Optimal time $T(\mathbf{x}_0)$			
0.0000	1.0000	2.0000	3.0000
2.5066	1.7841	2.1735	3.0547
3.5449	2.6831	2.5819	3.2088
4.3416	3.4328	3.0708	3.4392

$\rho_{i-1}(\mathbf{x}_0)$ slightly higher than ρ_i, when practically equal to the exact optimal value $T(\mathbf{x}_0)$. Again, good approximations are obtained with few moments for values of the initial state $\mathbf{x}_0$ corresponding to the right upper-triangle of Table 10.3.

10.4 Summary

We have presented the moment approach to OCP problems with state and/or control constraints. When the data of the OCP (dynamics, instantaneous costs, and constraints) are described by polynomials then one may devise a numerical scheme for approximating the global optimal value $J(0, \mathbf{x}_0)$ from below. It consists in a hierarchy of semidefinite relaxations of a weak formulation of the original OCP. Under some convexity conditions, one obtains convergence to the global optimum $J(0, \mathbf{x}_0)$.

Preliminary results on two minimum time examples with several values of the initial state $\mathbf{x}_0 \in \mathbf{X}$, are very encouraging as one often gets very close to the optimal value $T(\mathbf{x})$ with relatively few moments. Of course, even if not trivial, the two above examples are low-dimensional examples, and in view of the present status of semidefinite solvers, so far the appproach is limited to small problems.

In addition, and under appropriate convexity conditions like (10.10), one only gets convergence to the optimal value $J(0, \mathbf{x}_0)$ of the OCP. Therefore, this semidefinite programming approach should be viewed as a complement to existing methods for solving the OCP; in particular, as it may provide good approximations of the optimal value, it could be used to evaluate the efficiency of other approaches that compute also feasible controls $\mathbf{u} \in \mathscr{U}$. Using the semidefinite relaxations to obtain approximations of an optimal control $\mathbf{u}^* \in \mathscr{U}$ is an important topic of further research.

10.5 Notes and Sources

Most of the material of this chapter is from Lasserre *et al.* (2008a) and Theorem 10.1 is from Vinter (1993) (proved in the more general framework of differential inclusions $\dot{\mathbf{x}}(t) \in F(t, \mathbf{x}(t))$). Proposition 10.3 is from Prieur and Trélat (2005). In Henrion *et al.* (2008) is described a first attempt on how to use optimal solutions $w^* \in \mathbb{R}[t, \mathbf{x}]$ of the dual relaxation (10.18), for computing an approximate control feedback for the OCP. Preliminary results on a simple academic problem are encouraging.

For a detailed account of HJB theory in the case of state constraints, the interested reader is referred to Capuzzo-Dolcetta and Lions (1990) and Soner (1986). There exist many numerical methods to compute the solution of a given optimal control problem; for instance, *multiple shooting* techniques which solve two-point boundary value problems as described, e.g., in Pesch (1994), Stoer and Bulirsch (2002), or *direct methods*, as, e.g., in von Stryk and Bulirsch (1992), Fletcher (1980), Gill *et al.* (1981), which use, among others, descent or gradient-like algorithms. To deal with optimal control problems with state constraints, some adapted versions of the maximum principle have been developed (see Jacobson *et al.* (1971), Maurer (1977), and see Hartl *et al.* (1995) for a survey of this theory), but happen to be very hard to implement in general.

The LP approach for deterministic optimal control problems is described in e.g. Hernandez-Hernandez *et al.* (1996) whereas Vinter (1993) considers the even more general context of differential inclusions. General LP-approximation schemes based on grids have been proposed in e.g. Hernández-Lerma and Lasserre (1998a), and more recently in Gaitsgory and Rossomakhine (2006) in the control context.

This LP approach with occupation measures has also been used in the context of discrete-time Markov control processes, and is dual to Bellman's optimality principle. For more details the interested reader is referred to Borkar (2002), Hernández-Lerma and Lasserre (1996, 1999, 2003) and the many references therein. For some continuous-time stochastic control problems (e.g., modeled by diffusions) and optimal stopping problems, the LP approach has also been used with success to prove existence of stationary optimal policies; see for instance Jung and Stockbridge (2002), Helmes and Stockbridge (2000), Helmes *et al.* (2001), Kurtz and Stockbridge (1998), and also Fleming and Vermes (1989). In some of these works, the moment approach is also used to approximate the resulting infinite-dimensional LP.

Finally, let us mention the work of Prajna and Rantzer (2007) where the notion of occupation measures associated with trajectories of an autonomous dynamical system is used to certify some temporal properties.

Chapter 11

Convex Envelope and Representation of Convex Sets

This chapter considers the following problem: Given a rational function f on a basic semi-algebraic set $\mathbb{K}$, evaluate $\widehat{f}(\mathbf{x})$ at a particular point $\mathbf{x}$ in the domain of the convex envelope $\widehat{f}$ of f. We then consider the semidefinite representation of the convex hull $\text{co}(\mathbb{K})$ of $\mathbb{K}$. That is, finding a set defined by linear matrix inequalities in a lifted space, such that $\text{co}(\mathbb{K})$ is a projection of that set.

11.1 The Convex Envelope of a Rational Function

Recall that the convex envelope $\widehat{f}$ of a real-valued function $f : \mathbb{R}^n \to \mathbb{R}$ is the largest convex function majorized by f. Computing the convex envelope $\widehat{f}$ of f is a difficult problem, and so far, there is no efficient algorithm that approximates $\widehat{f}$ by convex functions (except for the simpler univariate case).

In this section, we consider the class of *rational fractions* f on a compact basic semi-algebraic set $\mathbb{K} \subset \mathbb{R}^n$ (and $f = +\infty$ outside $\mathbb{K}$). We view the problem as a particular instance of the generalized moment problem (1.1), and we provide an algorithm for computing *convex* and *uniform* approximations of its convex envelope $\widehat{f}$. More precisely, with $\mathbf{D} := \text{co}(\mathbb{K})$ being the convex hull of $\mathbb{K}$:

(a) We provide a sequence of *convex* functions $(f_i)_i$ that converges to $\widehat{f}$ **uniformly** on any compact subset of $\mathbf{D}$ where $\widehat{f}$ is continuous, as i increases.

(b) At each point $\mathbf{x} \in \mathbb{R}^n$, computing $f_i(\mathbf{x})$ reduces to solving a semidefinite program $\mathbf{Q}_{i\mathbf{x}}$.

(c) For every $\mathbf{x} \in \text{int}\,\mathbf{D}$, the semidefinite dual $\mathbf{Q}^*_{i\mathbf{x}}$ has an optimal solution, and any optimal solution provides an element of the subgradient $\partial f_i(\mathbf{x})$ at the point $\mathbf{x} \in \text{int}\,\mathbf{D}$.

11.1.1 *Convex envelope and the generalized moment problem*

Let $\mathbb{K} \subset \mathbb{R}^n$ be compact, and denote by $\mathbf{D}$ its convex hull. Hence, by a theorem of Caratheodory, $\mathbf{D}$ is convex and compact; see Rockafellar (1970).

Also, recall that $C(\mathbb{K})$ is the Banach space of real-valued continuous functions on $\mathbb{K}$ (hence bounded), equipped with the sup-norm $\|f\| := \sup_{\mathbf{x}\in\mathbb{K}} |f(\mathbf{x})|$, whereas the Banach space $\mathscr{M}(\mathbb{K})$ of finite *signed* Borel measures on $\mathbb{K}$, equipped with the norm of total variation, is identified with $C(\mathbb{K})^*$, the topological dual of $C(\mathbb{K})$. Its positive cone $\mathscr{M}(\mathbb{K})_+$ is the set of finite Borel measures on $\mathbb{K}$. Finally, let $\mathscr{P}(\mathbb{K}) \subset \mathscr{M}(\mathbb{K})_+$ be the set of Borel probability measures on $\mathbb{K}$. The spaces $\mathscr{M}(\mathbb{K})$ and $C(\mathbb{K})$ form a *dual pair* of vector spaces, with duality bracket

$$\langle \mu, f\rangle := \int_{\mathbb{K}} f\, d\mu, \qquad \mu \in \mathscr{M}(\mathbb{K}), f \in C(\mathbb{K}),$$

and the associated weak $\star$ topology on $\mathscr{M}(\mathbb{K})$ is the coarsest topology for which $\mu \to \langle \mu, f\rangle$ is continuous for every function f in $C(\mathbb{K})$.

Denote by $\widetilde{f}$, the natural extension to $\mathbb{R}^n$ of $f \in C(\mathbb{K})$, that is

$$\mathbf{x} \mapsto \widetilde{f}(\mathbf{x}) := \begin{cases} f(\mathbf{x}) \text{ on } \mathbb{K} \\ +\infty \text{ on } \mathbb{R}^n \setminus \mathbb{K}. \end{cases} \tag{11.1}$$

Note that $\widetilde{f}$ is lower-semicontinuous (l.s.c.), admits a minimum and its effective domain $\mathbb{K}$ is non-empty and compact (denoted by $\operatorname{dom} \tilde{f}$ in the sequel). With $f \in C(\mathbb{K})$, the convex envelope $\widehat{f}$ of $\widetilde{f}$ is the largest convex function majorized by $\widetilde{f}$.

With $f \in C(\mathbb{K})$, and $\mathbf{x} \in \mathbf{D}$ fixed, arbitrary, consider the infinite-dimensional linear program:

$$\begin{aligned} \rho^1_{\text{mom}}(\mathbf{x}) = \inf_{\mu\in\mathscr{M}(\mathbb{K})_+} \; & \langle \mu, f\rangle \\ \text{s.t. } & \langle \mu, z_i\rangle = x_i, \quad i = 1, \ldots, n \\ & \langle \mu, 1\rangle \;= 1 \end{aligned} \tag{11.2}$$

(where recall that $\langle \mu, z_i\rangle = \int_{\mathbb{K}} z_i\, d\mu$). Notice that (11.2) is a particular instance of the generalized moment problem (1.1).

Lemma 11.1. *Let* $\mathbf{D} = \mathrm{co}(\mathbb{K})$, $f \in C(\mathbb{K})$ *and* $\widetilde{f}$ *be as in (11.1). Then the convex envelope* $\widehat{f}$ *of* $\widetilde{f}$ *is given by:*

$$\widehat{f}(\mathbf{x}) = \begin{cases} \rho^1_{\mathrm{mom}}(\mathbf{x}), & \text{if } \mathbf{x} \in \mathbf{D}, \\ +\infty, & \text{otherwise}, \end{cases} \tag{11.3}$$

and so, $\mathbf{D} = \mathrm{dom}\,\widehat{f}$.

Proof. With $\mathbf{x} \in \mathbf{D}$ fixed, let $\Delta_{\mathbf{x}}(\mathbb{K}) \subset \mathscr{P}(\mathbb{K})$ be the set of probability measures μ on $\mathbb{K}$ centered at $\mathbf{x}$ (that is $\langle \mu, z_i \rangle = x_i$ for $i = 1, \ldots, n$), and let $\Delta^*_{\mathbf{x}}(\mathbb{K}) \subset \Delta_{\mathbf{x}}(\mathbb{K})$ be its subset of probability measures that have a finite support. Then:

$$\widehat{f}(\mathbf{x}) = \inf_{\mu \in \Delta^*_{\mathbf{x}}(\mathbb{K})} \langle \mu, f \rangle, \quad \forall \mathbf{x} \in \mathbf{D}.$$

(See Choquet (1969).) Next, since $\Delta^*_{\mathbf{x}}(\mathbb{K})$ is dense in $\Delta_{\mathbf{x}}(\mathbb{K})$ with respect to the weak $\star$ topology, and $\mathscr{P}(\mathbb{K})$ is metrizable and compact with respect to the same topology (see Choquet (1969)),

$$\widehat{f}(\mathbf{x}) = \min_{\mu \in \Delta_{\mathbf{x}}(\mathbb{K})} \langle \mu, f \rangle = \rho^1_{\mathrm{mom}}(\mathbf{x}),$$

for every $\mathbf{x} \in \mathbf{D}$. If $\mathbf{x} \notin \mathbf{D}$, there is no probability measure on $\mathbb{K}$, with finite support, and centered in $\mathbf{x}$; therefore $\widehat{f}(\mathbf{x}) = +\infty$. □

Next, let $p, q \in \mathbb{R}[\mathbf{x}]$, with $q > 0$ on $\mathbb{K}$, and let $f \in C(\mathbb{K})$ be defined as

$$\mathbf{x} \mapsto f(\mathbf{x}) = p(\mathbf{x})/q(\mathbf{x}), \qquad \mathbf{x} \in \mathbb{K}. \tag{11.4}$$

For every $\mathbf{x} \in \mathbf{D}$ fixed, consider the infinite-dimensional LP,

$$\begin{aligned} \rho_{\mathrm{mom}}(\mathbf{x}) = \inf_{\mu \in \mathscr{M}(\mathbb{K})_+} \; & \langle \mu, p \rangle \\ \text{s.t.} \; & \langle \mu, z_i\, q \rangle = x_i, \quad i = 1, \ldots, n \\ & \langle \mu, q \rangle \;\; = 1. \end{aligned} \tag{11.5}$$

The dual of (11.5) is the infinite-dimensional LP

$$\rho_{\mathrm{pop}}(\mathbf{x}) = \sup_{\gamma \in \mathbb{R}, \boldsymbol{\lambda} \in \mathbb{R}^n} \{\gamma + \langle \boldsymbol{\lambda}, \mathbf{x} \rangle : p(\mathbf{y}) - q(\mathbf{y})\langle \boldsymbol{\lambda}, \mathbf{y} \rangle \geq \gamma q(\mathbf{y}), \quad \forall \mathbf{y} \in \mathbb{K}\}, \tag{11.6}$$

Equivalently, as $q > 0$ everywhere on $\mathbb{K}$, and $f = p/q$ on $\mathbb{K}$,

$$\rho_{\text{pop}}(\mathbf{x}) = \sup_{\gamma\in\mathbb{R},\boldsymbol{\lambda}\in\mathbb{R}^m} \{\gamma + \langle\boldsymbol{\lambda},\mathbf{x}\rangle : \quad f(\mathbf{y}) - \langle\boldsymbol{\lambda},\mathbf{y}\rangle \geq \gamma, \quad \forall\,\mathbf{y}\in\mathbb{K}\}. \quad (11.7)$$

In view of the definition of $\tilde{f}$, notice that

$$f(\mathbf{y}) - \langle\boldsymbol{\lambda},\mathbf{y}\rangle \geq \gamma, \quad \forall\mathbf{y}\in\mathbb{K} \quad \Leftrightarrow \quad \tilde{f}(\mathbf{y}) - \langle\boldsymbol{\lambda},\mathbf{y}\rangle \geq \gamma, \quad \forall\mathbf{y}\in\mathbb{R}^n.$$

Hence (11.7) is just the dual of (11.2), for every $\mathbf{x}\in\mathbf{D}$, and so $\rho_{\text{pop}}(\mathbf{x}) \leq \rho^1_{\text{mom}}(\mathbf{x})$, for every $\mathbf{x}\in\mathbf{D}$. In fact we have the following well known result (that holds of course in a more general framework).

Theorem 11.2. *Let $p, q \in \mathbb{R}[\mathbf{x}]$ with $q > 0$ on $\mathbb{K}$, and let f be as in (11.4). Let $\mathbf{x}\in\mathbf{D}\,(=\text{co}(\mathbb{K}))$ be fixed, arbitrary, and let $\rho_{\text{mom}}(\mathbf{x})$ and $\rho_{\text{pop}}(\mathbf{x})$ be as in (11.5) and (11.7), respectively.*

Then both (11.5) and (11.7) have an optimal solution and there is no duality gap, i.e.,

$$\rho_{\text{pop}}(\mathbf{x}) = \rho_{\text{mom}}(\mathbf{x}) = \rho^1_{\text{mom}}(\mathbf{x}) = \widehat{f}(\mathbf{x}), \quad \forall\,\mathbf{x}\in\mathbf{D}. \qquad (11.8)$$

In fact, the above result can be completed. Let $f^* : \mathbb{R}^n \to \mathbb{R}$ be the Legendre-Fenchel conjugate of $\widetilde{f}$, i.e.,

$$\boldsymbol{\lambda} \mapsto f^*(\boldsymbol{\lambda}) := \sup_{\mathbf{y}\in\mathbb{R}^n} : \{\langle\boldsymbol{\lambda},\mathbf{y}\rangle - \widetilde{f}(\mathbf{y})\}.$$

Let $\partial\widehat{f}(\mathbf{x})$ denote the *subdifferential* of $\widehat{f}$ at the point $\mathbf{x}$. From Theorem 23.4 in Rockafellar (1970), $\partial\widehat{f}(\mathbf{x}) \neq \emptyset$ at least for every $\mathbf{x}$ in the relative interior of $\mathbf{D}$.

Corollary 11.3. *Let $\mathbf{x}\in\mathbf{D}$ be fixed, arbitrary, and let $\rho_{\text{pop}}(\mathbf{x})$ be as in (11.7).*

(a) *(11.7) is solvable if and only if $\partial\widehat{f}(\mathbf{x}) \neq \emptyset$, in which case any optimal solution $(\boldsymbol{\lambda}^*, \gamma^*)$ satisfies:*

$$\boldsymbol{\lambda}^* \in \partial\widehat{f}(\mathbf{x}), \quad \textit{and} \quad \gamma^* = -f^*(\boldsymbol{\lambda}^*). \qquad (11.9)$$

(b) *If f is as in (11.4), then $\partial\widehat{f}(\mathbf{x}) \neq \emptyset$ for every $\mathbf{x}$ in $\mathbf{D}$, and so, (11.7) is solvable and (a) holds for every $\mathbf{x}$ in $\mathbf{D}$.*

11.1.2 *Semidefinite relaxations*

So let $\mathbb{K} \subset \mathbb{R}^n$ be the basic closed semi-algebraic set defined in (4.1). The infinite dimensional LP (11.5) is an obvious instance of the generalized moment problem (1.1). Therefore, in the present context, the semidefinite relaxations (4.5) associated with (11.5) read:

$$\begin{array}{rll} \rho_i(\mathbf{x}) = \inf\limits_{\mathbf{y}} & L_{\mathbf{y}}(p) & \\ \text{s.t.} & L_{\mathbf{y}}(z_k\, q) \;=\; x_k, \quad k=1,\ldots,n & \\ & \mathbf{M}_i(\mathbf{y}) \succeq \mathbf{0}, & \\ & \mathbf{M}_{i-v_j}(g_j\mathbf{y}) \succeq \mathbf{0},\; j=1,\ldots,m & \\ & L_{\mathbf{y}}(q) = 1, & \end{array} \tag{11.10}$$

while its dual reads:

$$\begin{array}{rl} \rho_i^*(\mathbf{x}) = \sup\limits_{\gamma,\boldsymbol{\lambda},(\sigma_j)} & \gamma + \langle \boldsymbol{\lambda}, \mathbf{x}\rangle \\ \text{s.t.} & p(\mathbf{z}) - (\langle \boldsymbol{\lambda}, \mathbf{z}\rangle + \gamma) q(\mathbf{z}) = \sigma_0(\mathbf{z}) + \displaystyle\sum_{j=1}^m \sigma_j(\mathbf{z})\, g_j(\mathbf{z}), \quad \forall \mathbf{z} \\ & \sigma_j \in \Sigma[\mathbf{z}], \quad j=0,\ldots,m \\ & \deg \sigma_0,\ \deg \sigma_j\, g_j \;\leq 2i. \end{array} \tag{11.11}$$

Theorem 11.4. *Let $\mathbb{K}$ be as in (4.1), and let Assumption 2.1 hold. Let f be as in (11.4) with $p, q \in \mathbb{R}[\mathbf{x}]$, and with $q > 0$ on $\mathbb{K}$. Let $\widehat{f}$ be as in (11.3), and with $\mathbf{x} \in \mathbf{D}$ ($= \mathrm{co}(\mathbb{K})$) fixed, consider the semidefinite relaxations (11.10)-(11.11). Then:*
(a) *The function $f_i : \mathbb{R}^n \to \mathbb{R} \cup \{+\infty\}$ defined by*

$$\mathbf{x} \mapsto f_i(\mathbf{x}) := \rho_i(\mathbf{x}), \qquad \mathbf{x} \in \mathbb{R}^n, \tag{11.12}$$

is convex, and as $i \to \infty$, $f_i(\mathbf{x}) \uparrow \widehat{f}(\mathbf{x})$ pointwise, for all $\mathbf{x} \in \mathbb{R}^n$.
(b) *If $\mathbf{D}$ has a nonempty interior $\mathrm{int}\,\mathbf{D}$, then (11.11) is solvable and:*

$$\rho_i^*(\mathbf{x}) \;=\; \rho_i(\mathbf{x}) \;=\; f_i(\mathbf{x}), \qquad \mathbf{x} \in \mathrm{int}\,\mathbf{D}, \tag{11.13}$$

and for every optimal solution $(\boldsymbol{\lambda}_i^*, \gamma_i^*)$ of (11.11),

$$f_i(\mathbf{z}) - f_i(\mathbf{x}) \;\geq\; \langle \boldsymbol{\lambda}_i^*, \mathbf{z} - \mathbf{x}\rangle, \qquad \forall\, \mathbf{z} \in \mathbb{R}^n,$$

that is, $\boldsymbol{\lambda}_i^* \in \partial f_i(\mathbf{x})$.

So, Theorem 11.13 not only provides a convex approximation f_i of $\widehat{f}$ with monotone pointwise convergence $f_i(\mathbf{x}) \uparrow \widehat{f}(\mathbf{x})$ as $i \to \infty$, but it also provides a subgradient $\boldsymbol{\lambda} \in \partial f_i(\mathbf{x})$ at every point $\mathbf{x} \in \mathbf{D}$. We even get the stronger uniform convergence:

Corollary 11.5. *Let $\mathbb{K}$ be as in (4.1), and let Assumption 2.1 hold. Let f and $\widehat{f}$ be as in (11.4) and (11.3) respectively, and let $f_i : \mathbf{D} \to \mathbb{R}$, be as in Theorem 11.13.*

Then f_i is lower-semicontinuous. Moreover, for every compact $\Omega \subset \mathbf{D}$ on which $\widehat{f}$ is continuous,

$$\lim_{i \to \infty} \sup_{\mathbf{x} \in \Omega} |\widehat{f}(\mathbf{x}) - f_i(\mathbf{x})| = 0, \tag{11.14}$$

that is, the monotone nondecreasing sequence $\{f_i\}$ converges to $\widehat{f}$, uniformly on every compact on which $\widehat{f}$ is continuous.

Example 11.1. Consider the bivariate rational function $f : [-1,1]^2 \to \mathbb{R}$:

$$\mathbf{x} \mapsto f(\mathbf{x}) := \frac{x_1 x_2}{1 + x_1^2 + x_2^2}, \qquad \mathbf{x} \in [-1,1]^2,$$

on $[-1,1]^2$, displayed in Figure 11.1, with f_3 as well. In Figure 11.2 we

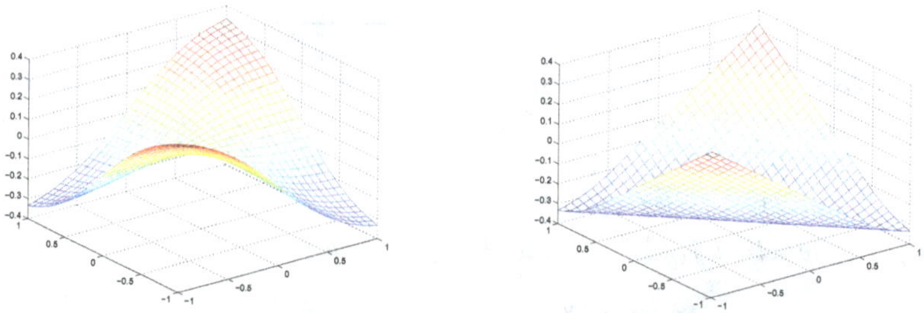

Fig. 11.1 Example 11.1, f and f_3 on $[-1,1]^2$.

have displayed $(f_3 - f_2)$ which is of the order 10^{-9} (which explains why for a few values of $x \in [-1,1]^2$ one may sometimes have $f_3(\mathbf{x}) \leq f_2(\mathbf{x})$ as we are at the limit of machine precision). It also means that again f_2 provides a very good approximation of the convex envelope $\widehat{f}$, that is, a very good approximation is already obtained at the first relaxation (i.e., here with $i = 2$) !

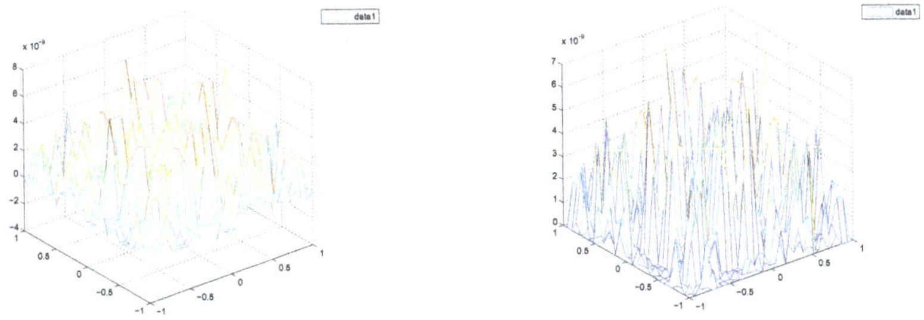

Fig. 11.2 Example 11.1, $f_3 - f_2$ and $(f_3 - f_2)^+$ on $[-1, 1]^2$.

11.2 Semidefinite Representation of Convex Sets

Semidefinite programming is a powerful optimization tool with a growing number of various applications. Therefore, an important issue is to characterize convex sets that have a *semidefinite representation* (SDr), and called SDr sets.

Definition 11.1. A convex set $\Omega \subset \mathbb{R}^n$ is SDr if there exist integers m, p and real $p \times p$ symmetric matrices $(\mathbf{A}_i)_{i=0}^n$, $(\mathbf{B}_j)_{j=1}^m$ such that:

$$\Omega = \{\mathbf{x} \in \mathbb{R}^n : \exists \mathbf{y} \in \mathbb{R}^m \text{ s.t. } \mathbf{A}_0 + \sum_{i=1}^n \mathbf{A}_i x_i + \sum_{j=1}^m \mathbf{B}_j y_j \succeq 0\}. \quad (11.15)$$

In other words, Ω is the linear projection on $\mathbb{R}^n$ of the convex set

$$\Omega' := \{(\mathbf{x}, \mathbf{y}) \in \mathbb{R}^n \times \mathbb{R}^m : \mathbf{A}_0 + \sum_{i=1}^n \mathbf{A}_i x_i + \sum_{j=1}^m \mathbf{B}_j y_j \succeq 0\}$$

of the *lifted* space $\mathbb{R}^{n+m}$, and Ω' is called a *semidefinite* representation (SDr) of Ω.

The set Ω' is also called a *lifted* SDr or a lifted LMI (Linear Matrix Inequality) because one sometimes needs additional variables $\mathbf{y} \in \mathbb{R}^m$ to obtain a description of Ω via appropriate LMIs. Here are examples of SDr sets:

- The intersection of half-spaces, i.e., a polyhedron $\{\mathbf{x} \in \mathbb{R}^n : \mathbf{A}\mathbf{x} \leq \mathbf{b}\}$, is a trivial example of convex sets whose SDr is readily available without lifting. Indeed $\mathbf{A}\mathbf{x} \leq \mathbf{b}$ is an LMI with diagonal matrices $\mathbf{A}_i$ in (11.15).
- The intersection of ellipsoids

$$\Omega := \{\mathbf{x} \in \mathbb{R}^n : \mathbf{x}'\mathbf{Q}_j\mathbf{x} + \mathbf{b}'\mathbf{x} + c_j \geq 0, \; j = 1, \ldots, m\}$$

(where $-\mathbf{Q}_j \succeq 0$ for all $j =, \ldots, m$) is a SDr set with lifted LMI representation in $\mathbb{R}^{(n+1)(n+2)/2-1}$:

$$\Omega' = \left\{ (\mathbf{x}, \mathbf{Y}) : \begin{array}{l} \begin{bmatrix} 1 & \mathbf{x}' \\ \mathbf{x} & \mathbf{Y} \end{bmatrix} \succeq 0 \\ \\ \mathrm{trace}\,(\mathbf{Q}_j \mathbf{Y}) + \mathbf{b}'\mathbf{x} + c_j \geq 0, \; j = 1, \ldots, m. \end{array} \right\}$$

- The epigraph of a univariate convex polynomial is a SDr set.
- Convex sets of $\mathbb{R}^2$ described from genus-zero plane curves are SDr sets; see Parrilo (2007).
- *Hyperbolic* cones obtained from 3-variables hyperbolic homogeneous polynomials are SDr sets; see the proof of the Lax conjecture in Lewis *et al.* (2005).

In this section, we consider semidefinite representations for the convex hull $\mathrm{co}(\mathbb{K})$ of compact basic semi-algebraic sets $\mathbb{K} \subset \mathbb{R}^n$ as described in (4.1). We show that $\mathrm{co}(\mathbb{K})$ has a SDr if the defining polynomials $(g_j) \subset \mathbb{R}[\mathbf{x}]$ of $\mathbb{K}$ satisfy what we call the *Putinar's Bounded Degree Representation* (P-BDR) of *affine* polynomials.

However, for general basic semi-algebraic sets $\mathbb{K}$, one cannot expect that the P-BDR property holds (if it ever holds) for *nice* values of the order r. Indeed otherwise one could minimize *any* affine polynomial on $\mathbb{K}$ *efficiently*. So from a practical point of view, the most interesting case is essentially the convex case, that is, when $\mathbb{K}$ is convex, and even more, when the defining polynomials g_j in (2.10) are concave, in which case one may hope that the P-BDR property holds for nice values of r.

11.2.1 *Semidefinite representation of* $\mathrm{co}(\mathbb{K})$

Let $\mathbb{K} \subset \mathbb{R}^n$ be the basic closed semi-algebraic set defined in (4.1) for some polynomials $g_j \in \mathbb{R}[\mathbf{x}]$, $j = 1, \ldots, m$. Recall that $Q(g) = \{\sum_{j=0}^m \sigma_j \, g_j : \sigma_j \in \Sigma[\mathbf{x}]\}$ is the *quadratic module* generated by the g_j's (with the convention

$g_0 = 1$), and given $r \in \mathbb{N}$, define $Q_r(g) \subset Q(g)$ to be the set

$$Q_r(g) := \{\sum_{j=0}^{m} \sigma_j\, g_j : \quad \sigma_j \in \Sigma[\mathbf{x}],\ \deg \sigma_j + \deg\, g_j \leq 2r\}. \tag{11.16}$$

For an affine polynomial $\mathbf{x} \mapsto f_0 + \sum_{i=1}^{n} f_i x_i$, let $(f_0, \mathbf{f}) \in \mathbb{R} \times \mathbb{R}^n$ be its vector of coefficients.

Definition 11.2 (P-BDR property). Given a compact set $\mathbb{K} \subset \mathbb{R}^n$ defined as in (4.1), the *Putinar's Bounded Degree Representation* (P-BDR) of affine polynomials holds for $\mathbb{K}$ if there exists $r \in \mathbb{N}$ such that

$$[\ f \text{ affine and positive on } \mathbb{K}\ ] \quad \Rightarrow \quad f \in Q_r(g), \tag{11.17}$$

except perhaps for a set of vectors $\mathbf{f}$ in $\mathbb{R}^n$ with Lebesgue measure zero. Call r its order.

Recall that if Assumption 2.1 holds for the g_j's that define $\mathbb{K}$ then by Theorem 2.14, if $f \in \mathbb{R}[\mathbf{x}]$ is positive on $\mathbb{K}$ then $f \in Q_r(g)$ for some $r = r_f$ that *depends* on f. In Definition 11.2, the P-BDR property requires that $r_f < r$ for almost all affine $f \in \mathbb{R}[\mathbf{x}]$, positive on $\mathbb{K}$.

Theorem 11.6. *Let $\mathbb{K} \subset \mathbb{R}^n$ be compact and defined as in (4.1) and let $s(2r) := \binom{n+2r}{n}$.*

If the P-BDR property holds for $\mathbb{K}$ with order r, then $\mathrm{co}(\mathbb{K})$ is a SDr set and the convex set

$$\Omega_r := \left\{ (\mathbf{x}, \mathbf{y}) \in \mathbb{R}^n \times \mathbb{R}^{s(2r)} : \begin{array}{ll} \mathbf{M}_{r-r_j}(g_j\, \mathbf{y}) \succeq 0, & j = 0, 1, \ldots, m \\ L_{\mathbf{y}}(z_i) & = x_i,\ i = 1, \ldots, n \\ y_0 & = 1 \end{array} \right\} \tag{11.18}$$

is a semidefinite representation of $\mathrm{co}(\mathbb{K})$.

Notice that in Theorem 11.6, the SDr (11.18) of $\mathrm{co}(\mathbb{K})$ is given *explicitly* in terms of the data (g_j) that define $\mathbb{K}$.

We have already seen that the intersection $\mathbb{K}$ of half-spaces and/or ellipsoids is a SDr set. It is relatively easy to show that in both cases the P-BDR property holds for $\mathbb{K}$ with $r = 1$. We next consider two other nontrivial examples.

Example 11.2. Let $n = 2$ with g_j concave and $\deg g_j = 2$ or 4, for all $j = 1, \ldots, m$, so that $\mathbb{K}$ is convex. It is known that in general $\mathbb{K}$ is not representable by a LMI in the variables x_1 and x_2 only. For instance take $m = 1$ and $g_1(\mathbf{x}) = 1 - x_1^4 - x_2^4$. The rigid convexity condition of Helton and Vinnikov (2007) is violated, and so there is no SDr of $\mathbb{K}$ involving only $\mathbf{x}$. But on the other hand, $\mathbb{K}$ is known to be SDr.

Assume $\mathbb{K}$ is compact, Slater's condition holds[1] and let $f \in \mathbb{R}[\mathbf{x}]$ be affine and nonnegative on $\mathbb{K}$ with global minimum $f^* \geq 0$ on $\mathbb{K}$. Convexity along with Slater's condition implies that the KKT optimality conditions (5.25) hold at any global minimizer $\mathbf{x}^* \in \mathbb{K}$; see Section C.1. And so there exist nonnegative Lagrange multipliers $\boldsymbol{\lambda} \in \mathbb{R}^m_+$ such that the (convex) Lagrangian $L_f := f - f^* - \sum_{j=1}^m \lambda_j g_j$ has optimal value 0 and, in addition, $\mathbf{x}^* \in \mathbb{K}$ is a global minimizer of L_f on $\mathbb{R}^2$. Therefore, the polynomial L_f being nonnegative on $\mathbb{R}^2$ and being quadratic or quartic in 2 variables, is s.o.s. That is, $L_f = \sigma$ for some $\sigma \in \Sigma[\mathbf{x}]$ and $\deg \sigma \leq 4$. But then

$$f \,=\, f^* + L_f + \sum_{j=1}^m \lambda_j\, g_j \,\in\, Q_2(g)$$

because as $f^* \geq 0$, $f^* + L_f \in \Sigma[\mathbf{x}]$. That is, the P-BDR property holds for $\mathbb{K}$ with order $r = 2$. Hence, $\mathbb{K}$ has the SDr (11.18).

Example 11.3. Let $m = 2$ with

$$g_i(\mathbf{x}) \,=\, \mathbf{x}'\mathbf{A}_i\mathbf{x} + c_i\,, \qquad i = 1, 2, \tag{11.19}$$

for some real symmetric matrices $\mathbf{A}_i$, and vector $\mathbf{c} = (c_1, c_2) \in \mathbb{R}^2$.

Given a linear polynomial $f \in \mathbb{R}[\mathbf{x}]$ with coefficient vector $\mathbf{f} = (f_i)_{i=1}^n \in \mathbb{R}^n$, consider the semidefinite program:

$$\rho = \min_{\mathbf{y}} \{\, L_\mathbf{y}(f) \;:\; \mathbf{M}_1(\mathbf{y}) \succeq 0;\; L_\mathbf{y}(g_i) \geq 0;\; i = 1, 2;\; y_0 = 1\,\}. \tag{11.20}$$

Proposition 11.7. *Let $\mathbb{K} \subset \mathbb{R}^n$ be defined as in (4.1) with $m = 2$ and g_j as in (11.19), and let $\mathbf{Q}$ be as in (11.20). Assume that $\mathbb{K}$ is compact with nonempty interior and*

$$\lambda_1\mathbf{A}_1 + \lambda_2\mathbf{A}_2 \prec 0 \tag{11.21}$$

---

[1] Recall that Slater's condition holds if there is some $\mathbf{x}_0 \in \mathbb{K}$ such that $g_j(\mathbf{x}_0) > 0$ for all $j = 1, \ldots, m$.

for some $\boldsymbol{\lambda} = (\lambda_1, \lambda_2) \geq 0$*. Then* $\rho = f^*$ *for almost all* $\mathbf{f} \in \mathbb{R}^n$*, and so the P-BDR property holds for* $\text{co}(\mathbb{K})$ *with order* $r = 1$*. That is,* $\text{co}(\mathbb{K})$ *has the semidefinite representation (11.18).*

Outer approximation of $\text{co}(\mathbb{K})$ *by an SDr set*

Let $\|\mathbf{x}\|$ denote the Euclidean norm of $\mathbf{x} \in \mathbb{R}^n$. With $\mathbf{B}_0 := \{\mathbf{x} \in \mathbb{R}^n : \|\mathbf{x}\| \leq 1\}$ and given a compact set $\Omega \subset \mathbb{R}^n$ and $\rho > 0$,

$$\Omega + \rho\,\mathbf{B}_0 = \{\mathbf{x} \in \mathbb{R}^n : \inf_{\mathbf{y} \in \Omega} \|\mathbf{x} - \mathbf{y}\| \leq \rho\}.$$

Theorem 11.8. *Let* $\mathbb{K} \subset \mathbb{R}^n$ *be a compact set as defined in (4.1), and let Assumption 2.1 hold. For every fixed* $\epsilon > 0$ *there is an integer* $r_\epsilon \in \mathbb{N}$ *such that*

$$\text{co}(\mathbb{K}) \subseteq \mathbb{K}_\epsilon \subset \text{co}(\mathbb{K}) + \epsilon\mathbf{B}_0,$$

where $\mathbb{K}_\epsilon$ *is the SDr convex set*

$$\left\{\mathbf{x} \in \mathbb{R}^n : \exists\, \mathbf{y} \in \mathbb{R}^{s(2r_\epsilon)} \text{ s.t. } \begin{bmatrix} \mathbf{M}_{r_\epsilon - r_j}(g_j\,\mathbf{y}) \succeq 0, & j = 0, \ldots, m \\ L_{\mathbf{y}}(z_i) & = x_i,\ i = 1, \ldots, n \\ y_0 & = 1. \end{bmatrix}\right\} \tag{11.22}$$

In addition, bounds on r_ϵ *are available.*

Hence Theorem 11.8 shows that even if $\text{co}(\mathbb{K})$ is not SDr, an arbitrarily close approximation $\mathbb{K}_\epsilon$ of $\text{co}(\mathbb{K})$ is SDr. Moreover, the semidefinite representation (11.22) is explicit in terms of the polynomials g_j that define $\mathbb{K}$.

11.2.2 *Semidefinite representation of convex basic semi-algebraic sets*

As already mentioned, for a general non convex basic semi-algebraic set $\mathbb{K} \subset \mathbb{R}^n$, one cannot expect the P-BDR property to hold (if it ever holds) for *nice* values of the order r. Indeed otherwise one could minimize any affine polynomial on $\mathbb{K}$ *efficiently*. So from a practical point of view, the most interesting case is essentially when $\mathbb{K}$ is convex, and even more, when the defining polynomials (g_j) in (2.10) are concave.

Therefore in this section, $\mathbb{K} \subset \mathbb{R}^n$ defined in (4.1) is compact and convex. We will use the Lagrangian L_f (see (5.26) and (C.9)) associated with the polynomials (g_j) and a linear $f \in \mathbb{R}[\mathbf{x}]$, already encountered in Example 11.2.

Lemma 11.9. *Let $\mathbb{K} \subset \mathbb{R}^n$ in (4.1) be compact, and assume that the g_j's that define $\mathbb{K}$ are all concave and Slater's condition holds. Given $f \in \mathbb{R}[\mathbf{x}]$, let $f^* := \min_{\mathbf{x}\in\mathbb{K}} f(\mathbf{x})$.*

For every linear $f \in \mathbb{R}[\mathbf{x}]$ with $\|\mathbf{f}\| = 1$, there exists $\boldsymbol{\lambda}^f \in \mathbb{R}^m_+$ such that the Lagrangian

$$\mathbf{x} \mapsto L_f(\mathbf{x}) := f(\mathbf{x}) - f^* - \sum_{j=1}^{m} \lambda_j^f \, g_j(\mathbf{x}) \tag{11.23}$$

is nonnegative on $\mathbb{R}^n$. In addition

$$|f^*| \leq \tau_{\mathbb{K}}; \quad \lambda_j^f \leq M_{\mathbb{K}}, \quad \forall j = 1, \ldots, m, \tag{11.24}$$

where $\tau_{\mathbb{K}}$ and $M_{\mathbb{K}}$ are independent of f.

Theorem 11.10. *Let $\mathbb{K} \subset \mathbb{R}^n$ be compact and defined as in (4.1). Assume that the (g_j) that define $\mathbb{K}$ are all concave and Slater's condition holds. Given a linear polynomial $f \in \mathbb{R}[\mathbf{x}]$, let L_f be the Lagrangian defined in (11.23).*

If L_f is s.o.s. for every linear $f \in \mathbb{R}[\mathbf{x}]$, then the P-BDR property holds for $\mathbb{K}$ with order $r = \max_{i=1,\ldots,m} \lceil \deg g_j/2 \rceil$, and $\mathbb{K}$ is a SDr set. In addition, the convex set

$$\left\{ (\mathbf{x}, \mathbf{y}) \in \mathbb{R}^n \times \mathbb{R}^{s(2r)} : \begin{bmatrix} \mathbf{M}_r(\mathbf{y}) \succeq 0 \\ L_{\mathbf{y}}(g_j) \geq 0, \;\; j = 1, \ldots, m \\ L_{\mathbf{y}}(z_i) \;\; = x_i, \; i = 1, \ldots, n \\ y_0 \qquad = 1 \end{bmatrix} \right\} \tag{11.25}$$

is a SDr of $\mathbb{K}$.

The semidefinite representation (11.25) of the convex set $\mathbb{K}$ has two important advantages over the semidefinite representation (11.18) of co($\mathbb{K}$) in the non convex case:

- The index r in Theorem 11.10 is known in advance.

• Moreover, the semidefinite constraint $\succeq 0$ is only concerned with the moment matrix $\mathbf{M}_r(\mathbf{y})$, that is, it does not depend on the data (g_j) that define the set $\mathbb{K}$!

Therefore, in view of Theorem 11.10, an important issue to find sufficient conditions to ensure that the Lagrangian L_f is s.o.s., and if possible, conditions that can be checked directly from the data g_j. For instance, we have seen in Section 2.1. a sufficient conditions on the coefficients of a polynomial to ensure it is s.o.s. Another condition is concerned with the Hesssian of L_f.

Recall from Definition 2.3 that a polynomial $f \in \mathbb{R}[\mathbf{x}]$ is s.o.s.-convex if its Hessian $\nabla^2 f$ is s.o.s., i.e., $\nabla^2 f = \mathbf{W}\,\mathbf{W}'$ for some possibily non square matrix polynomial $\mathbf{W} \in \mathbb{R}[\mathbf{x}]^{n \times k}$.

Theorem 11.11. *Let $\mathbb{K} \subset \mathbb{R}^n$ be as in (4.1) and let Assumption 2.1 hold. Assume that the g_j's that define $\mathbb{K}$ are all concave and Slater's condition holds.*
(a) *If $-g_j$ is s.o.s.-convex for every $j = 1, \ldots, m$, then $\mathbb{K}$ is SDr with semidefinite representation (11.25).*
(b) *If every $-g_j$ is either s.o.s.-convex or satisfies $-\nabla^2 g_j \succ 0$ on $\mathbb{K} \cap \{\mathbf{x} : g_j(\mathbf{x}) = 0\}$, then there exists $r \in \mathbb{N}$ such that the set Ω_r in (11.18) is a semidefinite representation of $\mathbb{K}$.*

Proof. Let $f \in \mathbb{R}[\mathbf{x}]$ be linear and as $\mathbb{K}$ is compact, let $\mathbf{x}^*$ be a minimizer of f on $\mathbb{K}$. Then from the definition (11.23) of L_f, it follows that $L_f(\mathbf{x}^*) = 0$ and $\nabla L_f(\mathbf{x}^*) = 0$.

(a) As every $-g_j$ is s.o.s.-convex then so is L_f. Therefore, by Lemma 2.30, $L_f \in \Sigma[\mathbf{x}]$, and the conclusion follows from Theorem 11.10.

(b) As $L_f(\mathbf{x}^*) = 0$ and $\nabla L_f(\mathbf{x}^*) = 0$,

$$L_f(\mathbf{x}) = (\mathbf{x} - \mathbf{x}^*)' \underbrace{\int_0^1 \int_0^t \nabla^2 L_f(\mathbf{x}^* + s(\mathbf{x} - \mathbf{x}^*))dsdt}_{\mathbf{F}(\mathbf{x},\mathbf{x}^*)} (\mathbf{x} - \mathbf{x}^*).$$

Observe that $\nabla^2 L_f = -\sum_{j \in J} \lambda_j^f \nabla^2 g_j$ where $J := \{j \in \{1, \ldots, m\} : \lambda_j^f >$

$0\} \neq \emptyset$. If $-\nabla^2 g_j(\mathbf{u}) \succ 0$ on $\mathbb{K} \cap \{\mathbf{u} : g_j(\mathbf{u}) = 0\}$ then the matrix polynomial

$$(\mathbf{x}, \mathbf{u}) \mapsto \mathbf{F}_j(\mathbf{x}, \mathbf{u}) := \int_0^1 \int_0^t -\nabla^2 g_j(\mathbf{u} + s(\mathbf{x} - \mathbf{u}))\, dsdt$$

is positive definite on $\Delta_j := \{(\mathbf{x}, \mathbf{u}) \in \mathbb{K} \times \mathbb{K} : g_j(\mathbf{u}) = 0\}$. Indeed, it is obviously positive semidefinite on Δ_j because $-\nabla^2 g_j \succeq 0$ and $\mathbb{K}$ is convex. Next, with $(\mathbf{x}, \mathbf{u}) \in \Delta_j$ fixed, suppose that

$$\xi' \mathbf{F}_j(\mathbf{x}, \mathbf{u}) \xi := \int_0^1 \int_0^t -\xi' \nabla^2 g_j(\mathbf{u} + s(\mathbf{x} - \mathbf{u})) \xi \, dsdt = 0$$

for some $\xi \in \mathbb{R}^n$. Then necessarily $-\xi' \nabla^2 g_j(\mathbf{u} + s(\mathbf{x} - \mathbf{u})) \xi = 0$ for almost all $s \in [0, 1]$, in contradiction with $-\nabla^2 g_j(\mathbf{u}) \succ 0$. Hence $\mathbf{F}_j(\mathbf{x}, \mathbf{u}) \succ 0$ on Δ_j. As the smallest eigenvalue of $\mathbf{F}_j$ is continuous in $(\mathbf{x}, \mathbf{u})$ and Δ_j is compact, its minimum on Δ_j is positive, and so $\mathbf{F}_j(\mathbf{x}, \mathbf{u}) \succeq \delta_j \mathbf{I}_n$ for all $(\mathbf{x}, \mathbf{u}) \in \Delta_j$ and some $\delta_j > 0$ (where $\mathbf{I}_n$ is the $n \times n$ identity matrix). In particular, $\mathbf{F}_j(\mathbf{x}, \mathbf{x}^*) \succeq \delta_j \mathbf{I}_n$ for all $j \in J$ and all $\mathbf{x} \in \mathbb{K}$ because $(\mathbf{x}, \mathbf{x}^*) \in \Delta_j$. Hence by Theorem 2.22, for every $j \in J$,

$$\mathbf{x} \mapsto \mathbf{F}_j(\mathbf{x}, \mathbf{x}^*) = \mathbf{F}_{j0}(\mathbf{x}) + \sum_{k=1}^m \mathbf{F}_{jk}(\mathbf{x})\, g_k(\mathbf{x}), \tag{11.26}$$

for some s.o.s. matrix polynomials $(\mathbf{F}_{jk})_{k=0}^m$. Similarly, if $-g_j$ is s.o.s.-convex then by Lemma 2.29, $\mathbf{x} \mapsto \mathbf{F}_j(\mathbf{x}, \mathbf{x}^*)$ is s.o.s. and so can be written as in (11.26) (take $\mathbf{F}_{jk} = 0$, $k \geq 1$). Hence (with $g_0 = 1$)

$$\begin{aligned}
\mathbf{x} \mapsto L_f(\mathbf{x}) &= (\mathbf{x} - \mathbf{x}^*)' \left(\sum_{j \in J} \lambda_j^f \mathbf{F}_j(\mathbf{x}, \mathbf{x}^*) \right) (\mathbf{x} - \mathbf{x}^*) \\
&= \sum_{k=0}^m g_k(\mathbf{x}) \left(\sum_{j \in J} \lambda_j^f (\mathbf{x} - \mathbf{x}^*)' \mathbf{F}_{jk}(\mathbf{x}) (\mathbf{x} - \mathbf{x}^*) \right) \\
&= \sum_{k=0}^m \sigma_j(\mathbf{x})\, g_j(\mathbf{x}),
\end{aligned}$$

for some s.o.s. polynomials $(\sigma_j) \in \Sigma[\mathbf{x}]$. Importantly, notice that the degrees of the s.o.s. matrix polynomial weights $(\mathbf{F}_{jk})$ do *not* depend on f. They only depend on $\delta_j, \deg g_j$, and $\|\mathbf{F}_j\|$ on the compact set Δ_j; see Theorem 2.22. Hence the degree of the s.o.s. polynomials $(\sigma_j) \subset \Sigma[\mathbf{x}]$ is bounded uniformly in f by say $2r \in \mathbb{N}$. Therefore, for every linear polynomial $f \in \mathbb{R}[\mathbf{x}]$,

$$f(\mathbf{x}) - f^* = L_f(\mathbf{x}) + \sum_{j\in J} \lambda_j^f g_j(\mathbf{x}) = \sum_{k=0}^{m} \sigma_j'(\mathbf{x})\, g_j(\mathbf{x}),$$

for some s.o.s. $(\sigma_j') \subset \Sigma[\mathbf{x}]$ whose degree is bounded uniformly in f by $2r+2$. This implies that the P-BDR property holds for $\mathbb{K}$ with order $r+1$, and so, by Theorem 11.6, Ω_{r+1} in (11.18) is a SDr of $\mathbb{K}$. $\square$

Example 11.4. Consider the class of convex sets $\mathbb{K} \subset \mathbb{R}^n$ where $g_j \in \mathbb{R}[\mathbf{x}]$ is concave and separable for every $j = 1, \ldots, m$. That is, for every $j = 1, \ldots, m$, g_j is concave and $g_j = \sum_{k=1}^n g_{jk}(x_k)$ for some univariate polynomials $g_{jk} \in \mathbb{R}[x_k]$, $k = 1, \ldots, n$. We have seen in Example 2.5 that $-g_j$ is s.o.s.-convex. Therefore by Theorem 11.11, $\mathbb{K}$ has the SDr (11.25).

In particular, letting $\|\mathbf{x}\|_d := (\sum_{i=1}^n x_i^{2d})^{1/2d}$, the d-Euclidean ball $\mathbb{K} := \{\mathbf{x} \in \mathbb{R}^n : \|\mathbf{x}\|_d^{2d} \leq 1\}$ is SDr for all $d \geq 1$.

Outer approximation of $\mathbb{K}$ by an SDr set

In case where we cannot prove that $\mathbb{K}$ is SDr, we next provide a convex SDr set $\mathbb{K}_\epsilon$ which is an outer approximation of $\mathbb{K}$, arbitrarily close to $\mathbb{K}$.

As $\mathbb{K}$ is compact, let $\tau_{\mathbb{K}} \in \mathbb{R}$ be such that:

$$\mathbf{x} \in \mathbb{K} \quad \Rightarrow \quad \|\mathbf{x}\| \leq \tau_{\mathbb{K}}. \tag{11.27}$$

Theorem 11.12. *Let $\mathbb{K} \subset \mathbb{R}^n$ as in (4.1) be compact, with $\tau_{\mathbb{K}}$ as in (11.27). Assume that the polynomials (g_j) that define $\mathbb{K}$ are all concave and Slater's condition holds. With $r \in \mathbb{N}$, $r \geq \lceil \deg g_j/2 \rceil$, $j = 1, \ldots, m$, let $\mathbf{x} \mapsto \Theta_r(\mathbf{x}) := \sum_{i=1}^n (x_i/\tau_{\mathbb{K}})^{2r}$. Then for every $\epsilon > 0$, there exists $r \in \mathbb{N}$ such that*

$$\mathbb{K} \subseteq \mathbb{K}_{\mathbf{r}} \subseteq \mathbb{K} + \epsilon\,\mathbf{B_0}, \tag{11.28}$$

where $\mathbb{K}_r$ is the SDr convex set

$$\left\{ \mathbf{x} \in \mathbb{R}^n : \exists\, \mathbf{y} \in \mathbb{R}^{s(2r)} \text{ s.t. } \begin{bmatrix} \mathbf{M}_r(\mathbf{y}) & \succeq 0 \\ L_{\mathbf{y}}(g_j) & \geq 0, \;\; j = 1, \ldots, m \\ L_{\mathbf{y}}(\Theta_r) & \leq 1 \\ L_{\mathbf{y}}(z_i) & = x_i, \; i = 1, \ldots, n \\ y_0 & = 1. \end{bmatrix} \right\} \tag{11.29}$$

Notice that like (11.25), the semidefinite representation (11.29) is explicit in terms of the g_j's and moreover, the semidefinite constraint $\succeq 0$ is only concerned with the moment matrix $\mathbf{M}_r(\mathbf{y})$, that is, it does *not* depend on the data (g_j) that define $\mathbb{K}$.

11.3 Algebraic Certificates of Convexity

In this final section we consider the problem of detecting whether some given basic semi-algebraic set $\mathbb{K}$ is convex. By detecting, we mean that if $\mathbb{K}$ is convex then one may obtain a *certificate* (or a proof) of convexity by some algorithm. The geometric characterization of convexity

$$\lambda \mathbf{x} + (1-\lambda)\mathbf{y} \in \mathbb{K}, \qquad \forall \mathbf{x}, \mathbf{y} \in \mathbb{K},\ \lambda \in (0,1) \tag{11.30}$$

is not a certificate because when this property holds, it cannot be checked by an algorithm.

Given the basic semi-algebraic set $\mathbb{K}$ defined in (2.10), let $\widehat{\mathbb{K}} \subset \mathbb{R}^n \times \mathbb{R}^n \times \mathbb{R}$ be the associated basic semi-algebraic set:

$$\widehat{\mathbb{K}} := \{(\mathbf{x}, \mathbf{y}, \lambda) \,:\, \hat{g}_j(\mathbf{x}, \mathbf{y}, \lambda) \geq 0,\ j = 1, \ldots, 2m+1\,\} \tag{11.31}$$

where:

$$\begin{aligned}
(\mathbf{x}, \mathbf{y}, \lambda) &\mapsto \hat{g}_j(\mathbf{x}, \mathbf{y}, \lambda) := g_j(\mathbf{x}), \quad j = 1, \ldots, m \\
(\mathbf{x}, \mathbf{y}, \lambda) &\mapsto \hat{g}_j(\mathbf{x}, \mathbf{y}, \lambda) := g_{j-m}(\mathbf{y}), \quad j = m+1, \ldots, 2m \\
(\mathbf{x}, \mathbf{y}, \lambda) &\mapsto \hat{g}_j(\mathbf{x}, \mathbf{y}, \lambda) := \lambda(1-\lambda), \quad j = 2m+1,
\end{aligned}$$

and let $P(\hat{g}) \subset \mathbb{R}[\mathbf{x}, \mathbf{y}, \lambda]$ be the preordering associated with the polynomials $\hat{g}_j$ that define $\widehat{\mathbb{K}}$ in (11.31), i.e.,

$$P(\hat{g}) := \left\{ \sum_{J \subseteq \{1,\ldots,2m+1\}} \sigma_J \left(\prod_{k \in J} \hat{g}_k \right) \,:\, \sigma_J \in \Sigma[\mathbf{x}, \mathbf{y}, \lambda] \right\}. \tag{11.32}$$

Our algebraic certificate of convexity is a follows.

Theorem 11.13. *Let $\mathbb{K} \subset \mathbb{R}^n$ be the basic semi-algebraic set defined in (2.10). Then $\mathbb{K}$ is convex if and only if for every $j = 1, \ldots, m$ and all $(\mathbf{x}, \mathbf{y}, \lambda) \in \mathbb{R}^n \times \mathbb{R}^n \times \mathbb{R}$:*

$$\theta_j(\mathbf{x}, \mathbf{y}, \lambda)\, g_j(\lambda\,\mathbf{x} + (1-\lambda)\,\mathbf{y}) = g_j(\lambda\,\mathbf{x} + (1-\lambda)\,\mathbf{y})^{2p_j} + h_j(\mathbf{x}, \mathbf{y}, \lambda) \tag{11.33}$$

for some polynomials $\theta_j, h_j \in P(\hat{g})$ and some integer $p_j \in \mathbb{N}$.

Proof. The set $\mathbb{K}$ is convex if and only if (11.30) holds, that is, if and only if for every $j = 1, \ldots, m$,

$$g_j(\lambda\mathbf{x} + (1-\lambda)\mathbf{y}) \geq 0, \qquad \forall \mathbf{x}, \mathbf{y} \in \mathbb{K},\ \lambda \in [0,1],$$

or, equivalently, if and only if for every $j = 1, \ldots, m$,

$$g_j(\lambda\mathbf{x} + (1-\lambda)\mathbf{y}) \geq 0, \qquad \forall (\mathbf{x}, \mathbf{y}, \lambda) \in \widehat{\mathbb{K}}. \tag{11.34}$$

But then (11.33) is just an application of Stengle's Nichtnegativstellensatz, i.e., Theorem 2.12(a), applied to (11.34). $\square$

The polynomials $\theta_j, h_j \in P(\hat{g})$, $j = 1, \ldots, m$, obtained in (11.33) indeed provide an obvious algebraic certificate of convexity for $\mathbb{K}$. This is because if (11.33) holds then for every $\mathbf{x}, \mathbf{y} \in \mathbb{K}$ and every $\lambda \in [0,1]$ one has $\theta_j(\mathbf{x}, \mathbf{y}, \lambda) \geq 0$ and $h_j(\mathbf{x}, \mathbf{y}, \lambda) \geq 0$ because $\theta_j, h_j \in P(\hat{g})$; and so $\theta_j(\mathbf{x}, \mathbf{y}, \lambda) g_j(\lambda\mathbf{x} + (1-\lambda)\mathbf{y}) \geq 0$. Therefore if $\theta_j(\mathbf{x}, \mathbf{y}, \lambda) > 0$ then $g_j(\lambda\mathbf{x}+(1-\lambda)\mathbf{y}) \geq 0$ whereas if $\theta_j(\mathbf{x}, \mathbf{y}, \lambda) = 0$ then $g_j(\lambda\mathbf{x}+(1-\lambda)\mathbf{y})^{2p_j} = 0$ which in turn implies $g_j(\lambda\mathbf{x} + (1-\lambda)\mathbf{y}) = 0$. Hence for every $j = 1, \ldots, m$, $g_j(\lambda\mathbf{x}+(1-\lambda)\mathbf{y}) \geq 0$ for every $\mathbf{x}, \mathbf{y} \in \mathbb{K}$ and every $\lambda \in [0,1]$, that is, (11.30) holds and so $\mathbb{K}$ is convex.

In principle, the algebraic certificate can be obtained numerically. Indeed, there is a bound on $p_j \in \mathbb{N}$, and the degrees of the s.o.s. weights σ_J in the representation (11.32) of the polynomial certificates $\theta_j, h_j \in P(\hat{g})$ in (11.33); see the discussion just after Theorem 2.12. Therefore, checking whether (11.33) holds reduces to checking whether some semidefinite program has a feasible solution. However the bound is so large that for practical implementation one should proceed as follows. Fix an *a priori* bound M on $p_j \in \mathbb{N}$ and on the degrees of the s.o.s. polynomial weights σ_J that define $h_j, \theta_j \in P(\hat{g})$; then check whether (11.33) holds true by solving the associated semidefinite program. If $\mathbb{K}$ is convex and the degrees of the certificates are small, then by increasing M one eventually finds a feasible solution. However, in practice such a certificate can be obtained only up to some machine precision, because of numerical inaccuracies inherent to semidefinite programming solvers.

Notice that the certificate in Theorem 11.13 involves polynomials in $\mathbb{R}[\mathbf{x}, \mathbf{y}, \lambda]$. We next provide another algebraic certificate that does not involve the variable λ, but for the class of basic semi-algebraic sets $\mathbb{K}$ whose defining polynomials (g_j) satisfy the following non degeneracy assumption on the boundary $\partial\mathbb{K}$.

Assumption 11.1 (Non-degeneracy). *For a basic semi-algebraic set $\mathbb{K} \subset \mathbb{R}^n$ as in (2.10), the polynomials $(g_j) \subset \mathbb{R}[\mathbf{x}]$ satisfy the nondegeneracy property if for every $j = 1, \ldots, m$, $\nabla g_j(\mathbf{x}) \neq 0$ whenever $\mathbf{x} \in \mathbb{K}$ and $g_j(\mathbf{x}) = 0$.*

We first have the following characterization of convexity:

Lemma 11.14. *Let $\mathbb{K}$ be as in (2.10) and let both Slater's condition and Assumption 11.1 hold. Then $\mathbb{K}$ is convex if and only if for every $j = 1, \ldots, m$,*

$$\langle \nabla g_j(\mathbf{y}), \mathbf{x} - \mathbf{y} \rangle \geq 0, \qquad \forall \mathbf{x}, \mathbf{y} \in \mathbb{K} \ \text{ with } g_j(\mathbf{y}) = 0. \tag{11.35}$$

Proof. The *only if part* is obvious. Indeed if $\langle \nabla g_j(\mathbf{y}), \mathbf{x} - \mathbf{y} \rangle < 0$ for some $\mathbf{x} \in \mathbb{K}$ and $\mathbf{y} \in \mathbb{K}$ with $g_j(\mathbf{y}) = 0$, then there is some $\bar{t} > 0$ such that $g_j(\mathbf{y} + t(\mathbf{x} - \mathbf{y})) < 0$ for all $t \in (0, \bar{t})$ and so the point $\mathbf{x}' := t\mathbf{x} + (1 - t)\mathbf{y}$ does not belong to $\mathbb{K}$, which in turn implies that $\mathbb{K}$ is not convex.

For the *if part*, (11.35) implies that at every point of the boundary, there exists a supporting hyperplane for $\mathbb{K}$. As $\mathbb{K}$ is closed with nonempty interior, the result follows from (Schneider, 1994, Theor. 1.3.3). □

As a consequence we obtain the following algebraic certificate of convexity.

Corollary 11.15 (Algebraic certificate of convexity). *Let $\mathbb{K}$ be as in (2.10), and let both Slater's condition and Assumption 11.1 hold. Then $\mathbb{K}$ is convex if and only if for every $j = 1, \ldots, m$,*

$$h_j \langle \nabla g_j(\mathbf{y}), \mathbf{x} - \mathbf{y} \rangle = \langle \nabla g_j(\mathbf{y}), \mathbf{x} - \mathbf{y} \rangle^{2l} + \theta_j + \varphi_j \, g_j \tag{11.36}$$

for some integer $l \in \mathbb{N}$, some polynomial $\varphi_j \in \mathbb{R}[\mathbf{x}, \mathbf{y}]$ and some polynomials h_j, θ_j in the preordering of $\mathbb{R}[\mathbf{x}, \mathbf{y}]$ generated by the family of polynomials $(g_k(\mathbf{x}), g_p(\mathbf{y}))$, $k, p \in \{1, \ldots, m\}$, $p \neq j$.

Proof. By Lemma 11.14, $\mathbb{K}$ is convex if and only if for every $j = 1, \ldots, m$, the polynomial $(\mathbf{x}, \mathbf{y}) \mapsto \langle \nabla g_j(\mathbf{y}), \mathbf{x} - \mathbf{y} \rangle$ is nonnegative on the set Ω_j defined by:

$$\Omega_j := \{(\mathbf{x}, \mathbf{y}) \in \mathbb{K} \times \mathbb{K} \, : \, g_j(\mathbf{y}) = 0 \}. \tag{11.37}$$

Then (11.36) follows from Theorem 2.12(a). □

So if $\mathbb{K}$ is convex then $(l, h_j, \theta_j, \varphi_j)$ provides us with the desired certificate of convexity, which in principle, can be obtained numerically as for the algebraic certificate (11.33). However, in practice such a certificate can be obtained only up to some machine precision, because of numerical inaccuracies inherent to semidefinite programming solvers. See the discussion after Theorem 11.13.

Observe that in Corollary 11.15, $\mathbb{K}$ is not necessarily compact. For compact basic semi-algebraic sets $\mathbb{K}$ that satisfy Assumption 2.1, one provides the following certificate.

Assumption 11.2 (Certificate of convexity). *Both Slater's condition and Assumption 11.1 hold for the set $\mathbb{K} \subset \mathbb{R}^n$ in (2.10). In addition, for every $j = 1, \ldots, m$, the polynomial g_j satisfies*

$$(\mathbf{x}, \mathbf{y}) \mapsto \langle \nabla g_j(\mathbf{y}), \mathbf{x} - \mathbf{y} \rangle = \sum_{k=0}^{m} \sigma_{jk}\, g_k(\mathbf{x}) + \sum_{k=0, k\neq j}^{m} \psi_{jk}\, g_k(\mathbf{y}) + \psi_j\, g_j(\mathbf{y}), \tag{11.38}$$

for some s.o.s. polynomials (σ_{jk}), $(\psi_{jk})_{k\neq j} \subset \Sigma[\mathbf{x}, \mathbf{y}]$, and some $\psi_j \in \mathbb{R}[\mathbf{x}, \mathbf{y}]$.

If Assumption 11.2 hold then $\mathbb{K}$ is convex because obviously (11.38) implies (11.35), which in turn, by Lemma 11.14, implies that $\mathbb{K}$ is convex.

By fixing an *a priori* bound $2d_j$ on the polynomial $(\sigma_{jk} g_k, \psi_{jk} g_k, \psi_j g_j)$, checking whether (11.38) holds reduces to solving a semidefinite program, simpler than for checking whether (11.36) holds. For instance, for every $j = 1, \ldots, m$, it suffices to solve the semidefinite program (recall that $r_k = \lceil (\deg g_k)/2 \rceil$, $k = 1 \ldots, m$)

$$\left\{ \begin{array}{ll} \rho_j := & \min_{\mathbf{z}} \; L_{\mathbf{z}}(\langle \nabla g_j(\mathbf{y}), \mathbf{x} - \mathbf{y} \rangle) \\ \text{s.t.} & \mathbf{M}_{d_j}(\mathbf{z}) \succeq 0 \\ & \mathbf{M}_{d_j - r_k}(g_k(\mathbf{x})\,\mathbf{z}) \succeq 0, \quad k = 1, \ldots, m \\ & \mathbf{M}_{d_j - r_k}(g_k(\mathbf{y})\,\mathbf{z}) \succeq 0, \quad k = 1, \ldots, m;\; k \neq j \\ & \mathbf{M}_{d_j - r_j}(g_j(\mathbf{y})\,\mathbf{z}) \;= 0 \\ & y_0 = 1. \end{array} \right. \tag{11.39}$$

If $\rho_j = 0$ for every $j = 1, \ldots, m$, then Assumption 11.2 holds. However, again because of the numerical inaccuracies inherent to the SDP solvers, one would only get $\rho_j \approx 0$; and so, this certificate of convexity is valid only up to machine precision.

Example 11.5. Consider the following simple illustrative example in $\mathbb{R}^2$:

$$\mathbb{K} := \{\mathbf{x} \in \mathbb{R}^2 : x_1x_2 - 1/4 \geq 0; 0.5 - (x_1 - 0.5)^2 - (x_2 - 0.5)^2 \geq 0\} \quad (11.40)$$

Obviously $\mathbb{K}$ is convex but its defining polynomial $\mathbf{x} \mapsto g_1(\mathbf{x}) := x_1x_2 - 1/4$ is not concave whereas $\mathbf{x} \mapsto g_2(\mathbf{x}) = 0.5 - (x_1 - 0.5)^2 - (x_2 - 0.5)^2$ is.

With $d_1 = 3$, solving (11.39) using GloptiPoly 3 yields the optimal value $\rho_1 \approx -4.58.10^{-11}$ which, in view of the machine precision for the SDP solvers used in GloptiPoly, could be considered to be zero, but of course with no guarantee. For $j = 2$ there is no test to perform because $-g_2$ being quadratic and convex yields

$$\langle \nabla g_2(\mathbf{y}), \mathbf{x} - \mathbf{y} \rangle = g_2(\mathbf{x}) - g_2(\mathbf{y}) + \underbrace{(\mathbf{x} - \mathbf{y})'(-\nabla^2 g_2(\mathbf{y}))(\mathbf{x} - \mathbf{y})}_{s.o.s.} \quad (11.41)$$

which is in the form (11.38) with $d_2 = 1$.

The next result states that under Assumption 11.2, the convex set $\mathbb{K}$ has a semidefinite representation. Let $d_j \in \mathbb{N}$ be such that $2d_j$ is larger than the maximum degree of the polynomials $\sigma_{jk}g_k, \psi_{jk}g_k, \psi_j g_j \in \mathbb{R}[\mathbf{x}, \mathbf{y}]$ in (11.38), $j = 1, \ldots, m$.

Theorem 11.16. *Let $\mathbb{K}$ be as in (2.10) and let both Assumption 2.1 and Assumption 11.2 hold. Then $\mathbb{K}$ is convex and with $r := \max_j d_j$, the set Ω_r in (11.18) is a semidefinite representation of $\mathbb{K}$.*

Example 11.6. Consider the convex set $\mathbb{K}$ of Example 11.5 for which the defining polynomial g_1 of $\mathbb{K}$ is not concave. We have seen that Assumption 11.2 holds (up to $\rho_1 \approx 10^{-11}$, close to machine precision) and $\max[d_1, d_2] = 3$. By Theorem 11.16, if ρ_1 would be exactly 0, the set

$$\Omega := \left\{ (\mathbf{x}, \mathbf{y}) \in \mathbb{R}^n \times \mathbb{R}^{s(6)} : \begin{cases} \mathbf{M}_3(\mathbf{y}) & \succeq 0 \\ \mathbf{M}_2(g_j\,\mathbf{y}) \succeq 0, & j = 1, 2 \\ L_{\mathbf{y}}(z_i) & = x_i, \quad i = 1, 2 \\ y_0 & = 1 \end{cases} \right. \quad (11.42)$$

would be a semidefinite representation of $\mathbb{K}$.

At least in practice, for every linear polynomial $f \in \mathbb{R}[\mathbf{x}]$, minimizing $L_{\mathbf{y}}(f)$ over Ω yields the desired optimal value $f^* := \min\{f(\mathbf{x}) : \mathbf{x} \in \mathbb{K}\}$, up to precision $M\rho_1 \approx -M10^{-11}$ for some constant M.

On the other hand, if $\mathbb{K}$ is convex there is no guarantee that (11.38) holds true and so, one may wonder how restrictive is Assumption 11.2. The following result provides some insight.

Corollary 11.17. *Let $\mathbb{K}$ in (2.10) be convex and let Assumption 2.1 and Slater's condition both hold. Assume that for every $j = 1, \ldots, m$, either $-g_j$ is s.o.s.-convex or $-g_j$ is convex on $\mathbb{K}$ and $-\nabla^2 g_j \succ 0$ on $\mathbb{K} \cap \{\mathbf{x} : g_j(\mathbf{x}) = 0\}$. Then Assumption 11.2 holds and so Theorem 11.16 applies.*

Hence from Corollary 11.17, it follows that Theorem 11.11 is a special case of Theorem 11.16.

11.4 Summary

In this chapter we have considered the problem of evaluating $\widehat{f}(\mathbf{x})$ at a point $\mathbf{x}$ for the convex envelope $\widehat{f}$ of a given rational function $f : \mathbb{K} \to \mathbb{R}$ on a compact basic semi-algebraic set $\mathbb{K}$. It turns out that for every $\mathbf{x}$ in the domain $\mathbf{D}$ of $\widehat{f}$, a hierarchy of semidefinite relaxations provide a monotone sequence $f_i(\mathbf{x}) \uparrow \widehat{f}(\mathbf{x})$ (in fact the convergence is even uniform on every compact subset of $\mathbb{K}$ where $\widehat{f}$ is continuous).

Then we have addressed the problem of finding an explicit semidefinite representation for co($\mathbb{K}$) (or $\mathbb{K}$ itself if convex) and provided several sufficient conditions for this to happen. Central is the so-called P-BDR property of the polynomials that define the set $\mathbb{K}$. Finally, we have also provided an algebraic certificate of convexity for arbitrary convex basic semi-algebraic sets, as well as two other certificates for convex basic semi-algebraic sets that satisfy a nondegeneracy property on their boundary. In principle they all can be obtained *numerically* via semidefinite programming, and so only up to machine precision, because of numerical inaccuracies inherent to SDP solvers.

11.5 Exercises

Exercise 11.1. Let $(s_{ji}) \subset \mathbb{N}$, $(g_{ji}) \subset \mathbb{R}_+$, and let $\mathbb{K} \subset \mathbb{R}^n$ be as in (2.10) with

$$\mathbf{x} \mapsto g_j(\mathbf{x}) := -\sum_{i=1}^{n} g_{ji} x_i^{2s_{ji}} + h_j(\mathbf{x}), \quad j = 1, \ldots, m,$$

where $h_j \in \mathbb{R}[\mathbf{x}]$ is a an affine polynomial, for every $j = 1, \ldots, m$. If $\mathbb{K}$ is compact, show that $\mathbb{K}$ has the semidefinite representation (11.25).

Exercise 11.2. Let Ω be as in (11.42) of Example 11.6. Show that for every linear polynomial $f \in \mathbb{R}[\mathbf{x}]$, minimizing $L_{\mathbf{y}}(f)$ over Ω yields the desired optimal value $f^* := \min \{f(\mathbf{x}) \,:\, \mathbf{x} \in \mathbb{K}\}$, up to $\rho_1 \approx -M10^{-11}$ for some constant M. (Hint: use a bound on the Lagrange KKT multipliers.)

11.6 Notes and Sources

11.1. The material in this section is from Laraki and Lasserre (2008a). Determining an index $i \in \mathbb{N}$ such that f_i approximates $\widehat{f}$ up to some prescribed error $\varepsilon > 0$ is an open problem.

For computing the convex envelope of a function f on a bounded domain, Brighi and Chipot (1994) propose triangulation methods and provide piecewise degree-1 polynomial approximations $f_h \geq \widehat{f}$, and derive estimates of $f_h - \widehat{f}$ (where h measures the size of the mesh). Another possibility is to view the problem as a particular instance of the general moment problem (like we do here), and use geometrical approaches as described in e.g. Anastassiou (1993) or Kemperman (1987); but, as acknowledged in the latter references, this approach is only practical for say, the univariate or bivariate cases.

11.2. Most of the material of that section is from Lasserre (2009a) and Lasserre (2009b). Theorem 11.11 is from Helton and Nie (2010). Theorem 11.13 is from Lasserre (2010) where algebraic certificates of convexity are discussed. The issue of characterizing SDr convex sets was raised in e.g. Ben-Tal and Nemirovski (2001) and Helton and Vinnikov (2007) who first show that *rigid convexity* is a necessary condition, also sufficient in dimension 2; this latter property is related to the Lax conjecture proved in Lewis *et al.* (2005). Parrilo (2007) exhibited specific semidefinite representations for genus-0 plane curves. In Helton and Nie (2010) and Helton and Nie (2009a) one may find several sufficient conditions (typically second order positive curvature conditions) for existence of semidefinite representations for basic semi-algebraic sets $\mathbb{K}$ (as well as for co($\mathbb{K}$) when $\mathbb{K}$ is not convex).

They also provide a nice technique to construct an explicit semidefinite representation for the convex hull $\mathrm{co}(\cup_{i=1}^{p} W_i)$ where each W_i already has a semidefinite representation. The authors even conjecture that every basic semi-algebraic convex set $\mathbb{K} \subset \mathbb{R}^n$ has a semidefinite representation. Finally, Chua and Tuncel (2008) consider even more general lifted conic representations of convex sets, called lifted G-representations (SDr being a special case) and discuss various geometric properties of convex sets admitting such lifted G-representations, as well as measures of "goodness" for such representations.

Chapter 12

Multivariate Integration

This chapter considers the problem of approximating multivariate integrals of rational functions or exponentials of a polynomial, on a compact basic semi-algebraic set. We also describe the moment approach as a tool for evaluating gradients or Hesssians in the maximum entropy approach for estimating an unknown density based on the knowledge of its moments.

Multivariate integrals arise in statistics, physics, and engineering and finance applications among other areas. For instance, such integrals are needed to compute probabilities over compact sets for multivariate normal random variables, and therefore, it is important to be able to approximate such integrals. In this chapter we are interested in the following two problems:

- Given a basic semi-algebraic set $\mathbb{K} \subset \mathbb{R}^n$, a finite Borel measure μ on $\mathbb{K}$ with all moments available, compute (or approximate) the integral $\int_{\mathbb{K}} f\, d\mu$ where f is a rational function p/q, with $p, q \in \mathbb{R}[\mathbf{x}]$ and $q > 0$ on $\mathbb{K}$.
- Given a polynomial $p \in \mathbb{R}[\mathbf{x}]$, compute (or approximate) the integral $\int_{\mathbf{B}} \exp p(\mathbf{x})\, d\mathbf{x}$ where $\mathbf{B} \subset \mathbb{R}^n$ is a simple set like e.g., the box $[-1, 1]^n$ or a simplex of $\mathbb{R}^n$.

As an application, we then apply the moment approach described for the latter problem to compute gradient and Hessian data needed in maximum entropy estimation.

12.1 Integration of a Rational Function

Let $\mathbb{K} \subset \mathbb{R}^n$ be a compact basic semi-algebraic set as defined in (4.1), and let μ be a given finite Borel measure on $\mathbb{K}$. As μ is supported on a compact set, it is moment determinate,[1] and so we assume that we are are given all its moments $\mathbf{z} = (z_{\boldsymbol{\alpha}})_{\boldsymbol{\alpha}\in\mathbb{N}^n}$. Typically, μ is a convex combination of standard distributions (uniform, (truncated) exponential, ...) on $\mathbb{K}$, so that in principle its moments can be obtained relatively easily.

With $f := p/q$ where $p, q \in \mathbb{R}[\mathbf{x}]$ and $q > 0$ on $\mathbb{K}$, one goal is to compute (or approximate) the integral $\int_{\mathbb{K}} f\,d\mu$, and more generally, to approximate finitely many moments of the new measure $d\nu := q^{-1}\,d\mu$. In the univariate case, a classical related problem is the computation of orthonormal polynomials[2] of $d\nu$ which is called the *modified measure.*

Lemma 12.1. *Let $\mathbb{K} \subset \mathbb{R}^n$ be compact, and let $q \in \mathbb{R}[\mathbf{x}]$ with $q > 0$ on $\mathbb{K}$. Let μ be a given finite Borel measure with support contained in $\mathbb{K}$. Then the only finite Borel measure φ with support in $\mathbb{K}$, solving the system of linear equations*

$$\int_{\mathbb{K}} \mathbf{x}^{\boldsymbol{\alpha}}\, q\, d\varphi = \int_{\mathbb{K}} \mathbf{x}^{\boldsymbol{\alpha}}\, d\mu, \qquad \forall\, \boldsymbol{\alpha} \in \mathbb{N}^n, \tag{12.1}$$

is $d\varphi = q^{-1}\,d\mu$. In particular, q^{-1} is the Radon-Nikodym derivative $[\frac{d\varphi}{d\mu}]$ of φ with respect to μ.

Proof. The linear system (12.1) has the particular solution $d\varphi = q^{-1}d\mu$ which has all its moments well defined because $q > 0$ on $\mathbb{K}$ and μ is finite with compact support. Moreover,

$$\int_{\mathbb{K}} \mathbf{x}^{\boldsymbol{\alpha}}\, q\, d\varphi = \int_{\mathbb{K}} \mathbf{x}^{\boldsymbol{\alpha}} q q^{-1} d\mu = \int_{\mathbb{K}} \mathbf{x}^{\alpha}\, d\mu, \qquad \forall \boldsymbol{\alpha} \in \mathbb{N}^n,$$

and so (12.1) holds.

We next prove that $d\varphi = q^{-1}d\mu$ is the only finite Borel measure solution of (12.1). So let (12.1) hold for some finite Borel measure φ with support in $\mathbb{K}$, and define ψ to be the Borel measure $q d\varphi$ (recall that $q > 0$ on $\mathbb{K}$), with all moments finite because $\mathbb{K}$ is compact. We thus have

$$\int_{\mathbb{K}} \mathbf{x}^{\boldsymbol{\alpha}}\, d\psi = \int_{\mathbb{K}} \mathbf{x}^{\boldsymbol{\alpha}}\, d\mu, \qquad \forall \boldsymbol{\alpha} \in \mathbb{N}^n.$$

Recall that Borel measures with compact support are moment determinate because by Weierstrass theorem, the space of polynomials is dense (for the

[1] See Definition 3.2.

[2] A family of univariate polynomials $(p_0, \ldots, p_d) \subset \mathbb{R}[x]$ is orthonormal with respect to some measure ν on $\mathbb{R}$ if $\int p_k p_t\, d\nu = \delta_{k=t}$, $\int x^k p_t d\nu = 0$ if $k < t$, and $\int x^t p_t d\nu > 0$, for all $0 \leq k, t \leq d$. This definition extends to the multivariate case; see Helton *et al.* (2008).

sup-norm) in the space of continuous functions on $\mathbb{K}$; therefore $\psi = \mu$. As ψ has compact support, let us define $d\nu := q^{-1}d\psi$, also well-defined because $q^{-1} > 0$ on $\mathbb{K}$, and ψ has its suppport contained in $\mathbb{K}$. Therefore, on the one hand,

$$\int_{\mathbb{K}} \mathbf{x}^{\boldsymbol{\alpha}}\, d\nu = \int_{\mathbb{K}} \mathbf{x}^{\boldsymbol{\alpha}} q^{-1} d\psi = \int_{\mathbb{K}} \mathbf{x}^{\boldsymbol{\alpha}} q^{-1} q\, d\varphi = \int_{\mathbb{K}} \mathbf{x}^{\boldsymbol{\alpha}}\, d\varphi, \qquad \forall \boldsymbol{\alpha} \in \mathbb{N}^n$$

(so that $\nu = \varphi$), whereas on the other hand, from $\psi = \mu$,

$$\int_{\mathbb{K}} \mathbf{x}^{\boldsymbol{\alpha}}\, d\nu = \int_{\mathbb{K}} \mathbf{x}^{\boldsymbol{\alpha}} q^{-1}\, d\psi = \int_{\mathbb{K}} \mathbf{x}^{\boldsymbol{\alpha}} q^{-1}\, d\mu, \quad \forall \boldsymbol{\alpha} \in \mathbb{N}^n.$$

Again, as Borel measures with compact support are moment determinate, it follows that $d\varphi = q^{-1}d\mu$, the desired result. □

We first consider the multivariable case $n > 1$ and then specialize to the one-dimensional case $n = 1$.

12.1.1 *The multivariable case*

Recalling that $\mathbf{z} = (z_{\boldsymbol{\alpha}})$ is the vector of moments of μ, consider the infinite-dimensional linear programs:

$$\begin{array}{rll} \rho^1_{\text{mom}} := & \displaystyle\sup_{\varphi \in \mathscr{M}(\mathbb{K})_+} & \displaystyle\int_{\mathbb{K}} p\, d\varphi \\ & \text{s.t.} & \displaystyle\int_{\mathbb{K}} q\, \mathbf{x}^{\boldsymbol{\alpha}}\, d\varphi = z_{\boldsymbol{\alpha}}, \qquad \forall \boldsymbol{\alpha} \in \mathbb{N}, \end{array} \tag{12.2}$$

and

$$\begin{array}{rll} \rho^2_{\text{mom}} := & \displaystyle\inf_{\varphi \in \mathscr{M}(\mathbb{K})_+} & \displaystyle\int_{\mathbb{K}} p\, d\varphi \\ & \text{s.t.} & \displaystyle\int_{\mathbb{K}} q\, \mathbf{x}^{\boldsymbol{\alpha}}\, d\varphi = z_{\boldsymbol{\alpha}}, \qquad \forall \boldsymbol{\alpha} \in \mathbb{N}. \end{array} \tag{12.3}$$

By Lemma 12.1, both linear programs (12.2) and (12.3) have the unique solution $d\varphi = q^{-1}\, d\mu$ and so,

$$\rho^1_{\text{mom}} = \rho^2_{\text{mom}} = \int_{\mathbb{K}} \frac{p}{q}\, d\mu.$$

Depending on parity, let $\deg q = 2v_0$ or $2v_0 - 1$, and with $\mathbb{K} \subset \mathbb{R}^n$ as in (4.1) let $\deg g_j = 2v_j$ or $2v_j - 1$, for all $j = 1, \ldots, m$. Consider the semidefinite programs:

$$\begin{aligned} \rho_i^1 := \sup_{\mathbf{y}} \; & L_{\mathbf{y}}(p) \\ \text{s.t.} \; & \mathbf{M}_i(\mathbf{y}) \succeq 0 \\ & \mathbf{M}_{i-v_j}(g_j\,\mathbf{y}) \succeq 0, \; j = 1, \ldots, m \\ & L_{\mathbf{y}}(q\,\mathbf{x}^{\boldsymbol{\alpha}}) \; = \; z_{\boldsymbol{\alpha}}, \quad |\boldsymbol{\alpha}| \leq 2i - \deg q, \end{aligned} \tag{12.4}$$

and

$$\begin{aligned} \rho_i^2 := \inf_{\mathbf{y}} \; & L_{\mathbf{y}}(p) \\ \text{s.t.} \; & \mathbf{M}_i(\mathbf{y}) \succeq 0 \\ & \mathbf{M}_{i-v_j}(g_j\,\mathbf{y}) \succeq 0, \; j = 1, \ldots, m \\ & L_{\mathbf{y}}(q\,\mathbf{x}^{\boldsymbol{\alpha}}) \; = \; z_{\boldsymbol{\alpha}}, \quad |\boldsymbol{\alpha}| \leq 2i - \deg q. \end{aligned} \tag{12.5}$$

The sequence (ρ_i^1) (resp. (ρ_i^2)) is monotone non increasing (resp. monotone non decreasing).

Theorem 12.2. *Let $\mathbb{K} \subset \mathbb{R}^n$ be as in (4.1) and Assumption 2.1 hold. Consider the semidefinite programs (12.4) and (12.5). Then, as $i \to \infty$,*

$$\rho_i^1 \downarrow \int_{\mathbb{K}} \frac{p}{q}\, d\mu \qquad \textit{and} \qquad \rho_i^2 \uparrow \int_{\mathbb{K}} \frac{p}{q}\, d\mu. \tag{12.6}$$

Proof. The semidefinite program (12.4) (resp. (12.5)) is the semidefinite relaxation (4.22) associated with the generalized moment problem (12.2) (resp. (12.3)), with countably many moment constraints. The assumptions of Theorem 4.3 are satisfied, and so (12.6) holds. □

12.1.2 *The univariate case*

We now consider the case $n = 1$. Let s be the degree of the polynomial $q \in \mathbb{R}[x]$, and let $I \subset \mathbb{R}[x]$ be the ideal generated by q. So, for every $k = 0, 1, \ldots$, write

$$x \mapsto x^k \; = \; r_k(x) + h_k(x) q(x), \quad k = 0, 1, \ldots \tag{12.7}$$

for some $h_k \in \mathbb{R}[x]$ and $r_k \in \mathbb{R}[x]/I$. Recall that $\mathbb{R}[x]/I$ is a $\mathbb{R}$-vector space of dimension s, with basis $1, x, \ldots, x^{s-1}$. Therefore, r_k has degree less than or equal to $s-1$, and reads

$$x \mapsto r_k(x) = \sum_{j=0}^{s-1} r_{kj}\, x^j,$$

for some vector of coefficients $\mathbf{r}_k \in \mathbb{R}^s$.

Next, let μ be a finite mesure on a compact basic semi-algebraic set $\mathbb{K} \subset \mathbb{R}$, with sequence of moments $\mathbf{z}$, and consider the sequence $\mathbf{y} = (y_k)$ defined as follows:

$$y_k = L_{\mathbf{y}}(r_k) + L_{\mathbf{y}}(h_k\, q) = \sum_{j=0}^{s-1} r_{kj} y_j + L_{\mathbf{z}}(h_k), \qquad k = 0, 1, \ldots, \tag{12.8}$$

with $r_k, h_k \in \mathbb{R}[x]$ as in (12.7) and where we use the constraint $L_{\mathbf{y}}(h_k q) = L_{\mathbf{z}}(h_k)$. Hence the semidefinite relaxations (12.4) and (12.5) simplify because they involve only s variables, namely the unkown moments $y_0, y_1, \ldots, y_{s-1}$. Indeed, the other moments y_k with $k \geq s$, are obtained via (12.8) as a linear combination of the first s moments $y_0, \ldots, y_{s-1}$, plus the constant term $L_{\mathbf{z}}(h_k) = \sum_j h_{kj}\, z_j$.

With $\mathbb{K} \subset \mathbb{R}$ being the interval $[a, b]$ (with $a, b \in \mathbb{R}$), let $g_1, g_2 \in \mathbb{R}[x]$ be the polynomials $x \mapsto g_1(x) := b - x$ and $x \mapsto g_2(x) := x - a$. With $i \geq \deg p$, consider the univariate semidefinite relaxation analogues of (12.4)-(12.5):

$$\begin{aligned} \rho_i^1 := \sup_{y_0, \ldots, y_{s-1}} \quad & L_{\mathbf{y}}(p) \\ \text{s.t. } & \mathbf{M}_i(\mathbf{y}) \succeq 0, \\ & \mathbf{M}_{i-1}(g_1 \mathbf{y}),\, \mathbf{M}_{i-1}(g_2 \mathbf{y}) \succeq 0 \end{aligned} \tag{12.9}$$

and

$$\begin{aligned} \rho_i^2 := \inf_{y_0, \ldots, y_{s-1}} \quad & L_{\mathbf{y}}(p) \\ \text{s.t. } & \mathbf{M}_i(\mathbf{y}) \succeq 0, \\ & \mathbf{M}_{i-1}(g_1\, \mathbf{y}),\, \mathbf{M}_{i-1}(g_2\, \mathbf{y}) \succeq 0, \end{aligned} \tag{12.10}$$

where in the Hankel (moment) matrix $\mathbf{M}_i(\mathbf{y})$ and localizing matrix $\mathbf{M}_{i-1}(g_j\, \mathbf{y})$, $j = 1, 2$, each entry y_k with $k \geq s$, is replaced with the expression (12.8), affine in $y_0, \ldots, y_{s-1}$. We then get the more specialized result:

Theorem 12.3. *Let $\mathbb{K} \subset \mathbb{R}$ be the interval $[a, b]$, and let $f := p/q$ with $p, q \in \mathbb{R}[x]$, and $q > 0$ on $\mathbb{K}$. Let μ be a finite Borel measure on $\mathbb{K}$, with sequence of moments $\mathbf{z}$, and consider the semidefinite relaxations (12.9) and (12.10). Then as $i \to \infty$,*

$$\rho_i^1 \downarrow \int_a^b f\,d\mu \quad \text{and} \quad \rho_i^2 \uparrow \int_a^b f\,d\mu.$$

It is important to emphasize that in the generic case where q^{-1} has only simple poles, one may decompose q^{-1} into a sum of elementary real fractions with denominators of degree at most 2. In this generic quadratic case, the semidefinite relaxations defined in (12.9) and (12.10) have only two variables, namely y_0 and y_1!

Example 12.1. Let $\mathbb{K} \subset \mathbb{R}$ be the interval $[a, b]$, with $a, b > 0$, and let $p = 1$ and $q = x$, so that $f = 1/x$. Hence in (12.7) we have

$$x^k = 0 + x^{k-1}\,x, \qquad k = 1, \ldots,$$

and so $r_k = 0$, and $h_k = x^{k-1}$, for all $k = 1, \ldots$. Let μ be the uniform distribution on $[a, b]$, and so

$$y_k = \int_a^b x^k\,dx = (b^{k+1} - a^{k+1})/(k+1), \qquad k = 0, 1, \ldots$$

We want to approximate $\int f d\mu = L_{\mathbf{y}}(1) = \int_a^b x^{-1} dx = \ln b/a$. The equation $L_{\mathbf{y}}(qx^k) = L_{\mathbf{z}}(x^k)$ yields

$$y_{k+1} = (b^{k+1} - a^{k+1})/(k+1), \quad k = 0, 1, \ldots$$

and so, the semidefinite relaxations (12.9) and (12.10) contain the single variable y_0 only! For instance:

$$\mathbf{M}_1(\mathbf{y}) = \begin{bmatrix} y_0 & b-a \\ b-a & (b^2-a^2)/2 \end{bmatrix};$$

$$\mathbf{M}_2(\mathbf{y}) = \begin{bmatrix} y_0 & b-a & (b^2-a^2)/2 \\ b-a & (b^2-a^2)/2 & (b^3-a^3)/3 \\ (b^2-a^2)/2 & (b^3-a^3)/3 & (b^4-a^4)/4 \end{bmatrix}.$$

With $a = 1, b = 2$, we observe a very fast convergence of both upper and lower bounds to the exact value $L_{\mathbf{y}}(1) = \ln 2 \approx 0.693147$; see Table 12.1.

Table 12.1 Example 12.1 : Upper and lower bounds ρ_i^1, ρ_i^2.

i	1	2	3	4
ρ_i^1	0.7000	0.693333	0.693152	0.693147
ρ_i^2	0.6666	0.692307	0.693122	0.693146

Example 12.2. Here $p = 1$ and $q = 1+x^2$, so that $f = 1/(1+x^2)$. We want to approximate $\int f d\mu = L_{\mathbf{y}}(1) = \int_a^b (1+x^2)^{-1} dx = \arctan(b) - \arctan(a)$. Using (12.8), the equation $L_{\mathbf{y}}(qx^k) = L_{\mathbf{z}}(x^k)$ yields

$$y_{k+2} = L_{\mathbf{y}}(r_k) + L_{\mathbf{z}}(h_k), \quad k = 0, 1, \ldots$$

In particular, $r_{2k}(x) = (-1)^k$ and $r_{2k+1}(x) = (-1)^k x$ for all $k = 1, \ldots$, and so, for instance, the moment matrix $M_2(\mathbf{y})$ reads

$$\mathbf{M}_2(\mathbf{y}) = \begin{bmatrix} y_0 & y_1 & -y_0 + (b-a) \\ y_1 & -y_0 + (b-a) & -y_1 + (b^2 - a^2)/2 \\ -y_0 + (b-a) & -y_1 + (b^2 - a^2)/2 & y_0 + (b^3 - a^3)/3 + (a-b) \end{bmatrix}.$$

With, $a = 1, b = 2$ one has $\int_1^2 f dx = L_{\mathbf{y}}(1) = \arctan(2) - \arctan(1) \approx 0.3217505$, and we obtain again a very good approximation with few moments; see Table 12.2.

Table 12.2 Example 12.2 : Upper and lower bounds ρ_i^1, ρ_i^2.

i	1	2	3	4
ρ_i^1	1.000	0.325	0.321785	0.321757
ρ_i^2	0.000	0.300	0.321260	0.321746

Example 12.3. Consider the Chebyshev measure $d\mu = (1-x^2)^{-1/2}\, dx$ on the interval $[-1, 1]$, and suppose that we want to approximate $\int_{-1}^1 (1 + x^2)^{-1}\, d\mu$. Results are displayed in Table 12.3. Notice that we already obtain very good bound with $i = 6$, i.e., with only 12 moments. The largest

Table 12.3 Example 12.3 : Upper and lower bounds ρ_i^1, ρ_i^2.

i	6	7	8	9
ρ_i^1	2.222102	2.221461	2.221461	2.221442
ρ_i^2	2.221328	2.221328	2.221438	2.221438

Table 12.4 Example 12.3 : Gauss-Kronrod quadrature formula.

n=2	n=3	n=4	n=5
2.22529479	2.22155480	2.22144480	2.22144156

semidefinite relaxation (12.9) with $i = 9$ has 2 variables and an LMI matrix size of 9×9. On the other hand, we have also computed another estimate via the Gauss-Kronrod quadrature formula which has $2n+1$ nodes (including ± 1) and is exact for polynomials up to degree $4n-1$; see (Gautschi, 1997, p. 167). The estimate with $n = 5$ gives the value 2.22144156 which, as expected, is between our upper and lower bounds.

12.2 Integration of Exponentials of Polynomials

Consider the following class of multivariate exponential integrals:

$$\rho_{\text{mom}} = \int_{\mathbb{K}} g(\mathbf{x})\, e^{h(\mathbf{x})}\, d\mathbf{x}, \tag{12.11}$$

where $\mathbf{x} \in \mathbb{R}^n$, $g, h \in \mathbb{R}[\mathbf{x}]$, and $\mathbb{K} \subset \mathbb{R}^n$ is a simple set like e.g. the box $\prod_{i=1}^n [a_i, b_i]$, or a simplex of $\mathbb{R}^n$. For clarity of exposition, we will only describe the approach for simple two-dimensional integrals on a box $[a, b] \times [c, d] \subset \mathbb{R}^2$. The multivariate case $n \geq 3$ essentially uses the same machinery but with more complicated and tedious notation.

12.2.1 *The moment approach*

Suppose that one wants to approximate:

$$\rho_{\text{mom}} = \int_{\mathbb{K}} g(x,y)\, e^{h(x,y)} dy\, dx \tag{12.12}$$

where $g, h \in \mathbb{R}[x,y]$, and $\mathbb{K} = [a,b] \times [c,d] \subset \mathbb{R}^2$. Consider the measure μ on $\mathbb{R}^2$ defined by

$$\mu(B) = \int_{\mathbb{K}\cap B} e^{h(x,y)} dy\, dx \qquad \forall\, B \in \mathscr{B}(\mathbb{R}^2), \tag{12.13}$$

and its sequence of moments $\mathbf{z} = (z_{\alpha\beta})$:

$$z_{\alpha\beta} = \int_{\mathbb{K}} x^\alpha y^\beta\, d\mu(x,y) = \int_a^b \int_c^d x^\alpha y^\beta e^{h(x,y)} dy\, dx \tag{12.14}$$

for all $(\alpha,\beta) \in \mathbb{N}^2$. Clearly, $\rho_{\text{mom}} = L_{\mathbf{z}}(g)$, where $L_{\mathbf{z}} : \mathbb{R}[x,y] \to \mathbb{R}$ is the usual linear functional

$$f\,(= \sum_{\alpha,\beta} f_{\alpha\beta} x^\alpha y^\beta) \quad \mapsto \quad L_{\mathbf{z}}(f) := \sum_{\alpha,\beta} f_{\alpha\beta}\, z_{\alpha\beta}.$$

Therefore we can compute ρ_{mom} once we have all necessary moments $\mathbf{z} = (z_{\alpha\beta})$. Integration by parts yields:

$$\begin{aligned} z_{\alpha\beta} = {} & \frac{1}{\beta+1} \int_a^b x^\alpha \left[y^{\beta+1}\, e^{h(x,y)} \right]_{y=c}^{y=d} dx \\ & - \frac{1}{\beta+1} \int_a^b \int_c^d x^\alpha y^{\beta+1} \frac{\partial h(x,y)}{\partial y}\, e^{h(x,y)}\, dy\, dx. \end{aligned}$$

If one writes $h(x,y) = \sum_{\gamma,\delta} h_{\gamma\delta} x^\gamma y^\delta$ then

$$z_{\alpha\beta} = \frac{d^{\beta+1}}{\beta+1} v_\alpha - \frac{c^{\beta+1}}{\beta+1} w_\alpha - \sum_{(\gamma,\delta)\in\mathbb{N}^2} \frac{\delta h_{\gamma\delta}}{\beta+1} z_{(\alpha+\gamma)(\beta+\delta)} \tag{12.15}$$

where $\mathbf{v} = (v_\alpha)$ and $\mathbf{w} = (w_\alpha)$ are the moments of the measures $d\nu = e^{h(x,d)} dx$ and $d\xi := e^{h(x,c)} dx$ on $[a,b]$, respectively.

Let $k, l \in \mathbb{R}[x]$ be the univariate polynomials $x \mapsto k(x) := h(x,d)$ and $x \mapsto l(x) := h(x,c)$. Integration by parts for v_α yields:

$$v_\alpha = \int_a^b x^\alpha \, e^{h(x,d)} dx = \frac{1}{\alpha+1}\left[x^{\alpha+1}\, e^{k(x)}\right]_{x=a}^{x=b} - \frac{1}{\alpha+1}\int_a^b x^{\alpha+1} k'(x)\, e^{k(x)} dx \qquad \forall \alpha \in \mathbb{N},$$

or, equivalently,

$$v_\alpha = \frac{b^{\alpha+1}\, e^{k(b)}}{\alpha+1} - \frac{a^{\alpha+1}\, e^{k(a)}}{\alpha+1} - \sum_{t\in\mathbb{N}} \frac{t\, k_t}{\alpha+1} v_{\alpha+t} \qquad \forall \alpha \in \mathbb{N}, \tag{12.16}$$

where (k_t) is the coefficient vector of the polynomial $k \in \mathbb{R}[x]$ of degree k_x.

Similarly:

$$w_\alpha = \frac{b^{\alpha+1}\, e^{l(b)}}{\alpha+1} - \frac{a^{\alpha+1}\, e^{l(a)}}{\alpha+1} - \sum_{t\in\mathbb{N}} \frac{t\, l_t}{\alpha+1} w_{\alpha+t} \quad \forall \alpha \in \mathbb{N}, \tag{12.17}$$

where (l_t) is the coefficient vector of the polynomial $l \in \mathbb{R}[x]$ of degree l_x.

In view of (12.16) and (12.17), all moments v_α and w_α are affine functions of $v_0, \ldots, v_{k_x-1}$, and $w_0, \ldots, w_{l_x-1}$, respectively.

The moment approach

Notice that in (12.12), the quantity ρ_{mom} to approximate, is a linear combination $L_{\mathbf{z}}(g)$ of moments of μ. Then, to compute upper and lower bounds on ρ_{mom}, it suffices to build up *two* hierarchies of semidefinite programs $\mathbf{Q}_i^u$ and $\mathbf{Q}_i^l$, as follows. Consider the vectors of moments $\mathbf{v}, \mathbf{w}$ and $\mathbf{z}$, up to order $2i$, $i \in \mathbb{N}$.

- The linear constraints of both $\mathbf{Q}_i^u$ and $\mathbf{Q}_i^l$ are obtained from the equations (12.15), (12.16), and (12.17), that contain only moments of order up to $2i$.
- The linear matrix inequality constraints of both $\mathbf{Q}_i^u$ and $\mathbf{Q}_i^l$ state necessary conditions on $\mathbf{v}, \mathbf{w}$ and $\mathbf{z}$, to be *moments* of some measures supported on $[a,b]$, $[a,b]$ and $[a,b] \times [c,d]$, respectively.

Then the semidefinite program $\mathbf{Q}_i^u$ (resp. $\mathbf{Q}_i^l$) maximizes (resp. minimizes) the linear criterion $L_{\mathbf{z}}(g)$ under the above constraints. Both are semidefinite relaxations of the original problem (12.12), and so their respective optimal values ρ_i^1 and ρ_i^2 provide upper and lower bounds on ρ_{mom}. In addition, the quality of both bounds increases with i because more and more constraints (12.15), (12.16), and (12.17), are taken into account.

12.2.2 *Semidefinite relaxations*

The support of the measure μ is the semiagebraic set $\{(x,y) \in \mathbb{R}^2 : \theta_i(x,y) \geq 0, \quad i = 1,2\}$, where $\theta_1, \theta_2 \in \mathbb{R}[x,y]$ are the polynomials

$$(x,y) \mapsto \begin{cases} \theta_1(x,y) := (b-x)(x-a) \\ \theta_2(x,y) := (d-y)(y-c) \end{cases}.$$

As θ_1, θ_2 are both quadratic, the necessary conditions for moment and localizing matrices associated with $\mathbf{z}$ read:

$$\mathbf{M}_i(\mathbf{z}) \succeq 0, \quad \mathbf{M}_{i-1}(\theta_k\,\mathbf{z}) \succeq 0, \quad k = 1,2. \tag{12.18}$$

Next, both measures ν and ξ are supported on the set $\{x \in \mathbb{R} : \theta_3(x) \geq 0\}$, with $x \mapsto \theta_3(x) := (b-x)(x-a)$, and so, analogues of (12.18) can be derived for $\mathbf{v}$ and $\mathbf{w}$. In addition, obvious bounds are available for z_0, v_0 and w_0. Namely:

$$\left.\begin{array}{ll} v_0 & \leq M_1 := (b-a)\sup_{x\in[a,b]}\, e^{|h(x,d)|} \\ w_0 & \leq M_2 := (b-a)\sup_{x\in[a,b]}\, e^{|h(x,c)|} \\ z_0 & \leq M_3 := (b-a)(d-c) \displaystyle\sup_{(x,y)\in[a,b]\times[c,d]} e^{|h(x,y)|} \end{array}\right\}. \tag{12.19}$$

Combining these necessary conditions and the linear relations for $\mathbf{z}$, $\mathbf{v}$, and $\mathbf{w}$ in (12.15), (12.16), and (12.17), one obtains an upper bound ρ_i^1 and a lower bound ρ_i^2 for ρ_{mom} by solving the following semidefinite programs:

$$\begin{aligned} \rho_i^1\,(\text{resp.}\,\rho_i^2) = & \sup_{\mathbf{z},\mathbf{v},\mathbf{w}}\ (\text{resp.}\ \inf_{\mathbf{z},\mathbf{v},\mathbf{w}}\ )\ L_{\mathbf{z}}(\mathbf{g}) \\ & \text{s.t.}\quad v_0 \leq M_1;\ w_0 \leq M_2;\ z_0 \leq M_3 \\ & \qquad \mathbf{M}_i(\mathbf{z}) \succeq 0, \mathbf{M}_{i-1}(\theta_j\,\mathbf{z}) \succeq 0, \quad j = 1,2 \\ & \qquad \mathbf{M}_i(\mathbf{v}) \succeq 0, \mathbf{M}_{i-1}(\theta_3\,\mathbf{v}) \succeq 0, \\ & \qquad \mathbf{M}_i(\mathbf{w}) \succeq 0, \mathbf{M}_{i-1}(\theta_3\,\mathbf{w}) \succeq 0, \end{aligned}$$

$$z_{\alpha\beta} = \frac{d^{\beta+1}}{\beta+1} v_\alpha - \frac{c^{\beta+1}}{\beta+1} w_\alpha - \sum_{(\gamma,\delta)\in\mathbb{N}^2} \frac{\delta h_{\gamma\delta}}{\beta+1} z_{(\alpha+\gamma)(\beta+\delta)},$$
$$\forall(\alpha,\beta) \in \mathbb{N}^2 : \alpha+\gamma+\beta+\delta \leq 2i.$$

$$v_\alpha = \frac{b^{\alpha+1}\, e^{k(b)}}{\alpha+1} - \frac{a^{\alpha+1}\, e^{k(a)}}{\alpha+1} - \sum_{t\in\mathbb{N}} \frac{t\, k_t}{\alpha+1} v_{\alpha+t},$$
$$\forall\alpha \in \mathbb{N} : \alpha + k_x \leq 2i.$$

$$w_\alpha = \frac{b^{\alpha+1}\, e^{l(b)}}{\alpha+1} - \frac{a^{\alpha+1}\, e^{l(a)}}{\alpha+1} - \sum_{t\in\mathbb{N}} \frac{t\, l_t}{\alpha+1} w_{\alpha+t}, \tag{12.20}$$
$$\forall\alpha \in \mathbb{N} : \alpha + l_x \leq 2i.$$

The sequence (ρ_i^1) (resp. (ρ_i^2)) is monotone non increasing (resp. monotone non decreasing).

Theorem 12.4. *Let ρ_i^1 and ρ_i^2 be as in (12.20) with M_j as in (12.19), $j = 1, 2, 3$. Then as $i\to\infty$,*

$$\rho_i^1 \downarrow \int_a^b \int_c^d g(x,y)\, e^{h(x,y)} dy\, dx \quad \text{and} \quad \rho_i^2 \uparrow \int_a^b \int_c^d g(x,y)\, e^{h(x,y)} dy\, dx.$$

Theorem 12.4 states that one can obtain an arbitrarily close approximation of ρ_{mom} by solving two hierarchies of semidefinite programs.

12.2.3 *The univariate case*

In this section we want to approximate the one-dimensional integral

$$\rho_{\text{mom}} = \int_0^1 g(x)\, e^{h(x)}\, dx, \tag{12.21}$$

where $x \in \mathbb{R}$, $g, h \in \mathbb{R}[x]$. Write $x \mapsto h(x) = \sum_{j=0}^d h_j\, x^j$ (with $h_d \neq 0$), and let $\mathbf{y} = (y_k)$ be the sequence of moments of the measure μ on $[0,1]$ with density $e^{h(x)}$ with respect to the Lebesgue measure. Integration by parts yields:

$$\int_0^1 x^k\, h'(x)\, e^{h(x)}\, dx = \left[x^k\, e^{h(x)}\right]_0^1 - k \int x^{k-1}\, e^{h(x)} dx, \tag{12.22}$$

for all $k = 0, 1, \ldots$. Therefore,

$$d\,h_d\,y_{d-1} \;=\; e^{h(1)} - e^{h(0)} - \sum_{j=1}^{d-1} j h_j\, y_{j-1} \tag{12.23}$$

and for every $k = 0, \ldots$

$$d\,h_d\,y_{d+k} \;=\; e^{h(1)} - (k+1)y_k - \sum_{j=1}^{d-1} j h_j\, y_{j+k}, \tag{12.24}$$

which shows that every moment y_k with $k \geq d-1$ is a linear combination of $\mathbf{y}(d) := (y_0, \ldots, y_{d-2})$, i.e.,

$$y_k \;=\; \langle \mathbf{v}_k, \mathbf{y}(d)\rangle, \qquad k \geq d-1, \tag{12.25}$$

for some vector $\mathbf{v}_k \in \mathbb{R}^{d-1}$, obtained from (12.23)-(12.24). Moreover,

$$y_0 \;\leq\; M \;:=\; \sup\,\{\, e^{|h(x)|} \;:\; x \in [0,1]\,\}. \tag{12.26}$$

With $i \geq \max[\deg g, d]$, consider the semidefinite program:

$$\begin{array}{rcl} \rho_i^1 \text{ (resp. } \rho_i^2) &=& \sup\limits_{\mathbf{y}(d)} \text{ (resp. } \inf\limits_{\mathbf{y}(d)}) \; L_{\mathbf{y}}(g) \\ && \text{s.t. } y_0 \leq M \\ && \quad \mathbf{M}_i(\mathbf{y}) \succeq 0 \\ && \quad \mathbf{M}_{i-1}(x(1-x)\,\mathbf{y}) \succeq 0 \\ && \quad y_k \;=\; \langle \mathbf{v}_k, \mathbf{y}(d)\rangle, \quad k = d-1, \ldots, 2i. \end{array} \tag{12.27}$$

Corollary 12.5. *Let ρ_i^1 and ρ_i^2 be as in (12.27) with M as in (12.26).*
(a) *Then as $i \to \infty$,*

$$\rho_i^1 \downarrow \int_0^1 g(x)\, e^{h(x)}\, dx \quad \text{and} \quad \rho_i^2 \uparrow \int_0^1 g(x)\, e^{h(x)}\, dx.$$

(b) *Let $\mathbf{y}^1(i)$ (resp. $\mathbf{y}^2(i)$) be an arbitrary optimal solution of the semidefinite programs (12.27) (garanteed to exist). Then for every $k \in \mathbb{N}$, and as $i \to \infty$,*

$$y_k^1(i) \to \int_0^1 x^k\, e^{h(x)}\, dx. \quad \text{and} \quad y_k^2(i) \to \int_0^1 x^k\, e^{h(x)}\, dx.$$

Proof. We briefly sketch the proof. Let $\mathbf{y}$ be a feasible solution of (12.27). Because of $y_0 \leq M$ and $\mathbf{M}_i(\mathbf{y}), \mathbf{M}_{i-1}(x(1-x)\mathbf{y}) \succeq 0$, one has $|y_k| \leq y_0 \leq M$ for every $k = 0, \ldots, 2i$. Hence, the feasible set is compact (as bounded and closed) and so, there is an optimal solution that we denote by $\mathbf{y}^1(i)$ in the "sup" case and $\mathbf{y}^2(i)$ in the "inf" case. By completing with zeros, one may consider $\mathbf{y}^j(i)$ as an element of l_∞, $j = 1, 2$, and even as an element of the M-ball of l_∞, which is compact in the weak $\star$ topology $\sigma(l_\infty, l_1)$. Therefore, there are subsequences $(i_p) \subset \mathbb{N}$, $(i_q) \subset \mathbb{N}$, and vectors $\mathbf{y}^1, \mathbf{y}^2 \in l_\infty$ such that, as p and $q \to \infty$,

$$y_j^1(i_p) \to y_j^1 \quad \text{and} \quad y_j^2(i_q) \to y_j^2, \qquad \forall j = 0, 1, \ldots \tag{12.28}$$

From the above convergence it follows that for every $i \in \mathbb{N}$, $\mathbf{M}_i(\mathbf{y}^k) \succeq 0$ and $\mathbf{M}_{i-1}(x(1-x)\mathbf{y}^k) \succeq 0$, $k = 1, 2$, which in turn implies that $\mathbf{y}^1$ (resp. $\mathbf{y}^2$) is the moment sequence of a measure μ^1 (resp. μ^2) on $[0, 1]$. In addition, with $\mathbf{y}^k(d) := (y_0^k, \ldots, y_{d-2}^k) \in \mathbb{R}^{d-1}$, $k = 1, 2$, and from the convergence (12.28), on obtains:

$$y_j^1 = \langle \mathbf{v}_j, \mathbf{y}^1(d) \rangle; \qquad y_j^2 = \langle \mathbf{v}_j, \mathbf{y}^2(d) \rangle, \quad \forall j \in \mathbb{N},$$

that is, (12.25) holds for $\mathbf{y}^1$ and $\mathbf{y}^2$. However, one may show that $d\mu = e^{h(x)}dx$ is the only measure on $[0, 1]$ whose moments $\mathbf{y}$ satisfy (12.25). Hence $\mu^1 = \mu^2 = \mu$, which shows that in fact the whole sequences $\mathbf{y}^1(i)$ and $\mathbf{y}^2(i)$ both converge to the same limit $\mathbf{y}^1 = \mathbf{y}^2 = \mathbf{y}$, and so (b) is proved. Finally (a) is an immediate consequence of (b). □

Notice that in the univariate case, one solves two hierarchies of semidefinite relaxations which contain $d-1$ variables only, as opposed to the multivariate case where the number of variables increases in the hierarchy.

12.3 Maximum Entropy Estimation

As a particular application of the above methodology, consider the *maximum entropy estimation* which is concerned with the following problem:

Let $f \in L_1([0, 1])$[3] be a nonnegative function only known via the first $2d + 1$ moments of its associated measure $d\mu = f dx$ on $[0, 1]$. From that partial knowledge one wishes (a) to provide an estimate f_d of f such that the first $2d + 1$ moments of $f_d dx$ match those of $f dx$, and (b) analyze

[3] $L_1([0, 1])$ denote the Banach space of integrable functions on the interval $[0, 1]$ of the real line, equipped with the norm $\|f\| = \int_0^1 |f(x)|\, dx$.

the asymptotic behavior of f_d when $d \to \infty$. This problem has important applications in various areas of physics, engineering, and signal processing in particular.

An elegant methodology is to search for f_d in a (finitely) parametrized family $\{f_d(\boldsymbol{\lambda}, x)\}$ of functions, and optimize over the unknown parameters $\boldsymbol{\lambda}$ via a suitable criterion. For instance, one may wish to select an estimate f_d that maximizes some appropriate *entropy*. Several choices of entropy functional are possible as long as one obtains a convex optimization problem in the finitely many coefficients λ_i's.

In this section, one chooses the Boltzmann-Shannon entropy, in which case the optimal estimate f_d^* is known to be the exponential of some polynomial of degree $2d$, whose coefficient vector $\boldsymbol{\lambda} \in \mathbb{R}^{2d+1}$ is then an optimal solution of a *convex* optimization problem. Therefore, results of the previous section can be used in any maximizing entropy algorithm to provide a numerical method to evaluate both gradient and Hessian of the function $\boldsymbol{\lambda} \mapsto \int_0^1 f_d(\boldsymbol{\lambda}, x)dx$, at each current iterate $\boldsymbol{\lambda}$. Its distinguishing feature is to avoid computing orthonormal polynomials of the measure $f_d(\boldsymbol{\lambda}; x)dx$.

12.3.1 *The entropy approach*

Consider the problem of estimating an unknown density $f : [0,1] \to \mathbb{R}_+$, based on the knowledge of its first $2d+1$ *moments*, $\mathbf{y} = (y_0, \ldots, y_{2d})$ only. That is,

$$y_j = \int_0^1 x^j\, f(x)\, dx, \qquad j = 0, \ldots, 2d,$$

where in general $y_0 = 1$ (as f is a density of some probability measure μ on $[0,1]$). The entropy approach is to compute an estimate f_d that maximizes some appropriate entropy, e.g., the usual Boltzmann-Shannon entropy $\mathscr{H} : L_1([0,1]) \to \mathbb{R} \cup \{-\infty\}$:

$$g \mapsto \mathscr{H}[g] := -\int_0^1 g(x)\, \ln g(x)\, dx,$$

a strictly concave functional. Therefore, the problem reduces to:

$$\sup_g \left\{ \mathscr{H}[g] : \int_0^1 x^j\, g(x)\, dx = y_j, \quad j = 0, \ldots, 2d \right\}. \tag{12.29}$$

The structure of this infinite-dimensional convex optimization problem permits to search for an optimal solution $g = f_d^*$ of the form

$$x \mapsto f_d^*(x) = \exp \sum_{j=0}^{2d} \lambda_j^* x^j,$$

where $\boldsymbol{\lambda}^* \in \mathbb{R}^{2d+1}$ is unique when f_d^* has to satisfy the constraints of (12.29). This is because the Legendre-Fenchel conjugate[4] ψ^* of the function

$$x \mapsto \psi(x) := \begin{cases} x \ln x & \text{if } x > 0 \\ 0 & \text{if } x = 0 \\ +\infty & \text{if } x < 0 \end{cases}$$

is $z \mapsto \psi^*(z) := \exp(z-1)$. And so, the conjugate functional $\mathscr{H}^*$ of $-\mathscr{H}$ is

$$g \mapsto \mathscr{H}^*(g) = \int_0^1 \exp(g(x) - 1)dx.$$

Therefore, given $\boldsymbol{\lambda} \in \mathbb{R}^{2d+1}$, define $f_d(\boldsymbol{\lambda}, \cdot) : \mathbb{R}_+ \to \mathbb{R}$ to be:

$$x \mapsto f_d(\boldsymbol{\lambda}, x) := \exp \sum_{j=0}^{2d} \lambda_j x^j, \qquad x \in \mathbb{R}. \tag{12.30}$$

Problem (12.29) reduces to the concave finite-dimensional optimization problem:

$$\mathbf{P}: \quad \sup_{\boldsymbol{\lambda} \in \mathbb{R}^{2d+1}} \left\{ \langle \mathbf{y}, \boldsymbol{\lambda} \rangle - \int_0^1 f_d(\boldsymbol{\lambda}, x)\, dx \right\}.$$

Notice that solving $\mathbf{P}$ is just evaluating g_d^* at the point $\mathbf{y}$, where g_d^* is the Legendre-Fenchel transform of the function $g_d : \mathbb{R}^{2d+1} \to \mathbb{R}$,

$$\boldsymbol{\lambda} \mapsto g_d(\boldsymbol{\lambda}) := \int_0^1 f_d(\boldsymbol{\lambda}, x)\, dx, \tag{12.31}$$

i.e., the mass of the measure $d\mu_d := f_d(\boldsymbol{\lambda}, x)dx$ on $[0, 1]$.

Lemma 12.6. *Let f_d, g_d be defined as in (12.30) and (12.31) respectively, and let $\mathbf{z}(\boldsymbol{\lambda}) = (z_k(\boldsymbol{\lambda})) \subset \mathbb{R}$ be the sequence:*

---

[4]Let $f : \mathbb{R}^n \to \mathbb{R} \cup \{+\infty\}$ with $f \not\equiv +\infty$ be such that there is an affine function minorizing f on $\mathbb{R}^n$. Then the function $\boldsymbol{\lambda} \mapsto f^*(\boldsymbol{\lambda}) := \sup\{\boldsymbol{\lambda}'\mathbf{x} - f(\mathbf{x}) \,:\, \mathbf{x} \in \mathbb{R}^n\}$ is called the Legendre-Fenchel conjugate of f. See e.g. (Hiriart-Urruty and Lemarechal, 1993, p. 37).

$$z_k(\boldsymbol{\lambda}) := \int_0^1 x^k f_d(\boldsymbol{\lambda}, x)\, dx, \quad k = 0, 1, \ldots$$

Then g_d *is convex with gradient* $\nabla g_d = (\partial g_d / \partial \lambda_k)_k$ *given by*

$$\boldsymbol{\lambda} \mapsto \frac{\partial g_d(\boldsymbol{\lambda})}{\partial \lambda_k} = \int_0^1 x^k f_d(\lambda, x)\, dx = z_k(\boldsymbol{\lambda}), \quad \forall k = 0, \ldots, 2d, \tag{12.32}$$

and Hessian $\nabla^2 g_d = (\frac{\partial^2 g_d}{\partial \lambda_j \partial \lambda_k})_{j,k}$ *given by*

$$\boldsymbol{\lambda} \mapsto \frac{\partial^2 g_d(\boldsymbol{\lambda})}{\partial \lambda_j \partial \lambda_k} = \int_0^1 x^{j+k} f_d(\lambda, x)\, dx = z_{j+k}(\boldsymbol{\lambda}), \quad \forall j, k = 0, \ldots, 2d. \tag{12.33}$$

12.3.2 *Gradient and Hessian computation*

Notice that computing the first $2d + 1$ moments $\mathbf{z}(\boldsymbol{\lambda})$ of the measure $d\mu_d = f_d(\boldsymbol{\lambda}, x)dx$ yields the gradient $\nabla g_d(\boldsymbol{\lambda})$, which permits to implement a first-order minimization algorithm. Computing an additional $2d$ moments provides us with the Hessian $\nabla^2 g_d(\boldsymbol{\lambda})$ as well. For instance, to solve $\mathbf{P}$ one may wish to implement Newton's method, which in view of (12.32)-(12.33), yields the iterates

$$\boldsymbol{\lambda}^{(k+1)} = \boldsymbol{\lambda}^{(k)} + [\nabla^2 g_d(\boldsymbol{\lambda}^{(k)})]^{-1}\, \hat{\mathbf{z}}(\boldsymbol{\lambda}^{(k)}), \qquad k = 0, 1, \ldots \tag{12.34}$$

where $\hat{\mathbf{z}}(\boldsymbol{\lambda}^{(k)}) = \mathbf{y} - \mathbf{z}(\boldsymbol{\lambda}^{(k)})$.

In fact, one only needs compute the vector

$$\mathbf{z}^d(\boldsymbol{\lambda}) = (z_0(\boldsymbol{\lambda}), \ldots, z_{2d-2}(\boldsymbol{\lambda}))$$

of first $2d-1$ moments because any other moment $z_k(\boldsymbol{\lambda})$ with $k \geq 2d-1$, can be expressed from $\mathbf{z}^d(\boldsymbol{\lambda})$ at no additional cost. This follows from (12.23)-(12.24) in substituting $e^{h(x)}$ with $f_d(\boldsymbol{\lambda}, x)$.

One possibility is to run $2d-1$ semidefinite relaxations (12.27) with $x \mapsto g(x) := x^k$ to get an approximation of $z_k(\boldsymbol{\lambda})$ for every $k = 0, \ldots, 2d-2$. But it is better to run a single semidefinite relaxation (12.27) (which maximizes some criterion $L_{\mathbf{z}}(g)$) for some index i sufficiently large. Indeed by Corollary 12.5(b), for every $k = 0, 1, \ldots$,

$$z_k^1(i) \to \int_0^1 x^k e^{h(x)}\, dx = \int_0^1 x^k f_d(\boldsymbol{\lambda}, x)\, dx = z_k(\boldsymbol{\lambda}),$$

Table 12.5 Example 12.4 : $f(x) = (1+x)^{-1}$ and $d = 1$.

k	moments			coefficients $\lambda^{(k)} \in \mathbb{R}^3$		
0	111.5	98.24	88.69	1.0000	1.000	5.000
1	41.28	36.25	32.69	0.3323	0.2306	5.4424
2	15.48	13.45	12.10	0.0569	-1.4160	6.3745
8	.7023	.3149	.2005	0.0125	-1.0867	0.4837
9	.6935	.3071	.1934	-0.0049	-0.9257	0.2461
10	.6931	.3068	.1931	-0.0055	-0.9205	0.2383
⋆	.6931	.3069	.1931			

as $i \to \infty$ (where $\mathbf{z}^1(i)$ is an optimal solution of the semidefinite relaxation (12.27)).

Example 12.4. Let $x \mapsto f(x) := (1+x)^{-1}$ be the unknown density. With $d = 1$ (i.e. $\boldsymbol{\lambda} \in \mathbb{R}^3$) and $i = 5$ in (12.27), i.e., with at most 10 moments, results are displayed in Table 12.5. The last line with a "⋆" displays the exact moments. Very good results are obtained after 10 iterations only of Newton's method (12.34). Gradient components are about $O(10^{-5})$ or $O(10^{-6})$, and the first seven moments of the measure $d\mu = f_d(\boldsymbol{\lambda}, x)dx$ read:

$$(0.6931, 0.3068, 0.1931, 0.1402, 0.1099, 0.0903, 0.0766),$$

a fairly good approximation. Finally, results with $d = 2$ (i.e. with $\boldsymbol{\lambda} \in \mathbb{R}^5$), are displayed in Table 12.6, again with very good results. However, we had to set $i = 7$ (i.e. 14 moments) for the semidefinite relaxations (12.27). In case the Hessian would be ill-conditioned when close to an optimal solution, more sophisticated second-order methods, or even first-order methods might be preferable for larger r. At iteration 8 the gradient is $O(10^{-10})$. Figure 12.1 displays $f_d - f$ on $[0, 1]$ because both curves of f_d and f are almost indistinguishable. Indeed, the scale in Figure 12.1 is 10^{-4}.

12.4 Summary

In this chapter, we have considered the moment approach to solve (or at least approximate) two types of multivariate integrals $J = \int_{\mathbb{K}} f d\mu$: Namely,

- When $\mathbb{K} \subset \mathbb{R}^n$ is a basic compact semi-algebraic set, f is a rational

Table 12.6 Example 12.4 : $f(x) = (1+x)^{-1}$ and $d = 2$.

k	moments					coefficients $\lambda^{(k)} \in \mathbb{R}^5$				
0	8.2861	6.1806	5.1359	4.4543	3.9579	0.7500	0.7500	0.7500	0.7500	0.7500
1	3.3332	2.3940	1.9635	1.6918	1.4974	0.2198	-0.0129	0.9905	0.7967	0.7684
2	1.5477	1.0112	0.8010	0.6781	0.5939	0.0171	-0.6777	0.3247	1.2359	0.8959
4	0.7504	0.3588	0.2408	0.1842	0.1507	0.0195	-1.5589	4.0336	-8.0020	5.5820
8	0.6931	0.3069	0.1931	0.1402	0.1098	-0.0001	-0.9958	0.4658	-0.2191	0.0562
$\star$	0.6931	0.3069	0.1931	0.1402	0.1098					

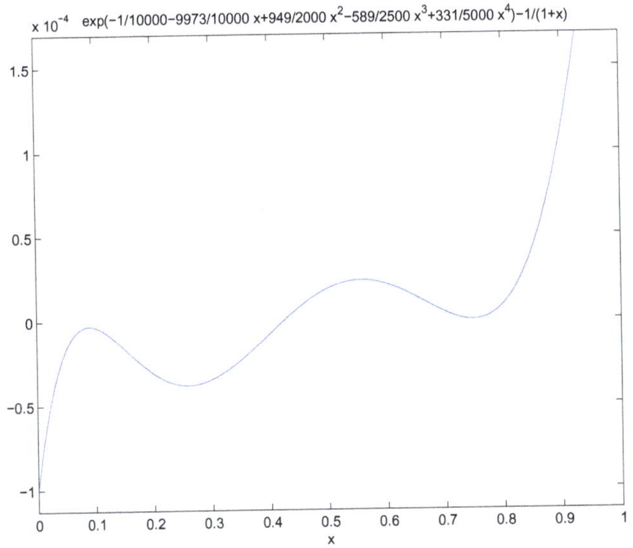

Fig. 12.1 Example 12.4, $d = 2$: $f_d - f$ on $[0, 1]$.

function p/q with $q > 0$ on $\mathbb{K}$, and μ is a given measure on $\mathbb{K}$ whose sequence of moments is known and available.

- When $\mathbb{K}$ is a simple set like e.g. the unit box $[-1, 1]^n \subset \mathbb{R}^n$, μ is the Lebesgue measure, and f is the exponential of some given polynomial.

In both cases we have defined a hierarchy of semidefinite relaxations that provide upper and lower bounds converging to the exact value J.

Finally, we have also considered the maximum entropy approach to estimate an unknown density $f \in L_1(\mathbb{K})$ from the knowledge of its moments.

When $\mathbb{K}$ is a simple set, then the approximation scheme developed earlier permits to evaluate gradient and Hessian data needed in an optimization algorithm for maximum entropy.

12.5 Exercises

Exercise 12.1. Let $n = 1$, $a > 0$, and with the Gaussian density $x \mapsto f(x) ::= \exp(-x^2/2)$, one wishes to approximate the integral $\int_0^a f(x)dx$. Apply the semidefinite relaxations (12.27) and compare with approximations already available in some Tables.

12.6 Notes and Sources

12.1 Most of the material in this section is from Lasserre (2009c). For a numerical approximation of $\int f d\mu$, the most popular methods are Monte Carlo schemes and (at least in the one-dimensional case $n = 1$) generalizations of Gauss quadrature formulas, i.e., formulas which integrate *exactly* in space of Laurent polynomials or more general rational functions. The former approach is described in Niederreiter (1992) and the latter procedure is standard. One has first to compute the recurrence coefficients for orthogonal polynomials associated with μ, and then compute the nodes and weights from these coefficients. However, as noted in Deun *et al.* (2005) even for the one-dimensional case, if the (recurrence) coefficients are not known *explicitly* then problems arise in their computation. This is also true even for polynomial quadrature formulae with arbitrary measures; see (Deun *et al.*, 2005, p. 1). An exception is precisely in (Deun *et al.*, 2005, §3) where explicit recurrence coefficients are provided for the case of so-called Chebyshev weight functions. On the other hand, if one uses standard Gaussian quadrature (or cubature if $n > 1$) formulas exact for polynomials, then one has only an estimate and some error occurs. In the one-dimensional case, another possibility is to use modification algorithms for the previously mentioned modified measure $f d\mu$, get orthogonal polynomials, and then the first moment of $f d\mu$.

12.2 Most of the material of this section and in particular, Theorem 12.4 and Corollary 12.5, are from Bertsimas *et al.* (2008) where in fact, $\mathbb{K} \subset \mathbb{R}^n$

can also be the more general basic semi-algebraic set:

$$\{\mathbf{x} \in \mathbb{R}^n : b_1^l \leq x_1 \leq b_1^u,\ b_i^l(\mathbf{x}[i-1]) \leq x_i \leq b_i^u(\mathbf{x}[i-1]), \quad \forall i = 2, \ldots, n\},$$

where $\mathbf{x}[i] \in \mathbb{R}^i$ is the vector of first i elements of $\mathbf{x}$ for all $i = 1, \ldots, n$, $b_i^l, b_i^u \in \mathbb{R}[\mathbf{x}[i-1]]$ for all $i = 2, \ldots, n$, and $b_1^l, b_1^u \in \mathbb{R}$.

12.3 Most of the material is from Lasserre (2007a). For more details on the maximum entropy approach, the interested reader is referred to e.g. Byrnes and Lindquist (2006), Borwein and Lewis (1991b,a, 1992), Georgiou (2006), Mead and Papanicolaou (1984), and Tagliani (2002a,b). As early as in Mead and Papanicolaou (1984), it was recognized that such entropy methods may outperform classical Padé-like approximations. Except for the homotopy approach developed in Georgiou (2006), optimization algorithms using first or second order methods to search for $\boldsymbol{\lambda}$, need evaluate the gradient $\nabla_{\boldsymbol{\lambda}} g_d$ and Hessian $\nabla_{\boldsymbol{\lambda}}^2 g_d$ at the current iterate $\boldsymbol{\lambda}$. Of course, this can be done via some quadrature formula and a typical example of this approach is the Newton method described in Mead and Papanicolaou (1984). In principle, the quadrature formula should be with respect to the weight function and so, requires to repeatedly compute orthogonal polynomials with respect to the measure $f_d(\boldsymbol{\lambda}, x)dx$ on $[0, 1]$, not an easy task in general. This is why one rather uses some Gaussian quadrature formula with respect to dx on $[0, 1]$ and incorporate $f_d(\boldsymbol{\lambda}, \bullet)$ in the integrand to evaluate. In doing so, one estimates both gradient and Hessian with some unknown error. However, very good numerical results may be obtained because Gaussian quadrature formula seem to perform well for integrating exponentials of polynomials, at least in the univariate case on $[0, 1]$.

Chapter 13

Min-Max Problems and Nash Equilibria

This chapter considers some min-max problems, as well as the minimization of the supremum of finitely many rational functions on a compact basic semi-algebraic set. Then this is used to compute (or approximate) the value of Nash equilibria for N-player finite games. We end up with applying the moment approach to polynomial games.

13.1 Robust Polynomial Optimization

Let $\mathbb{K} \subset \mathbb{R}^n$ be a basic semi-algebraic set and consider the following min-max optimization problem:

$$f^* := \inf_{\mathbf{x}\in\mathbb{K}} \sup_{\mathbf{y}\in\Omega(\mathbf{x})} f(\mathbf{x},\mathbf{y}) \tag{13.1}$$

where $f \in \mathbb{R}[\mathbf{x},\mathbf{y}]$, and for each $\mathbf{x} \in \mathbb{K}$, the set $\Omega(\mathbf{x}) \subset \mathbb{R}^p$ is a convex polytope.

Problem (13.1) is a robust version of the global polynomial optimization problem of Chapter 5. Indeed, one may interpret (13.1) as a *game* against nature "$\mathbf{y}$", where one wishes to take an optimal *robust* decision $\mathbf{x}$, where "robust" means against the worst possible outcome of the opponent $\mathbf{y} \in \Omega(\mathbf{x})$.

We consider the case where when $\mathbf{x}$ is fixed:

- The polynomial $\mathbf{y} \mapsto f_{\mathbf{x}}(\mathbf{y})\ (:= f(\mathbf{x},\mathbf{y})) \in \mathbb{R}[\mathbf{y}]$ is *affine*, and
- for every $\mathbf{x} \in \mathbb{K}$, the set $\Omega(\mathbf{x})$ is the convex polytope:

$$\Omega(\mathbf{x}) := \{\mathbf{y} \in \mathbb{R}^p \,:\, \mathbf{A}(\mathbf{x})\,\mathbf{y} = \mathbf{b}(\mathbf{x})\,;\ \mathbf{y} \geq 0\,\} \tag{13.2}$$

where for some $s \in \mathbb{N}$, the mapping $\mathbf{x} \mapsto \mathbf{A}(\mathbf{x}) \in \mathbb{R}^{s\times p}$ defines a matrix polynomial and the mapping $\mathbf{x} \mapsto \mathbf{b}(\mathbf{x})$ defines a vector polynomial, that is, $\mathbf{A} \in \mathbb{R}[\mathbf{x}]^{s\times p}$ and $\mathbf{b} \in \mathbb{R}[\mathbf{x}]^p$. Therefore, writing

$$(\mathbf{x},\mathbf{y}) \mapsto \quad f(\mathbf{x},\mathbf{y}) = f_0(\mathbf{x}) + \sum_{j=1}^{p} f_j(\mathbf{x})\, y_j,$$

Problem (13.1) reads:

$$f^* = \inf_{\mathbf{x}\in\mathbb{K}} \sup_{\mathbf{y}} \{f_0(\mathbf{x}) + \sum_{j=1}^{p} f_j(\mathbf{x})\, y_j \;:\; \mathbf{A}(\mathbf{x})\,\mathbf{y} = \mathbf{b}(\mathbf{x});\; \mathbf{y} \geq 0\}.$$

Let $\mathbf{A}_j(\mathbf{x})$ denote the jth column of $\mathbf{A}(\mathbf{x})$. Standard LP-duality in the inner maximization problem, yields:

$$\begin{aligned} f^* = \inf_{\mathbf{x},\mathbf{u}} \; & f_0(\mathbf{x}) + \mathbf{u}'\,\mathbf{b}(\mathbf{x}) \\ \text{s.t. } & \mathbf{A}_j(\mathbf{x})'\mathbf{u} - f_j(\mathbf{x}) \geq 0, \quad j = 1,\ldots,p. \\ & \mathbf{x} \in \mathbb{K},\; \mathbf{u} \in \mathbb{R}^s. \end{aligned} \tag{13.3}$$

Observe that for every $j = 1,\ldots,p$, the function

$$(\mathbf{x},\mathbf{u}) \longmapsto h_j(\mathbf{x},\mathbf{y}) := \mathbf{A}_j(\mathbf{x})'\mathbf{u} - f_j(\mathbf{x}),$$

is a polynomial in $\mathbf{x}$ and $\mathbf{u}$. And so is the function $(\mathbf{x},\mathbf{u}) \mapsto \mathbf{u}'\mathbf{b}(\mathbf{x})$. In addition, the set $\widehat{\mathbb{K}} \subset \mathbb{R}^n \times \mathbb{R}^s$ defined by

$$\widehat{\mathbb{K}} := \{(\mathbf{x},\mathbf{u}) \in \mathbb{R}^n \times \mathbb{R}^s \;:\; h_j(\mathbf{x},\mathbf{u}) \geq 0,\; j = 1,\ldots,p;\; \mathbf{x} \in \mathbb{K}\,\}$$

is a basic semi-algebraic set. Hence the min-max problem (13.1) reduces to solving the global polynomial optimization problem (13.3) in $\mathbb{R}^n \times \mathbb{R}^s$. Therefore, the semidefinite relaxations defined in Chapter 5 can be applied. In particular, if $\widehat{\mathbb{K}}$ is compact and Assumption 2.1 holds, their convergence is guaranteed by Theorem 5.6.

Therefore, solving the robust optimization problem (13.1) reduces to solving a polynomial optimization problem in $\mathbb{R}^{n+p}$, as opposed to $\mathbb{R}^n$ in the non robust setting.

In what precedes, the decision $\mathbf{x} \in \mathbb{R}^n$ lies in a basic semi-algebraic set $\mathbb{K} \subset \mathbb{R}^n$, whereas the "perturbation" $\mathbf{y}$ lies in a nice convex set $\Omega(\mathbf{x})$. We next see the reverse situation.

13.1.1 *Robust Linear Programming*

Consider now the following linear program:

$$\sup_{\mathbf{y}} \{ \mathbf{c}(\mathbf{x})' \mathbf{y} : \quad \mathbf{A}\,\mathbf{y} = \mathbf{b}; \quad \mathbf{y} \geq 0 \}, \tag{13.4}$$

where $\mathbf{y} \in \mathbb{R}^p$ is a decision vector while $\mathbf{x} \in \mathbb{R}^n$ is a perturbation (taking values in a basic semi-algebraic set $\mathbb{K} \subset \mathbb{R}^n$), unknown at the time where the decision $\mathbf{y}$ should be taken. Hence, now the decision $\mathbf{y}$ lies in a nice convex set $\Omega \subset \mathbb{R}^p$, whereas the perturbation $\mathbf{x}$ lies in a compact basic semi-algebraic set $\mathbb{K} \subset \mathbb{R}^n$. One further assumes that the reward mapping $\mathbf{x} \mapsto \mathbf{c}(\mathbf{x})$ is polynomial, that is, $\mathbf{c} \in \mathbb{R}[\mathbf{x}]^p$. Therefore, in this context, the robust LP problem reads:

$$\rho^* = \sup_{\mathbf{y}\in\Omega} \inf_{\mathbf{x}\in\mathbb{K}} \mathbf{c}(\mathbf{x})' \mathbf{y}, \tag{13.5}$$

where $\Omega := \{\mathbf{y} \in \mathbb{R}^p : \mathbf{A}\,\mathbf{y} = \mathbf{b};\, \mathbf{y} \geq 0\}$. Equivalently,

$$\rho^* = \sup_{\boldsymbol{\lambda},\mathbf{y}} \{ \lambda : \mathbf{y} \in \Omega, \quad \mathbf{x} \mapsto \mathbf{c}(\mathbf{x})' \mathbf{y} - \boldsymbol{\lambda} > 0 \text{ on } \mathbb{K} \}.$$

We next define appropriate semidefinite relaxations for problem (13.5). Let $\mathbb{K} \subset \mathbb{R}^n$ be the basic semi-algebraic set in (4.1), and for $2i \geq i_0 := \max[\deg \mathbf{c}, \deg g_j]$, consider the hierarchy of semidefinite programs:

$$\begin{aligned} \rho_i := \sup_{\boldsymbol{\lambda},\mathbf{y},\sigma_j} \; & \boldsymbol{\lambda} \\ \text{s.t. } & \mathbf{A}\,\mathbf{y} = \mathbf{b}; \quad \mathbf{y} \geq 0 \\ & \mathbf{c}(\mathbf{x})' \mathbf{y} - \boldsymbol{\lambda} = \sigma_0(\mathbf{x}) + \sum_{j=1}^{m} \sigma_j(\mathbf{x}) g_j(\mathbf{x}), \quad \forall \mathbf{x} \in \mathbb{R}^n \\ & (\sigma_j)_{j=0}^{m} \in \Sigma(\mathbf{x}) \\ & \deg \sigma_0,\, \deg \sigma_j + \deg g_j \leq 2i, \qquad j = 1, \ldots, m. \end{aligned} \tag{13.6}$$

That (13.6) is a semidefinite program is because writing $\mathbf{x} \mapsto c_k(\mathbf{x}) = \sum_{\boldsymbol{\alpha}} c_{k\boldsymbol{\alpha}} \mathbf{x}^{\boldsymbol{\alpha}}$ for every $k = 1, \ldots, p$, and recalling how the semidefinite relaxations introduced in Chapters 4 and 5 are formulated, the constraint

$$\mathbf{c}(\mathbf{x})' \mathbf{y} - \boldsymbol{\lambda} = \sigma_0(\mathbf{x}) + \sum_{j=1}^{m} \sigma_j(\mathbf{x}) g_j(\mathbf{x}), \quad \forall \mathbf{x} \in \mathbb{R}^n$$

is the same as stating (with $\mathbb{N}^n_{2i} = \{\boldsymbol{\alpha} \in \mathbb{N} : |\boldsymbol{\alpha}| \leq 2i\}$)

$$\sum_{k=1}^{p} c_{k\boldsymbol{\alpha}}\, y_k - \lambda 1_{\{0\}}(\boldsymbol{\alpha}) = \langle \mathbf{B}_{\boldsymbol{\alpha}}, \mathbf{X}\rangle + \sum_{j=1}^{m} \langle \mathbf{C}^j_{\boldsymbol{\alpha}}, \mathbf{Z}_j\rangle, \quad \forall \boldsymbol{\alpha} \in \mathbb{N}^n_{2i};\ \mathbf{X}, \mathbf{Z}_j \succeq 0,$$

for some appropriate real symmetric matrices $(\mathbf{B}_{\boldsymbol{\alpha}}, \mathbf{C}^j_{\boldsymbol{\alpha}})$, and where the unknowns are $\lambda \in \mathbb{R}$, $\mathbf{y} \in \mathbb{R}^p$ and the real symmetric matrices $\mathbf{X}, \mathbf{Z}_j \succeq 0$.

The sequence (ρ_i) is obviously monotone nondecreasing and in addition, $\rho_i \leq \rho^*$ for every $i \geq i_0$.

Theorem 13.1. *Let $\mathbb{K}$ be as in (4.1) and satisfy Assumption (2.1). Let ρ^* be as in (13.5) and assume that $\Omega = \{\mathbf{y} \in \mathbb{R}^p : \mathbf{A}\mathbf{y} = \mathbf{b}, \mathbf{y} \geq 0\}$ is compact. Consider the hierarchy of semidefinite programs defined in (13.6). Then $\rho_i \uparrow \rho^*$ as $i \to \infty$.*

Proof. For every $\mathbf{y} \in \Omega$, let $\mathbf{y} \mapsto \rho(\mathbf{y}) := \inf\{\mathbf{c}(\mathbf{x})'\mathbf{y} : \mathbf{x} \in \mathbb{K}\}$ so that $\rho^* = \sup\{\rho(\mathbf{y}) : \mathbf{y} \in \Omega\}$. Observe that necessarily, $\boldsymbol{\lambda} \leq \rho(\mathbf{y}) \leq \rho^*$ for every feasible solution $(\mathbf{y}, \boldsymbol{\lambda}, (\sigma_j))$ of (13.6). On the other hand, as $\mathbb{K}$ is compact and $\mathbf{c} \in \mathbb{R}[\mathbf{x}]^p$, $\rho(\mathbf{y})$ is continuous; hence $\rho^* = \rho(\mathbf{y}^*)$ for some $\mathbf{y}^* \in \Omega$ because Ω is compact. So with $\epsilon > 0$ fixed, arbitrary, by Theorem 2.14, there exists s.o.s. polynomials $(\sigma^\epsilon_j) \subset \Sigma[\mathbf{x}]$ such that

$$c(\mathbf{x})'\mathbf{y}^* - \rho^* + \epsilon = \sigma^\epsilon_0(\mathbf{x}) + \sum_{j=1}^{m} \sigma^\epsilon_j(\mathbf{x}) g_j(\mathbf{x}), \quad \mathbf{x} \in \mathbb{R}^n.$$

Hence for sufficiently large i, $(\mathbf{y}^*, \boldsymbol{\lambda}(:= \rho^* - \epsilon), (\sigma^\epsilon_j))$ is a feasible solution of (13.6), and so $\rho_i \geq \rho^* - \epsilon$. Therefore, as $\epsilon > 0$ was arbitrary, letting $\epsilon \to 0$ yields the desired result $\rho_i \uparrow \rho^*$. □

Observe that the semidefinite relaxations (13.6) are basically the same as those defined for a polynomial optimization problem on $\mathbb{K}$, with the additional variables $\mathbf{y} \in \mathbb{R}^p$ and linear constraints $\mathbf{A}\mathbf{y} = \mathbf{b}$ (that do not change with the index of the relaxation).

If Ω and $\mathbb{K}$ are compact then $\rho^* = \mathbf{c}(\mathbf{x}^*)'\mathbf{y}^*$ for some $\mathbf{y}^* \in \Omega$ and some $\mathbf{x}^* \in \mathbb{K}$. Indeed let $(\mathbf{y}^i)_i \subset \Omega$ be a maximizing sequence in (13.5). For every $\mathbf{y}^i \in \Omega$ there exists $\mathbf{x}^i \in \mathbb{K}$ that minimizes $\mathbf{x} \mapsto \mathbf{c}(\mathbf{x})'\mathbf{y}^i$ on $\mathbb{K}$ with value $\rho^i := \mathbf{c}(\mathbf{x}^i)'\mathbf{y}^i \to \rho^*$ as $i \to \infty$. For some subsequence (i_k) one gets

$(\mathbf{y}^{i_k}, \mathbf{x}^{i_k}) \to (\mathbf{y}^*, \mathbf{x}^*) \in \Omega \times \mathbb{K}$, and by continuity,

$$\rho^* = \lim_{k\to\infty} \rho^{i_k} = \lim_{k\to\infty} \mathbf{c}(\mathbf{x}^{i_k})'\mathbf{y}^{i_k} = \mathbf{c}(\mathbf{x}^*)'\mathbf{y}^*.$$

Corollary 13.2. *Let $\mathbb{K}$ be as in (4.1) and satisfy Assumption (2.1). Let ρ^* be as in (13.5) and assume that $\Omega = \{\mathbf{y} \in \mathbb{R}^p : \mathbf{A}\mathbf{y} = \mathbf{b}, \mathbf{y} \geq 0\}$ is compact. Let $(\mathbf{y}^*, \mathbf{x}^*) \in \Omega \times \mathbb{K}$ be an optimal solution of (13.5), and consider the hierarchy of semidefinite programs defined in (13.6).*

If the polynomial $\mathbf{x} \mapsto \mathbf{c}(\mathbf{x})'\mathbf{y}^* - \mathbf{c}(\mathbf{x}^*)'\mathbf{y}^*$ which is nonnegative on $\mathbb{K}$, has Putinar's representation (2.13), i.e., if

$$\mathbf{x} \mapsto \mathbf{c}(\mathbf{x})'\mathbf{y}^* - \rho^* = \sigma_0^* + \sum_{j=1}^{m} \sigma_j^* \, g_j, \tag{13.7}$$

for some s.o.s. polynomials $(\sigma_j^*) \subset \Sigma[\mathbf{x}]$, then $\rho^* = \rho_{i^*}$ for some index i^*, that is, finite convergence occurs.

Proof. Let $2i^* := \max[2i_0, \max_j \deg \sigma_j^* g_j]$ with (σ_j) as in (13.7). Then $(\mathbf{y}^*, \rho^*, (\sigma_j^*))$ is a feasible solution of the semidefinite program (13.6), with value $\rho_{i^*} = \rho^*$. □

So in this case, by solving the semidefinite program (13.6) for some index i^*, one obtains an optimal robust decision $\mathbf{y}^* \in \Omega$ and one also identifies an associated worst "move" $\mathbf{x}^* \in \mathbb{K}$ by nature.

13.1.2 *Robust Semidefinite Programming*

Consider now the following semidefinite program:

$$\sup_{\mathbf{y}} \{ \mathbf{c}(\mathbf{x})'\,\mathbf{y} : \sum_{j=1}^{p} \mathbf{A}_j \, y_j \preceq \mathbf{A}_0 \} \tag{13.8}$$

where $(\mathbf{A}_j) \subset \mathbb{R}^{t\times t}$ are real symmetric matrices and $\mathbf{y}$ is a decision vector of $\mathbb{R}^p$. As before, $\mathbf{x} \in \mathbb{R}^n$ is a perturbation taking values in a basic semi-algebraic set $\mathbb{K} \subset \mathbb{R}^n$, unknown at the time where the decision $\mathbf{y}$ should be taken, and again, one assumes that the reward mapping $\mathbf{x} \mapsto \mathbf{c}(\mathbf{x})$ is polynomial, i.e., $\mathbf{c} \in \mathbb{R}[\mathbf{x}]^p$.

Therefore, the robust semidefinite problem reads:

$$\rho^* = \sup_{\mathbf{y}\in\Omega} \inf_{\mathbf{x}\in\mathbb{K}} \mathbf{c}(\mathbf{x})'\,\mathbf{y},$$

which is the same as (13.5) but now with $\Omega := \{\mathbf{y} \in \mathbb{R}^p : \sum_{j=1}^p \mathbf{A}_j y_j \preceq \mathbf{A}_0\}$, still a convex set (but not a polytope).

Therefore, with $\mathbb{K} \subset \mathbb{R}^n$ be the basic semi-algebraic set defined in (4.1), and for $2i \geq \max[\deg \mathbf{c}, \deg g_j]$, consider the hierarchy of semidefinite programs:

$$
\begin{aligned}
\rho_i := \sup_{\boldsymbol{\lambda},\mathbf{y},\sigma_j} \ & \boldsymbol{\lambda} \\
\text{s.t.} \ & \sum_{j=1}^p \mathbf{A}_j\, y_j \preceq \mathbf{A}_0 \\
& \mathbf{c}(\mathbf{x})'\mathbf{y} - \boldsymbol{\lambda} = \sigma_0(\mathbf{x}) + \sum_{j=1}^m \sigma_j(\mathbf{x}) g_j(\mathbf{x}), \quad \forall \mathbf{x} \in \mathbb{R}^n \\
& (\sigma_j)_{j=0}^m \in \Sigma(\mathbf{x}),\ \deg \sigma_j + \deg g_j \leq 2i, \qquad j = 1,\ldots,m.
\end{aligned}
\tag{13.9}
$$

The sequence (ρ_i) is obviously monotone non decreasing.

Theorem 13.3. *Let $\mathbb{K}$ be as in (4.1) and satisfy Assumption (2.1). Let ρ^* be as in (13.5) and assume that $\Omega = \{\mathbf{y} \in \mathbb{R}^p : \sum_{i=1}^p \mathbf{A}_j y_j \preceq \mathbf{A}_0\}$ is compact. Consider the hierarchy of semidefinite programs defined in (13.9). Then $\rho_i \uparrow \rho^*$ as $i \to \infty$.*

The proof is a verbatim copy of that of Theorem 13.1 and is omitted. Finally, an obvious analogue of Corollary 13.2 holds with $\Omega = \{\mathbf{y} \in \mathbb{R}^p : \sum_{j=1}^p \mathbf{A}_j y_j \preceq \mathbf{A}_0\}$.

13.2 Minimizing the Sup of Finitely Many Rational Functions

In this section we consider the global optimization problem:

$$
\rho^* := \inf_{\mathbf{x} \in \mathbb{K}} \left[f_0(\mathbf{x}) + \max_{j=1,\ldots,p} f_j(\mathbf{x}) \right] \tag{13.10}
$$

where:

- $\mathbb{K}$ is a basic semi-algebraic set as in (4.1) and,

• $\mathbf{x} \mapsto f_j(\mathbf{x}) := p_j(\mathbf{x})/q_j(\mathbf{x})$ with $p_j, q_j \in \mathbb{R}[\mathbf{x}]$ and $q_j > 0$ on $\mathbb{K}$, for all $j = 0, 1, \ldots, p$.

Let $\widehat{\mathbb{K}} \subset \mathbb{R}^{n+1}$ be the basic semi-algebraic set

$$\{ (\mathbf{x}, z) \in \mathbb{R}^n \times \mathbb{R} \,:\, \mathbf{x} \in \mathbb{K}; \quad z\, q_j(\mathbf{x}) - p_j(\mathbf{x}) \geq 0,\ j = 1, \ldots, p \}, \quad (13.11)$$

and consider the new optimization problem

$$\rho_{\text{mom}} \;=\; \inf_{\mu \in \mathscr{M}(\widehat{\mathbb{K}})_+} \{ \int_{\widehat{\mathbb{K}}} (p_0 + z\, q_0)\, d\mu \,:\, \int_{\widehat{\mathbb{K}}} q_0\, d\mu = 1 \}, \qquad (13.12)$$

(where recall that $\mathscr{M}(\widehat{\mathbb{K}})_+$ is the set of finite Borel measures on $\widehat{\mathbb{K}}$). In Problem (13.12) one recognizes an instance of the generalized moment problem (1.1).

Proposition 13.4. *Let ρ^* and ρ_{mom} be as in (13.10) and (13.12) respectively. If $\rho^* > -\infty$ then $\rho^* = \rho_{\text{mom}}$*

Proof. For $\epsilon > 0$ fixed, arbitrary, let $\mathbf{x} \in \mathbb{K}$ be such that $f_0(\mathbf{x}) + \max_{i=1,\ldots,p} f_i(\mathbf{x}) \leq \rho^* + \epsilon$. Let $z := \max_{i=1,\ldots,p} f_i(\mathbf{x})$ so that $(\mathbf{x}, z) \in \widehat{\mathbb{K}}$ because $\mathbf{x} \in \mathbb{K}$ and $q_i > 0$ on $\mathbb{K}$ for every $i = 1, \ldots, p$. With $\delta_{(\mathbf{x},z)}$ being the Dirac measure at $(\mathbf{x}, z) \in \widehat{\mathbb{K}}$, let μ be the measure $q_0(\mathbf{x})^{-1}\delta_{(\mathbf{x},z)}$. Then $\mu \in \mathscr{M}(\widehat{\mathbb{K}})_+$ with $\int q_0\, d\mu = 1$, and $\int (p_0 + z q_0) d\mu = p_0(\mathbf{x})/q_0(\mathbf{x}) + z \leq \rho + \epsilon$. As $\epsilon > 0$ was arbitrary, it follows that $\rho_{\text{mom}} \leq \rho^*$.

On the other hand, let $\mu \in \mathscr{M}(\widehat{\mathbb{K}})_+$ be such that $\int q_0\, d\mu = 1$. As $f_0(\mathbf{x}) + \max_{i=1,\ldots,p} f_i(\mathbf{x}) \geq \rho^*$ for all $\mathbf{x} \in \mathbb{K}$, it follows that $p_0(\mathbf{x}) + z q_0(\mathbf{x}) \geq \rho^* q_0(\mathbf{x})$ for all $(\mathbf{x}, z) \in \widehat{\mathbb{K}}$ (as $q_0 > 0$ on $\mathbb{K}$). Integrating with respect to μ yields $\int (p_0 + z q_0) d\mu \geq \rho^* \int q_0 d\mu = \rho^*$, which proves that $\rho_{\text{mom}} \geq \rho^*$, and so, $\rho_{\text{mom}} = \rho^*$, the desired result. □

Semidefinite relaxations

If $\mathbb{K}$ is compact, and under Assumption 2.1, let

$$M_1 \;:=\; \max_{i=1,\ldots,p} \left\{ \frac{\max\{|p_i(\mathbf{x})| : \mathbf{x} \in \mathbb{K}\}}{\min\{q_i(\mathbf{x}) \,:\, \mathbf{x} \in \mathbb{K}\}} \right\}, \qquad (13.13)$$

and

$$M_2 \;:=\; \min_{i=1,\ldots,p} \left\{ \frac{\min\{p_i(\mathbf{x}) : \mathbf{x} \in \mathbb{K}\}}{\max\{q_i(\mathbf{x}) \,:\, \mathbf{x} \in \mathbb{K}\}} \right\}. \qquad (13.14)$$

Redefine the set $\widehat{\mathbb{K}}$ to be

$$\widehat{\mathbb{K}} := \{(\mathbf{x}, z) \in \mathbb{R}^n \times \mathbb{R} \,:\, h_j(\mathbf{x}, z) \geq 0, \quad j = 1, \ldots p + m + 2\} \tag{13.15}$$

with

$$\begin{cases} (\mathbf{x}, z) \mapsto h_j(\mathbf{x}, z) := g_j(\mathbf{x}) & j = 1, \ldots, p \\ (\mathbf{x}, z) \mapsto h_j(\mathbf{x}, z) := z\, q_{j-p}(\mathbf{x}) - p_{j-p}(\mathbf{x}) & j = p + 1, \ldots, p + m \\ (\mathbf{x}, z) \mapsto h_j(\mathbf{x}, z) := (M_1 - z)(z - M_2) & j = m + p + 1 \end{cases} . \tag{13.16}$$

Lemma 13.5. *Let $\mathbb{K} \subset \mathbb{R}^n$ be compact and let Assumption 2.1 hold. Then the set $\widehat{\mathbb{K}} \subset \mathbb{R}^{n+1}$ defined in (13.15) also satisfies Assumption 2.1.*

Proof. As Assumption 2.1 holds for $\mathbb{K}$, equivalently, the quadratic polynomial $\mathbf{x} \mapsto M - \|\mathbf{x}\|^2$ can be written in the form (2.13); see Theorem 2.15. So consider the quadratic polynomial $w \in \mathbb{R}[\mathbf{x}, z]$ defined by:

$$(\mathbf{x}, z) \mapsto w(\mathbf{x}, z) = M - \|\mathbf{x}\|^2 + (M_1 - z)(z - M_2).$$

Obviously, its level set $\{(\mathbf{x}, z) \,:\, w(\mathbf{x}, z) \geq 0\} \subset \mathbb{R}^{n+1}$ is compact and moreover, w can be written in the form

$$w(\mathbf{x}, z) = \sigma_0(\mathbf{x}) + \sum_{j=1}^{p} \sigma_j(\mathbf{x})\, g_j(\mathbf{x}) + h_{m+p+1}(\mathbf{x}, z)$$

for some s.o.s. polynomials $(\sigma_j) \subset \mathbb{R}[\mathbf{x}, z]$. Therefore Assumption 2.1 holds for $\widehat{\mathbb{K}}$, the desired result. □

We are now in position to define a hierarchy of semidefinite relaxations for solving (13.10). Let $\mathbf{y} = (y_{\boldsymbol{\alpha}})$ be a real sequence indexed in the monomial basis $(\mathbf{x}^{\boldsymbol{\beta}} z^k)$ of $\mathbb{R}[\mathbf{x}, z]$ (with $\boldsymbol{\alpha} = (\boldsymbol{\beta}, k) \in \mathbb{N}^n \times \mathbb{N}$).

Let $h_0(\mathbf{x}, z) := p_0(\mathbf{x}) + z q_0(\mathbf{x})$, and let $v_j := \lceil (\deg h_j)/2 \rceil$ for every $j = 0, \ldots, m + p + 1$. For $i \geq i_0 := \max_{j=0,\ldots,p+m+1} v_j$, introduce the hierarchy of semidefinite programs:

$$\begin{aligned} \rho_i := \inf_{\mathbf{y}} \; & L_{\mathbf{y}}(h_0) \\ \text{s.t. } & \mathbf{M}_i(\mathbf{y}) \succeq 0 \\ & \mathbf{M}_{i-v_j}(h_j\, \mathbf{y}) \succeq 0, \quad j = 1, \ldots, m + p + 1 \\ & L_{\mathbf{y}}(q_0) = 1. \end{aligned} \tag{13.17}$$

Theorem 13.6. *Let $\mathbb{K} \subset \mathbb{R}^n$ be compact and as in (4.1). Let Assumption 2.1 hold and assume that $\rho^* > -\infty$. Consider the semidefinite program (13.17) with $(h_j) \subset \mathbb{R}[\mathbf{x}, z]$ and M_1, M_2 defined in (13.16) and (13.13)-(13.14) respectively. Then:*
(a) $\rho_i \uparrow \rho^* = \rho_{\text{mom}}$ *as* $i \to \infty$.
(b) *Let $\mathbf{y}^i$ be an optimal solution of (13.17). If*

$$\operatorname{rank} \mathbf{M}_i(\mathbf{y}^i) = \operatorname{rank} \mathbf{M}_{i-i_0}(\mathbf{y}^i) =: t \tag{13.18}$$

then one may extract t points $(\mathbf{x}^*(k))_{k=1}^t \subset \mathbb{K}$, all global minimizers of problem (13.10).

Theorem 13.6 is just Theorem 4.1 applied to the generalized moment problem (13.12).

13.3 Application to Nash Equilibria

Nash equilibrium, a central concept in game theory, is a profile of mixed strategies (a strategy for each player) such that each player is best-responding to the strategies of the opponents. In this section we are concerned with algorithms to effectively compute (or at least approximate) Nash equilibria in the context of N-players non zero-sum finite games, and 2-players zero-sum polynomial games.

13.3.1 *N-player games*

A finite game is a tuple $(N, (S^i)_{i=1,\ldots,N}, (g^i)_{i=1,\ldots,N})$ where $N \in \mathbb{N}$ is the set of players, S^i is the finite set of pure strategies[1] and $g^i : S \to \mathbb{R}$ is the payoff function of player i, where $S := S^1 \times \cdots \times S^N$. The set

$$\Delta^i = \left\{ \mathbf{p}^i = \left(p^i(x_i)\right)_{x_i \in S^i} : \quad p^i(x_i) \geq 0, \quad \sum_{x_j \in S^i} p^i(x_i) = 1 \right\}$$

of probability distributions over S^i is called the set of *mixed strategies* of player i. Notice that Δ^i is a simplex of $\mathbb{R}^{|S^i|}$, hence a compact basic semi-

[1] When using a pure strategy, Player i chooses a single action $x_i \in S^i$ deterministically, whereas when using a mixed strategy he randomly selects an action $x_i \in S^i$, according to some probability distribution on S^i.

algebraic set. If each player i chooses some mixed strategy $\mathbf{p}^i(\cdot)$, the vector denoted $\mathbf{p} = (\mathbf{p}^1, ..., \mathbf{p}^N) \in \Delta = \Delta^1 \times ... \times \Delta^N$ is called the *profile*, and the expected payoff of player i is

$$\mathbf{g}^i(\mathbf{p}) = \sum_{\mathbf{x}\in S} p^1(x_1)\cdots p^N(x_N) g^i(\mathbf{x}).$$

For a player i, and a profile $\mathbf{p}$, let $\mathbf{p}^{-i}$ be the profile of the other players except i, that is, $\mathbf{p}^{-i} = (\mathbf{p}^1, ..., \mathbf{p}^{i-1}, \mathbf{p}^{i+1}, ..., \mathbf{p}^N)$, and let $S^{-i} = S^1 \times \cdots \times S^{i-1} \times S^{i+1} \cdots \times S^N$. Given a profile $\mathbf{p}$ and an action $x_i \in S^i$, define

$$g^i(x_i, \mathbf{p}^{-i}) := \sum_{\mathbf{x}^{-i}\in S^{-i}} p^1(x_1)\cdots p^{i-1}(x_{i-1}) p^{i+1}(x_{i+1}) \cdots p^N(x_N) g^i(\mathbf{x}),$$

where $\mathbf{x}^{-i} \in S^{-i}$ denotes the vector $(x_1, \ldots, x_{i-1}, x_{i+1}, \ldots, x_N)$.

The value of the game is given by:

$$\rho^* := \inf_{\mathbf{p}\in\Delta} \sup_{i\in N, x_i\in S^i} \left\{ g^i(x_i, \mathbf{p}^{-i}) - \mathbf{g}^i(\mathbf{p}) \right\}, \tag{13.19}$$

and a profile $\mathbf{p}_0$ is a Nash equilibrium (in mixed strategies) if and only if, for all $i \in N$ and all $x_i \in S^i$, $g^i(\mathbf{p}_0) \geq g^i(x_i, \mathbf{p}_0^{-i})$, or equivalently if

$$\mathbf{p}_0 \in \arg\min_{\mathbf{p}\in\Delta} \max_{i\in N, x_i\in S^i} \left\{ g^i(x_i, \mathbf{p}^{-i}) - \mathbf{g}^i(\mathbf{p}) \right\}. \tag{13.20}$$

As each S^i is finite, the min-max problem (13.19) is a particular instance of problem (13.10). Therefore, one can use the hierarchy of semidefinite relaxations (13.17) to approximate ρ^* as closely as desired. In addition, if (13.18) is satisfied at some step in the hierarchy, then one obtains an optimal strategy. Moreover, since the optimal value is zero, one knows when the algorithm should stop and if it does not stop, one has a bound on payoffs so that one knows which ϵ-equilibrium is reached.

Example 13.1. Consider the simple illustrative example of a 2×2 game with data

	x_2^1	x_2^2
x_1^1	(a, c)	$(0, 0)$
x_1^2	$(0, 0)$	(b, d)

for some scalars (a, b, c, d). For instance, when Player 1 selects action x_1^1 and Player 2 selects action x_2^1, the reward of Player 1 (resp. Player 2) is a

(resp. c). Let $\mathbf{p} := (p^1, p^2)$ where $p^1 \in [0,1]$ is the probability that player 1 plays x_1^1 and $p^2 \in [0,1]$ is the probability that player 2 plays x_2^1. Then one has to solve:

$$\inf_{\mathbf{p}\in[0,1]^2} \quad \sup \begin{cases} a\,p^1 - a\,p^1p^2 - b\,(1-p^1)(1-p^2) \\ b\,(1-p^2) - a\,p^1p^2 - b\,(1-p^1)(1-p^2) \\ c\,p^1 - c\,p^1p^2 - d\,(1-p^1)(1-p^2) \\ d\,(1-p^1) - c\,p^1p^2 - d\,(1-p^1)(1-p^2), \end{cases}$$

that is, one has to minimize the supremum of four bivariate bilinear polynomials on the box $[0,1]^2$. The first relaxation in the hierarchy of semidefinite programs (13.17) reads:

$$\begin{aligned} \rho_1 = \inf_{\mathbf{y}}\ & y_{001} \\ \text{s.t. } & \mathbf{M}_1(\mathbf{y}) \succeq 0 \\ & -y_{002} + (M_1+M_2)y_{001} - M_1M_2y_{000} \geq 0 \\ & y_{100} - y_{200} \geq 0,\ y_{010} - y_{020} \geq 0 \\ & y_{001} - ay_{100} + ay_{110} + b(y_0 - y_{100} - y_{010} + y_{110}) \geq 0 \\ & y_{001} - by_0 + by_{010} + ay_{110} + b(y_0 - y_{100} - y_{010} + y_{110}) \geq 0 \\ & y_{001} - cy_{100} + cy_{110} + d(y_0 - y_{100} - y_{010} + y_{110}) \geq 0 \\ & y_{001} - dy_0 + dy_{100} + cy_{110} + d(y_0 - y_{100} - y_{010} + y_{110}) \geq 0 \\ & y_{000} \ = 1 \end{aligned}$$

e.g. with $M_1 = \max[a,b,c,d]$ and $M_2 = -\max[a+b, c+d]$. Solving (13.17) with the GloptiPoly software (see Section D) and with $(a,b,c,d) = (0.05, 0.82, 0.56, 0.76)$, one obtains $\rho_3 \approx 3.93.10^{-11}$ and the three optimal solutions $\mathbf{p} \in \{(0,0), (1,1), (0.57575, 0.94253)\}$. With randomly generated $a,b,c,d \in [0,1]$ we also obtained a very good approximation of the global optimum 0 and 3 optimal solutions in most cases with $i = 3$ (i.e. with moments or order 6 only) and sometimes $i = 4$.

We have also considered 2-player non-zero sum $p \times q$ games with randomly generated reward matrices $\mathbf{A}, \mathbf{B} \in \mathbb{R}^{p\times q}$ and $p, q \leq 5$. We could solve 5×2 and $4 \times q$ (with $q \leq 3$) games exactly with the 4th (sometimes 3rd) semidefinite relaxation, i.e. $\rho_4 = O(10^{-10}) \approx 0$ and one could extract an optimal solution.[2] However, the size is relatively large and one is close to the limit of present public semidefinite solvers like SeDuMi. Indeed, for

[2] In fact GloptiPoly 3 extracts *all* solutions because most semidefinite solvers that one may call in GloptiPoly 3 (e.g. SeDuMi) use primal-dual interior points methods which find an optimal solution in the relative interior of the feasible set. In the present context of (13.17) this means that at an optimal solution $\mathbf{y}^*$, the moment matrix $\mathbf{M}_i(\mathbf{y}^*)$ has maximum rank and its rank corresponds to the number of solutions.

a 2-player 5×2 or 4×3 game, the third semidefinite relaxation has 923 variables and $\mathbf{M}_3(\mathbf{y}) \in \mathbb{R}^{84\times 84}$, whereas the fourth relaxation has 3002 variables and $\mathbf{M}_4(\mathbf{y}) \in \mathbb{R}^{210\times 210}$. For a 4×4 game the third relaxation with 1715 variables and $\mathbf{M}_3(\mathbf{y}) \in \mathbb{R}^{120\times 120}$ is still solvable, whereas the fourth relaxation has 6434 variables and $\mathbf{M}_4(\mathbf{y}) \in \mathbb{R}^{330\times 330}$.

13.3.2 *Two-player zero-sum polynomial games*

Consider now the min-max optimization problem:

$$\rho^* := \inf_{\mu\in\mathscr{P}(\mathbb{K}_1)} \sup_{\nu\in\mathscr{P}(\mathbb{K}_2)} \int_{\mathbb{K}_2}\int_{\mathbb{K}_1} p(\mathbf{x},\mathbf{z})\, d\mu(\mathbf{x})\, d\nu(\mathbf{z}) \tag{13.21}$$

where $\mathbb{K}_1 \subset \mathbb{R}^{n_1}, \mathbb{K}_2 \subset \mathbb{R}^{n_2}$ are basic semi-algebraic set as defined in (4.1), and $\mathscr{P}(\mathbb{K}_1)$ (resp. $\mathscr{P}(\mathbb{K}_2)$) denotes the set of probability measures on $\mathbb{K}_1$ (resp. $\mathbb{K}_2$).

Problem (13.21) models a two-player zero-sum *polynomial* game. A probability measure $\mu \in \mathscr{P}(\mathbb{K}_1)$ (resp. $\nu \in \mathscr{P}(\mathbb{K}_2)$) corresponds to a mixed strategy of player 1 (resp. player 2). In such a game, player 1 chooses an action $\mathbf{x} \in \mathbb{K}_1$ according to the probability distribution μ whereas player 2 chooses an action $\mathbf{z} \in \mathbb{K}_2$ according to the probability distribution ν. The payoff of player 1 is $-p(\mathbf{x},\mathbf{z})$ and $p(\mathbf{x},\mathbf{z})$ for player 2.

If $\mathbb{K}_1$ and $\mathbb{K}_2$ are compact, there exists an optimal solution (μ^*,ν^*) in which μ^* (resp. ν^*) corresponds to an optimal mixed strategy for player 1 (resp. player 2). In addition

$$\rho^* := \sup_{\nu\in\mathscr{P}(\mathbb{K}_2)} \inf_{\mu\in\mathscr{P}(\mathbb{K}_1)} \int_{\mathbb{K}_1}\int_{\mathbb{K}_2} p(\mathbf{x},\mathbf{z})\, d\nu(\mathbf{x})\, d\mu(\mathbf{z}) \tag{13.22}$$

We next show how to determine (or approximate) ρ^* and an optimal pair of strategies (μ^*,ν^*).

With $p \in \mathbb{R}[\mathbf{x},\mathbf{z}]$ as in (13.21), write

$$\begin{aligned} p(\mathbf{x},\mathbf{z}) &= \sum_{\boldsymbol{\alpha}\in\mathbb{N}^{n_2}} p_{\boldsymbol{\alpha}}(\mathbf{x})\,\mathbf{z}^{\boldsymbol{\alpha}} \qquad \text{with} \\ p_{\boldsymbol{\alpha}}(\mathbf{x}) &= \sum_{\boldsymbol{\beta}\in\mathbb{N}^{n_1}} p_{\boldsymbol{\alpha}\boldsymbol{\beta}}\,\mathbf{x}^{\boldsymbol{\beta}}, \qquad |\boldsymbol{\alpha}| \le d_{\mathbf{z}}, \end{aligned} \tag{13.23}$$

where $d_{\mathbf{z}}$ is the total degree of p when seen as a polynomial in $\mathbb{R}[\mathbf{z}]$. So, let $p_{\boldsymbol{\alpha}\boldsymbol{\beta}} := 0$ for every $\boldsymbol{\beta} \in \mathbb{N}^{n_1}$ whenever $|\boldsymbol{\alpha}| > d_{\mathbf{z}}$. Given a sequence $\mathbf{y} = (y_{\boldsymbol{\alpha}})$ indexed in the canonical basis $(\mathbf{x}^{\boldsymbol{\alpha}})$ of $\mathbb{R}[\mathbf{x}]$, let $p_{\mathbf{y}} \in \mathbb{R}[\mathbf{z}]$ be the polynomial defined by:

$$\mathbf{z} \mapsto p_{\mathbf{y}}(\mathbf{z}) := \sum_{\boldsymbol{\alpha}\in\mathbb{N}^{n_2}} \left(\sum_{\boldsymbol{\beta}\in\mathbb{N}^{n_1}} p_{\boldsymbol{\alpha}\boldsymbol{\beta}}\, y_{\boldsymbol{\beta}} \right) \mathbf{z}^{\boldsymbol{\alpha}}. \tag{13.24}$$

Let $\mathbb{K}_1 \subset \mathbb{R}^{n_1}$ (resp. $\mathbb{K}_2 \subset \mathbb{R}^{n_2}$) be the basic semi-algebraic set defined in (4.1) with m_1 polynomials $(g_j) \subset \mathbb{R}[\mathbf{x}]$ (resp. m_2 polynomials $(h_k) \subset \mathbb{R}[\mathbf{z}]$). Let $r_j := \lceil \deg g_j/2 \rceil$, for every $j = 1, \ldots, m_1$, and with $h_0 = 1$, consider the following hierarchy of semidefinite programs:

$$\begin{aligned} \rho_i := \ & \inf_{\mathbf{y},\lambda,\sigma_k} \ \lambda \\ \text{s.t.} \quad & \lambda - p_{\mathbf{y}} = \sum_{k=0}^{m_2} \sigma_k\, h_k \\ & \mathbf{M}_i(\mathbf{y}) \succeq 0 \\ & \mathbf{M}_{i-r_j}(g_j\, \mathbf{y}) \succeq 0, \quad j = 1, \ldots, m_1 \\ & y_0 = 1 \\ & \sigma_k \in \Sigma[\mathbf{z}];\ \deg \sigma_k + \deg h_k \le 2i, \quad k = 0, 1, \ldots, m_2. \end{aligned} \tag{13.25}$$

Observe that with $p_{\mathbf{y}}$ as in (13.24), for any admissible solution $(\mathbf{y}, \lambda)$ of (13.25),

$$\lambda \ge \sup_{\mathbf{z}} \{ p_{\mathbf{y}}(\mathbf{z}) \ :\ \mathbf{z} \in \mathbb{K}_2 \}. \tag{13.26}$$

Similarly, with p as in (13.21), write

$$\begin{aligned} p(\mathbf{x}, \mathbf{z}) &= \sum_{\boldsymbol{\alpha}\in\mathbb{N}^{n_1}} \hat{p}_{\boldsymbol{\alpha}}(\mathbf{z})\, \mathbf{x}^{\boldsymbol{\alpha}} \qquad \text{with} \\ \hat{p}_{\boldsymbol{\alpha}}(\mathbf{z}) &= \sum_{\boldsymbol{\beta}\in\mathbb{N}^{n_2}} \hat{p}_{\boldsymbol{\alpha}\boldsymbol{\beta}}\, \mathbf{z}^{\boldsymbol{\beta}}, \qquad |\alpha| \le d_{\mathbf{x}}, \end{aligned} \tag{13.27}$$

where $d_{\mathbf{x}}$ is the total degree of p when seen as a polynomial in $\mathbb{R}[\mathbf{x}]$. So, let $\hat{p}_{\boldsymbol{\alpha}\boldsymbol{\beta}} := 0$ for every $\boldsymbol{\beta} \in \mathbb{N}^{n_2}$ whenever $|\boldsymbol{\alpha}| > d_{\mathbf{x}}$.

Given a sequence $\mathbf{y} = (y_{\boldsymbol{\alpha}})$ indexed in the canonical basis $(\mathbf{z}^{\boldsymbol{\alpha}})$ of $\mathbb{R}[\mathbf{z}]$, let

$$\mathbf{x} \mapsto \hat{p}_{\mathbf{y}}(\mathbf{x}) := \sum_{\boldsymbol{\alpha}\in\mathbb{N}^{n_1}} \left(\sum_{\boldsymbol{\beta}\in\mathbb{N}^{n_2}} \hat{p}_{\boldsymbol{\alpha}\boldsymbol{\beta}}\, y_{\boldsymbol{\beta}} \right) \mathbf{x}^{\boldsymbol{\alpha}}. \tag{13.28}$$

Let $l_k := \lceil \deg h_k/2 \rceil$, for every $k = 1, \ldots, m_2$, and consider the following hierarchy of semidefinite programs (with $g_0 = 1$):

$$\begin{array}{rl} \rho^i := \sup\limits_{\mathbf{y},\gamma,\sigma_j} & \gamma \\ \text{s.t.} & \hat{p}_{\mathbf{y}} - \gamma = \sum\limits_{j=0}^{m_1} \sigma_j \, g_j \\ & \mathbf{M}_i(\mathbf{y}) \succeq 0 \\ & \mathbf{M}_{i-l_k}(h_k \, \mathbf{y}) \succeq 0, \quad k = 1, \ldots, m_2 \\ & y_0 = 1 \\ & \sigma_j \in \Sigma[\mathbf{x}]; \, \deg \sigma_j + \deg g_j \leq 2i, \quad j = 0, 1, \ldots, m_1. \end{array} \tag{13.29}$$

Observe that with $\hat{p}_{\mathbf{y}}$ as in (13.28), for any admissible solution $(\mathbf{y}, \gamma)$ of (13.29),

$$\gamma \leq \inf_{\mathbf{x}} \{ \hat{p}_{\mathbf{y}}(\mathbf{x}) \; : \; \mathbf{x} \in \mathbb{K}_1 \}.$$

Let $r_1 := \max_{j=1,\ldots,m_1} \lceil \deg g_j/2 \rceil$, and let $r_2 := \max_{k=1,\ldots,m_2} \lceil \deg h_k/2 \rceil$.

Theorem 13.7. *Let $\mathbb{K}_1 \subset \mathbb{R}^{n_1}$ and $\mathbb{K}_2 \subset \mathbb{R}^{n_2}$ both satisfy Assumption 2.1 and let ρ^* be the value of the game (13.21). Let ρ_i and ρ^i be the respective optimal values of the semidefinite programs (13.25) and (13.29). Then:*

(a) $\lim_{i\to\infty} \rho_i = \lim_{i\to\infty} \rho^i = \rho^*$.

(b) *Let $\mathbf{y}^1$ be part of an optimal solution of (13.25) with optimal value ρ_i, and let $\mathbf{y}^2$ be part of an optimal solution of (13.29) with optimal value ρ^k. If $\rho_i \leq \rho^k$ and if*

$$\begin{aligned} \operatorname{rank} \mathbf{M}_i(\mathbf{y}^1) &= \operatorname{rank} \mathbf{M}_{i-r_1}(\mathbf{y}^1) \; (=: s_1) \\ \operatorname{rank} \mathbf{M}_k(\mathbf{y}^2) &= \operatorname{rank} \mathbf{M}_{k-r_2}(\mathbf{y}^2) \; (=: s_2) \end{aligned}$$

then $\rho_i = \rho^k = \rho^*$. In addition, there is an optimal strategy $\mu^* \in \mathscr{P}(\mathbb{K}_1)$ for player 1 which is supported on s_1 points of $\mathbb{K}_1$ and an optimal strategy $\nu^* \in \mathscr{P}(\mathbb{K}_2)$ for player 2 which is supported on s_2 points of $\mathbb{K}_2$.

Proof. (a) Let $\mu^* \in \mathscr{P}(\mathbb{K}_1), \nu^* \in \mathscr{P}(\mathbb{K}_2)$ be optimal strategies of player 1 and player 2 respectively, and let $\mathbf{y}^* = (y^*_\alpha)$ be the sequence of moments of μ^* (well-defined because $\mathbb{K}_1$ is compact). Then

$$\begin{aligned}
\rho^* &= \sup_{\nu\in\mathscr{P}(\mathbb{K}_2)} \int_{\mathbb{K}_2} \left(\int_{\mathbb{K}_1} p(\mathbf{x},\mathbf{z})d\mu^*(\mathbf{x})\right) d\nu(\mathbf{z}) \\
&= \sup_{\nu\in\mathscr{P}(\mathbb{K}_2)} \int_{\mathbb{K}_2} \sum_{\boldsymbol{\alpha}\in\mathbb{N}^{n_2}} \left(\sum_{\boldsymbol{\beta}\in\mathbb{N}^{n_1}} p_{\boldsymbol{\alpha}\boldsymbol{\beta}} \int_{\mathbb{K}_1} \mathbf{x}^{\boldsymbol{\beta}}\, d\mu^*(\mathbf{x})\right) \mathbf{z}^{\boldsymbol{\alpha}} d\nu(\mathbf{z}) \\
&= \sup_{\nu\in\mathscr{P}(\mathbb{K}_2)} \int_{\mathbb{K}_2} \sum_{\boldsymbol{\alpha}\in\mathbb{N}^{n_2}} \left(\sum_{\boldsymbol{\beta}\in\mathbb{N}^{n_1}} p_{\boldsymbol{\alpha}\boldsymbol{\beta}} y^*_{\boldsymbol{\beta}}\right) \mathbf{z}^{\boldsymbol{\alpha}} d\nu(\mathbf{z}) \\
&= \sup_{\nu\in\mathscr{P}(\mathbb{K}_2)} \int_{\mathbb{K}_2} p_{\mathbf{y}^*}(\mathbf{z})\, d\nu(\mathbf{z}) \;=\; \sup_{\mathbf{z}} \{\, p_{\mathbf{y}^*}(\mathbf{z}) \;:\; \mathbf{z}\in\mathbb{K}_2 \,\} \\
&= \inf_{\lambda,\sigma_k} \{\, \lambda \,:\, \lambda - p_{\mathbf{y}^*} \;=\; \sigma_0 + \sum_{k=1}^{m_2} \sigma_k\, h_k; \quad (\sigma_j)_{j=0}^{m_2} \subset \Sigma[\mathbf{z}] \,\},
\end{aligned}$$

where recall that $\mathbf{z} \mapsto p_{\mathbf{y}^*}(\mathbf{z}) = \sum_{\boldsymbol{\alpha}\in\mathbb{N}^{n_2}} \left(\sum_{\boldsymbol{\beta}\in\mathbb{N}^{n_1}} p_{\boldsymbol{\alpha}\boldsymbol{\beta}}\, y^*_{\boldsymbol{\beta}}\right) \mathbf{z}^{\boldsymbol{\alpha}}$. Fix $\epsilon > 0$ arbitrary. Then

$$\rho^* - p_{\mathbf{y}^*} + \epsilon \;=\; \sigma_0^\epsilon + \sum_{k=1}^{m_2} \sigma_k^\epsilon\, h_k, \tag{13.30}$$

for some polynomials $(\sigma_k^\epsilon) \subset \Sigma[\mathbf{z}]$ of degree at most d_ϵ^1. Hence $(y^*, \rho^*+\epsilon, \sigma_k^\epsilon)$ is an admissible solution for the semidefinite program (13.25) whenever $2i \geq d_\epsilon := d_\epsilon^1 + \sup_k \deg h_k$, and so

$$\rho_i \;\leq\; \rho^* + \epsilon, \qquad \forall\, 2i \geq d_\epsilon. \tag{13.31}$$

Now, let $(\mathbf{y}^i, \lambda_i)$ be an admissible solution of the semidefinite program (13.25) with value $\lambda_i \;\leq\; \rho_i + 1/i$. By an argument already used several times in Chapter 4, there exists $\hat{\mathbf{y}} \in \mathbb{R}^\infty$ and a subsequence (i_k) such that the pointwise convergence $\mathbf{y}^{i_k} \to \hat{\mathbf{y}}$ holds, that is,

$$\lim_{k\to\infty} y^{i_k}_{\boldsymbol{\alpha}} \;=\; \hat{\mathbf{y}}_{\boldsymbol{\alpha}} \qquad \forall \boldsymbol{\alpha} \in \mathbb{N}^{n_1}. \tag{13.32}$$

But then, invoking (13.32) yields

$$\mathbf{M}_r(\hat{\mathbf{y}}) \succeq 0 \quad \text{and} \quad \mathbf{M}_r(g_j\, \hat{\mathbf{y}}) \succeq 0, \quad \forall j = 1,\ldots,m_1; \quad r = 0,1,\ldots$$

By Theorem 2.14, there exists $\hat{\mu} \in \mathscr{P}(\mathbb{K}_1)$ such that

$$\hat{y}_{\alpha} = \int_{\mathbb{K}_1} \mathbf{x}^{\alpha} \, d\hat{\mu}, \qquad \forall \alpha \in \mathbb{N}^{n_1}.$$

On the other hand,

$$\begin{aligned} \rho^* \le \sup_{\nu \in \mathscr{P}(\mathbb{K}_2)} \int_{\mathbb{K}_2} \left(\int_{\mathbb{K}_1} p(\mathbf{x}, \mathbf{z}) d\hat{\mu}(\mathbf{x}) \right) d\nu(\mathbf{z}) &= \sup_{\mathbf{z}} \{ p_{\hat{\mathbf{y}}}(\mathbf{z}) \,:\, \mathbf{z} \in \mathbb{K}_2 \} \\ = \inf \{ \lambda \,:\, \lambda - p_{\hat{\mathbf{y}}} = \sigma_0 + \sum_{k=1}^{m_2} \sigma_k \, h_k; \quad (\sigma_j)_{j=0}^{m_2} \subset \Sigma[\mathbf{z}] \}, \end{aligned}$$

with $\mathbf{z} \mapsto p_{\hat{\mathbf{y}}}(\mathbf{z}) = \sum_{\alpha \in \mathbb{N}^{n_2}} \left(\sum_{\beta \in \mathbb{N}^{n_1}} p_{\alpha\beta} \, \hat{y}_{\beta} \right) \mathbf{z}^{\alpha}$. Let $\rho := \sup_{\mathbf{z}} \{ p_{\hat{\mathbf{y}}}(\mathbf{z}) \,:\, \mathbf{z} \in \mathbb{K}_2 \}$ (hence $\rho \ge \rho^*$), and consider the polynomial

$$\mathbf{z} \mapsto p_{\mathbf{y}^i}(\mathbf{z}) := \sum_{\alpha \in \mathbb{N}^{n_2}} \left(\sum_{\beta \in \mathbb{N}^{n_1}} p_{\alpha\beta} \, y^i_{\beta} \right) \mathbf{z}^{\alpha}.$$

It has same degree as $p_{\hat{\mathbf{y}}}$, and by (13.32), $\| p_{\hat{\mathbf{y}}} - p_{\mathbf{y}^{i_k}} \| \to 0$ as $k \to \infty$ (for some norm on the (finite dimensional vector) of coefficients).

Hence, $\sup_{\mathbf{z}} \{ p_{\mathbf{y}^{i_k}}(\mathbf{z}) \,:\, \mathbf{z} \in \mathbb{K}_2 \} \to \rho \ge \rho^*$ as $k \to \infty$. Moreover, by (13.26) one obtains $\lambda_{i_k} \ge \sup_{\mathbf{z}} \{ p_{\mathbf{y}^{i_k}}(\mathbf{z}) \,:\, \mathbf{z} \in \mathbb{K}_2 \}$. Therefore, $\lambda_{i_k} \ge \rho - \epsilon \ge \rho^* - \epsilon$ for all sufficiently large k (say for all $i_k \ge d^2_{\epsilon}$).

This combined with $\lambda_{i_k} \le \rho^* + \epsilon + 1/i_k$ for all $2i_k \ge d^1_{\epsilon}$, yields the desired result that $\lim_{k \to \infty} \rho_{i_k} = \lim_{k \to \infty} \lambda_{i_k} = \rho^*$, because $\epsilon > 0$ (fixed) was arbitrary. Finally, as the converging subsequence (i_k) was arbitrary, we get that the entire sequence (ρ_i) converges to ρ^*. Similar arguments are used to prove $\rho^i \to \rho^*$.

(b) Let $\mathbf{y}^1, \mathbf{y}^2$ be as in Theorem 13.7. If the two rank conditions hold, by Theorem 3.11, $\mathbf{y}^1$ (resp. $\mathbf{y}^2$) is the moment sequence of some probability measure $\mu^* \in \mathscr{P}(\mathbb{K}_1)$ supported on s_1 points of $\mathbb{K}_1$ (resp. $\nu^* \in \mathscr{P}(\mathbb{K}_2)$ supported on s_2 points of $\mathbb{K}_2$). From the proof of (a) it follows immediately that $\rho_i \ge \sup_{\mathbf{z}} \{ p_{\mathbf{y}^1}(\mathbf{z}) \,:\, \mathbf{z} \in \mathbb{K}_2 \} \ge \rho^*$ and similarly $\rho^k \le \inf_{\mathbf{x}} \{ \hat{p}_{\mathbf{y}^2}(\mathbf{x}) \,:\, \mathbf{x} \in \mathbb{K}_1 \} \le \rho^*$. As $\rho_i \le \rho^k$ it follows that $\rho_i = \rho^k = \rho^*$. Hence, (μ^*, ν^*) is an optimal pair of strategies for the game (13.21). □

13.3.3 *The univariate case*

Let $\mathbb{K}_1, \mathbb{K}_2 \subset \mathbb{R}$ be intervals of the real line, say $\mathbb{K}_1 = [a, b]$ and $\mathbb{K}_2 = [c, d]$. Then the optimal value ρ^* of the game (13.7) is obtained by solving a *single* semidefnite program.

Namely, let d_1 be the maximum degree of the univariate polynomial $x \mapsto p_{\boldsymbol{\alpha}}(x)$ defined in (13.23), when $|\boldsymbol{\alpha}| \leq d_z$. Similarly, let d_2 be the maximum degree of the univariate polynomial $z \mapsto \hat{p}_{\boldsymbol{\alpha}}(z)$ defined in (13.27), when $|\boldsymbol{\alpha}| \leq d_x$. Consider the semidefinite programs:

$$\begin{array}{rl} \inf\limits_{\mathbf{y},\lambda,\sigma_k} & \lambda \\ \text{s.t.} & \lambda - p_{\mathbf{y}} = \sigma_0 + \sigma_1(d-z)(z-c) \\ & \mathbf{M}_{\lceil d_1/2\rceil}(\mathbf{y}) \succeq 0 \\ & \mathbf{M}_{\lceil d_1/2\rceil-1}((b-x)(x-a),\mathbf{y}) \succeq 0 \\ & y_0 = 1 \\ & \sigma_0,\sigma_1 \in \Sigma[z];\ \deg\sigma_j \leq d_z,\ j=0,1, \end{array} \tag{13.33}$$

and

$$\begin{array}{rl} \sup\limits_{\mathbf{y},\lambda,\sigma_k} & \lambda \\ \text{s.t.} & \hat{p}_{\mathbf{y}} - \lambda = \sigma_0 + \sigma_1(b-x)(x-a) \\ & \mathbf{M}_{\lceil d_2/2\rceil}(\mathbf{y}) \succeq 0 \\ & \mathbf{M}_{\lceil d_2/2\rceil-1}((d-z)(z-c),\mathbf{y}) \succeq 0 \\ & y_0 = 1 \\ & \sigma_0,\sigma_1 \in \Sigma[x];\ \deg\sigma_j \leq d_x,\ j=0,1. \end{array} \tag{13.34}$$

Corollary 13.8. *If $\mathbb{K}_1$ and $\mathbb{K}_2$ are intervals of the real line $\mathbb{R}$, then the value ρ^* of the game (13.21) is the optimal value of the semidefinite programs (13.33) and (13.34).*

13.4 Exercises

Exercise 13.1. Prove the following lemma:

Lemma 13.9. *Let $\mathbb{K} \subset \mathbb{R}^n$ be compact and let $p, q : \mathbb{R}^n \to \mathbb{R}$ be continuous with $q > 0$ on $\mathbb{K}$. Let $\mathscr{P}(\mathbb{K}) \subset \mathscr{M}(\mathbb{K})_+$ be the set of Borel probability measures on $\mathbb{K}$. Then*

$$\inf_{\mu \in \mathscr{P}(\mathbb{K})} \frac{\displaystyle\int_{\mathbb{K}} p\, d\mu}{\displaystyle\int_{\mathbb{K}} q\, d\mu} = \inf_{\varphi \in \mathscr{M}(\mathbb{K})_+} \{ \int_{\mathbb{K}} p\, d\varphi : \int_{\mathbb{K}} q\, d\varphi = 1 \}$$

$$= \inf_{\mu \in \mathscr{P}(\mathbb{K})} \int_{\mathbb{K}} \frac{p}{q}\, d\mu = \inf_{\mathbf{x} \in \mathbb{K}} \frac{p(\mathbf{x})}{q(\mathbf{x})}.$$

13.5 Notes and Sources

13.1. Most of this section is from Lasserre (2006d) and for a survey on algorithms and complexity of robust optimization the interested reader is referred to e.g. Ben-Tal *et al.* (2000). For recent various applications see Ben-Tal *et al.* (2006) and all papers in the same issue of the journal.

13.2 and 13.3. The material of these sections is from Laraki and Lasserre (2012). Corollary 13.8 was first proved in Parrilo (2006).

It is well known that any 2-player zero-sum *finite* game is reducible to a linear program and hence could be computed in polynomial time; see e.g. Dantzig (1963). Lemke and Howson (1964) provided a famous algorithm that computes a Nash equilibrium of any 2-player *non* zero-sum finite game. The algorithm has been extended to n-player finite games in Rosenmüller (1971), Wilson (1971) and Govindan and Wilson (2003).

An alternative to the Lemke-Howson algorithm for 2-player games is provided in van den Elzen and Talman (1991) and has been extended to n-player games in Herings and van den Elzen (2002). As shown in the recent survey Herings and Peeters (2010), all these algorithms are homotopy-based and converge under some regularity condition.

Recently, Savani and von Stengel (2006) proved that the Lemke-Howson algorithm for 2-player games may be exponential. One may expect that this result extends to all known homotopy methods. Daskalakis *et al.* (2006) proved that solving numerically 3-player finite games is hard.[3] The result has been extended to 2-player finite games by Chen and Deng (2006). For a recent survey on the complexity of computing equilibria on game theory, see Roughgarden (2010).

A different approach to solve the problem is to view the set of Nash

[3] More precisely, it is complete in the PPAD class of all search problems that are guaranteed to exist by means of a direct graph argument. This class was introduced in Papadimitriou (1994) and is between P and NP.

equilibria as the set of real nonnegative solutions to a system of polynomial equations. Methods of computational algebra (e.g. using Gröbner bases) can be applied as suggested and studied in e.g. Datta (2009), Lipton and Markakis (2004) and Sturmfels (2002). However, in this approach, one first computes *all* complex solutions to sort out all real nonnegative solutions afterwards. Interestingly, polynomial equations can also be solved via homotopy-based methods (see e.g. Verschelde (1999)). Notice that if one adopts the viewpoint of Nash equilibria as *real* solutions of polynomial equations, then the machinery of Chapter 6 can be applied so as to avoid computing complex solutions.

These above algorithms are concerned with finite games. On the other hand, in Polynomial games introduced by Dresher *et al.* (1950), the set of pure strategies $\mathbf{S}^i$ of Player i is a product of compact intervals. When the game is zero-sum and $\mathbf{S}^i = [0,1]$ for each player i, Parrilo (2006) showed that finding an optimal solution is equivalent to solving a single semidefinite program; see Corollary 13.8. Shah and Parrilo (2007) extended the methodology to discounted zero-sum stochastic games in which the transition is controlled by one player only. Finally, it is worth noticing recent algorithms designed to solve some specific classes of infinite games; for instance see Gürkan and Pang (2009).

Chapter 14

Bounds on Linear PDE

This chapter applies the moment approach to provide bounds on functionals of solutions of linear partial differential equations with boundary conditions and polynomial coefficients.

14.1 Linear Partial Differential Equations

In general, it is impossible to obtain analytical solutions of partial differential equations (PDEs) and so one has to invoke numerical methods. A typical approach is to first discretize the domain and then obtain an approximate solution by solving the resulting equations, and matching boundary values and initial conditions. As an immediate consequence, the computational complexity is exponential in the dimension because with $O(1/\epsilon)$ points in each dimension, one needs to solve a system of size $O((1/\epsilon)^n)$ for n-dimensional PDEs and a desired accuracy of $O(\epsilon)$.

In this chapter we provide a methodology to compute upper and lower bounds on some functional of the solution of a PDE with boundary conditions. Let L be the following partial differential operator with polynomial coefficients :

$$L = \sum_{\boldsymbol{\alpha} \in \mathbb{N}^n} L_{\alpha} \frac{\partial^{\boldsymbol{\alpha}}}{\partial \mathbf{x}^{\boldsymbol{\alpha}}}, \qquad L_{\alpha} \in \mathbb{R}[\mathbf{x}] \quad \forall \boldsymbol{\alpha} \in \mathbb{N}^n, \tag{14.1}$$

where for every $\boldsymbol{\alpha} \in \mathbb{N}^n$,

$$\frac{\partial^{\boldsymbol{\alpha}} u(\mathbf{x})}{\partial \mathbf{x}^{\boldsymbol{\alpha}}} = \frac{\partial^{|\boldsymbol{\alpha}|} u(\mathbf{x})}{\partial^{\alpha_1} x_1 \cdots \partial^{\alpha_n} x_n}.$$

Let $\Omega \subset \mathbb{R}^n$ with boundary $\partial\Omega$, $f : \mathbb{R}^n \to \mathbb{R}$, and consider the linear PDE:

$$Lu = f \qquad \text{on } \Omega, \tag{14.2}$$

with some additional appropriate boundary conditions on $\partial\Omega$.

The goal is to evaluate (or provide upper and lower bounds on) the functional

$$J := \int_{\Omega} (G\,u)\,d\mathbf{x}, \tag{14.3}$$

for some linear operator $G := \sum_{\alpha} G_{\alpha} \frac{\partial^{\alpha}}{\partial \mathbf{x}^{\alpha}}$ as in (14.1), and where u is a solution of the PDE (14.2).

The moment approach

Equation (14.2) should be understood in a weak sense, namely

$$Lu = f \Longleftrightarrow \int_{\Omega} (Lu)\phi\,d\mathbf{x} = \int_{\Omega} f\phi\,d\mathbf{x}, \quad \forall \phi \in \mathscr{D},$$

where $\mathscr{D}$ is some subset of the space of smooth functions $\mathscr{C}^{\infty}$. One also assumes that $\mathscr{D}$ has a countable dense subset generated by $\mathscr{F} = (\phi_1, \phi_2, \ldots)$, so that:

$$\begin{aligned} Lu = f \Longleftrightarrow & \int_{\Omega} (Lu)\phi\,d\mathbf{x} = \int_{\Omega} f\phi\,d\mathbf{x}, \quad \forall \phi \in \mathscr{D}, \\ \Longleftrightarrow & \int_{\Omega} (Lu)\phi_i\,d\mathbf{x} = \int_{\Omega} f\phi_i\,d\mathbf{x}, \quad \forall \phi_i \in \mathscr{F}. \end{aligned} \tag{14.4}$$

Assume that a solution u is bounded from below by u_0. Then the basic underlying idea is to consider the function $v := u - u_0$ as a nonnegative element of $L_1(\Omega)$, i.e. the density of a measure $d\mu = v dx$ on Ω, absolutely continuous with respect to the Lebesgue measure.

Integration by parts of (14.4) yields countably many linear conditions on the moments of μ and additional variables (moments of appropriate measures on the boundary $\partial\Omega$). Similarly, J in (14.3) becomes a linear expression in the same variables.

Then one computes a lower (resp. upper) bound by minimizing (resp. maximizing) the latter linear expression under:

- finitely many linear moment constraints obtained from (14.4), and
- appropriate moment conditions for support constraints of the involved measures.

By increasing the number of moment constraints, one obtains tighter lower and upper bounds. As the form of the moment constraints is problem dependent, the approach is better illustrated on the following simple one-dimensional example. Let the PDE (14.2) be the ordinary differential equation (ODE):

$$u'' + au' + bu \,=\, f \quad \text{or, equivalently,} \quad v'' + av' + bv \,=\, \tilde{f}$$

(with $\tilde{f} := f - bu_0$) on the interval $\Omega = [0,1]$. Let the boundary conditions on $\partial\Omega$ be $u'(0) = c_0$ and $u'(1) = c_1$. With $\phi \in \mathscr{F}$, integration by parts of (14.4) yields

$$\int_0^1 \phi \tilde{f}\, dx \,=\, v'\phi\,|_0^1 - v\phi'\,|_0^1 + av\phi\,|_0^1 + \int_0^1 (\phi'' - a\phi' + b\phi)\, v dx$$

that is,

$$\begin{aligned}\int_0^1 \phi \tilde{f}\, dx = {}& \phi(1)c_1 - \phi(0)c_0 \\ & + v(1)(a\phi(1) - \phi'(1)) + v(0)(\phi'(0) - a\phi(0)) \\ & + \int_0^1 (\phi'' - a\phi' + b\phi)\, v dx.\end{aligned}$$

Let $\mathbf{y} = (y_k)$ be the vector of moments of $d\mu = v dx$, and let $\gamma = (\gamma_j)$ with

$$\gamma_j \,:=\, \int_0^1 x^j\, \tilde{f}(x)\, dx, \qquad j = 0, 1, \ldots$$

With $x \mapsto \phi_j(x) := x^j$, one obtains the moment conditions

$$\begin{aligned}\gamma_j = {}& c_1 - c_0\delta_{j=0} \\ & + v(1)(a - j) + v(0)(\delta_{j=1} - a\delta_{j=0}) \\ & + j(j-1)\, y_{j-2} - aj\, y_{j-1} + b\, y_j,\end{aligned} \tag{14.5}$$

with the Kronecker symbol $\delta_{j=\cdot}$ and the convention $y_j = 0$ whenever $j < 0$.

Next, suppose that $G(v) := x^k v + x^p v' + x^q v"$ for some $k, p, q \in \mathbb{N}$, so that

$$\begin{aligned}\int_0^1 (Gv)\, dx = {}& y_k + v(1) - v(0)\delta_{p=0} - p\, y_{p-1} + c_1 - c_0\delta_{q=0} \\ & - q(v(1) - v(0)\delta_{q=1}) + q(q-1)\, y_{q-2}\end{aligned} \tag{14.6}$$

again with the convention $y_j = 0$ if $j < 0$. With $x \mapsto g(x) := x(1-x)$, and for $i \geq i_0 := \max[k, p-1, q-2]$, consider the semidefinite programs:

$$\rho_i^1 = \sup_{\mathbf{y},v(1),v(0)} \quad (14.6)$$

$$\text{s.t. } \mathbf{M}_i(\mathbf{y}) \succeq 0,\ \mathbf{M}_{i-1}(g\,\mathbf{y}) \succeq 0 \qquad (14.7)$$

$$(14.5), \quad 0 \leq j \leq 2i,$$

and

$$\rho_i^2 = \inf_{\mathbf{y},v(1),v(0)} \quad (14.6)$$

$$\text{s.t. } \mathbf{M}_i(\mathbf{y}) \succeq 0,\ \mathbf{M}_{i-1}(g\,\mathbf{y}) \succeq 0 \qquad (14.8)$$

$$(14.5), \quad 0 \leq j \leq 2i.$$

Of course $\rho_i^2 \leq J \leq \rho_i^1$ for every $i \geq i_0$ and so, as $i \to \infty$,

$$\rho_i^2 \uparrow \underline{\rho} \leq J \quad \text{and} \quad \rho_i^1 \downarrow \overline{\rho} \geq J. \qquad (14.9)$$

14.2 Notes and Sources

This chapter is inspired from Bertsimas and Caramanis (2006). The moment approach to PDE was already described in Dawson (1980) but with no effective computation as general multivariate moment conditions were not available at that time. The first effective application of the moment-approach to solving PDEs seems to date back to the "eigenvalue moment method" initiated in Handy and Bessis (1985) in the context of one-dimensional quantum systems with rational fraction potential functions. The motivation was to provide an alternative to eigenvalue methods for computing bounds on the quantum ground-state energy. Using a monomial basis of test functions, integration by parts of the associated Schrödinger equation yields linear constraints on the moments of $d\mu := u\,d\mathbf{x}$ where u is the solution of the PDE. At that time, semidefinite programming was not a widely spread out technique and so the authors used Hankel-Hadamard determinant constraints (from the Hamburger moment problem) as moment conditions. The latter can be expressed in terms of finitely many moment variables (the so-called missing moments) after elimination of the others via the linear moment equations from the Schrödinger PDE. As observed in several examples, the convergence of upper and lower bounds to the ground-state energy is fast. As the method becomes untractable for more than 3 missing moments, Handy *et al.* (1988) provided an improvement by using cutting planes to approximate the convex feasible region defined by the positive semidefiniteness of the Hankel-moment matrix; see also Handy *et al.* (1989) and Handy (2001) for refinements and extensions. A modern treatment would simply use a hierarchy of semidefinite relaxations in the spirit of (14.7)-(14.8).

More recently, Bertsimas and Caramanis (2006) applied this technique to Bessel and Helmholtz equations as well as on a queueing application where some queue length is approximated via a reflected Brownian motion. They also reported excellent numerical results.

Final Remarks

We hope to have convinced the reader that the moment approach developed in this book can be useful to help solve or approximate the Generalized Moment Problem and some of its various applications. As already mentioned, and since most of the addressed problems are difficult, the reader will have understood that this methodology has practical limitations mainly due to the size of the initial problem to solve. Indeed so far, and in view of the present status of the semidefinite programming solvers, the approach is limited to problems of modest size. Fortunately, many problems of larger size exhibit some sparsity patterns or symmetries that can be exploited as we have also indicated.

In the book, we have not touched upon the (new) theory of positive polynomials in the non commutative case, i.e., polynomials in non commuting variables. Several beautiful results have already been obtained in *Non Commutative Real Algebraic Geometry*, a new and promising domain of Mathematics. Among these results, one finds some non commutative Positivstellensatze analogues of the Positivstellenstaze presented in Chapter 2 of this book. For an introduction to these concepts the interested reader is referred to the survey Helton and Putinar (2007) as well as Schmüdgen (2008) and the many references therein. Potential applications arise in e.g. control as well as in quantum information. For control applications the interested reader is referred to e.g. de Oliveira *et al.* (2008). Also, a recent non commutative version of the moment approach for polynomial optimization presented in Chapter 5 of this book, has been defined in Pironio *et al.* (2009) and tested successfully in a quantum information application described in Navascués *et al.* (2008).

Appendix A

Background from Algebraic Geometry

A main concern in algebraic geometry is the analysis of the set of solutions for systems of polynomial equations. In this section, we review some of the basic ingredients of algebraic geometry with emphasis on the strong interplay between geometric notions (the set of solution "points"), and their algebraic representation (the polynomials equations).

A.1 Fields and Cones

We denote by $x+y$ and $x \cdot y\,(= xy)$ the addition and multiplication respectively of two elements x, y of a field.

If both F, F_1 are fields and $F \subseteq F_1$, then F_1 is a **field extension** of F, denoted F_1/F. A field extension F_1/F is called **algebraic** if every element of F_1 is a root of some nonzero polynomial with coefficients in F. A field extension which is not algebraic is called **transcendental**. For instance $\mathbb{C}/\mathbb{R}$ is an algebraic field extension, whereas $\mathbb{R}/\mathbb{Q}$ is transcendental. If F_1 is regarded as a vector space over F, its dimension is called the degree of the extension. According to this degree, an algebraic field extension is further classified as finite or infinite. All finite field extensions are algebraic, but the converse is not true.

A field F is **algebraically closed** if every univariate polynomial of degree at least 1, and with coefficients in F, has a root in F. Every field F is contained in a field $\overline{F}$ which is algebraically closed and such that every element of $\overline{F}$ is the root of a nonzero univariate polynomial with coefficients in F. This field (unique up to isomorphism) is called the **algebraic closure** of F. For instance, the field $\mathbb{C}$ is the algebraic closure of the field $\mathbb{R}$.

We begin with the definitions of an **ordered** field, a **cone** and a **real** field.

Definition A.1. An **ordered field** $(F, \leq)$ is a field F, equipped with an ordering $\leq$, i.e., a total order relation $\leq$ such that for any $x, y \in F$:
(a) $x \leq y \Rightarrow x + z \leq y + z$ for all $z \in F$.
(b) If $0 \leq x$ and $0 \leq y$ then $xy \geq 0$.

The ordering $\leq$ on F is called **Archimedean** and F is called an **Archimedean ordered field** if for each $x \in F$ there is $n \in \mathbb{N}$ such that $x \leq n$. For example, with their usual ordering, both $\mathbb{Q}$ and $\mathbb{R}$ are Archimedean.

Definition A.2. A **cone** of a field F is a subset $P \subset F$ such that
(a) If $x, y \in P$ then $x + y \in P$ and $xy \in P$.
(b) $x^2 \in P$ for every $x \in F$.

A cone P with the property that $-1 \notin P$ is called **proper**.

The set $\{x \in F : x \geq 0\}$ is called the positive cone of $(F, \leq)$, and the set $\Sigma F^2 \subset F$ of finite sums of squares of elements of F, is a cone contained in every cone of F.

Theorem A.1. *Let F be a field. Then the following are equivalent:*
(a) *F can be ordered.*
(b) *F has a proper cone.*
(c) $-1 \notin \Sigma F^2$.
(d) *For every $x_1, \ldots, x_n$ in F*

$$\sum_{i=1}^{n} x_i^2 = 0 \quad \Rightarrow x_1 = \cdots = x_n = 0.$$

A field that satifies the above four properties in Theorem A.1 is called a **real** field. A **real closed** field F is a field that has *no* nontrivial real algebraic extension $F_1 \supset F$, $F_1 \neq F$. For instance, the field $\mathbb{Q}$ of rationals is a real field and its extension field $\mathbb{R}_{\text{alg}}$ of real algebraic numbers (which is its algebraic closure), is a real closed field. Of course, $\mathbb{R}$ is also a real closed field.

A.2 Ideals

Let k be an arbitrary field and $k[\mathbf{x}] = k[x_1, \ldots, x_n]$ be the (commutative) ring of polynomials in n variables with coefficients in k.

Definition A.3. An **ideal** $I \subset k[\mathbf{x}]$ is a subspace of $k[\mathbf{x}]$ that satisfies

$$\forall f \in k[\mathbf{x}], \quad g \in I \Rightarrow fg \in I.$$

Given a finite family $F = \{f_1, \ldots, f_m\} \subset k[\mathbf{x}]$, we denote by $I = \langle f_1, \ldots, f_n \rangle$ the ideal generated by F, that is,

$$\langle f_1, \ldots, f_n \rangle := \left\{ \sum_{i=1}^{m} g_i f_i \quad : \quad g_i \in k[\mathbf{x}],\, i = 1, \ldots, m \right\}.$$

The following theorem is a key result on ideals.

Theorem A.2. (The Hilbert Basis Theorem) *Every ideal I of $k[\mathbf{x}]$ is* **finitely** *generated, that is, there exists a finite family $F = \{f_1, \ldots, f_s\} \subset k[\mathbf{x}]$ such that*

$$f \in I \Rightarrow f = \sum_{i=1}^{s} h_i f_i, \quad h_i \in k[\mathbf{x}],\ i = 1, \ldots, s.$$

As every ideal of $k[\mathbf{x}]$ is finitely generated, one may wonder whether some representations are better than others. Reduced Gröbner bases are precisely representations that are interesting in the sense that if a finite subset $F \subset k[\mathbf{x}]$ is a reduced Gröbner basis of an ideal $I \subset k[\mathbf{x}]$ then the **ideal membership problem** of detecting whether or not a given polynomial $f \in k[\mathbf{x}]$ is in I, is solvable by Buchberger's algorithm (the multi-dimensional analogue of the Euclidean algorithm in the univariate case).

An important notion is the **quotient ring** $k[\mathbf{x}]/I$ associated with an ideal I of $k[\mathbf{x}]$. Its elements, of the form $f + I$, are called **cosets** of I. Namely, $k[\mathbf{x}]/I$ consists of equivalence classes modulo I of polynomials $f \in k[\mathbf{x}]$. Equivalently, given two polynomials $f, g \in k[\mathbf{x}]$, f is **congruent** to g modulo I if and only if $f - g \in I$. It follows that $k[\mathbf{x}]/I$ is a commutative ring with the usual addition and multiplication inherited from $k[\mathbf{x}]$.

Definition A.4. The **radical** ideal $\sqrt{I} \subset k[\mathbf{x}]$ of an ideal $I \subset k[\mathbf{x}]$ is defined as follows

$$\sqrt{I} := \{f \in k[\mathbf{x}] : \quad \text{there exists } e \in \mathbb{N} \text{ such that } f^e \in I\}.$$

For instance, with $n = 1$ and $a \in \mathbb{R}$, let $I = \langle (x-a)^2 \rangle \subset \mathbb{R}[x]$ be the ideal of $\mathbb{R}[x]$ generated by the polynomial $x \mapsto x - a$. Then, $\sqrt{I} = \langle x - a \rangle$.

A.3 Varieties

An important geometric notion is that of a variety.

Definition A.5. Given a subset $S \subset k[\mathbf{x}]$, and a field $K \supseteq k$, the set $V_K(S) \subset K^n$ defined by

$$V_K(S) := \{\mathbf{x} \in K^n : f(\mathbf{x}) = 0, \ \forall f \in S\}$$

is called a **variety** of K^n.

In particular, given an ideal $I \subset k[\mathbf{x}]$, $V_K(I)$ denotes the variety of K^n associated with I, and if $I = \langle f_1, \ldots, f_m \rangle \subset k[\mathbf{x}]$,

$$V_K(I) = \{\mathbf{x} \in K^n : f_i(\mathbf{x}) = 0, \ i = 1, \ldots, m\}.$$

Note that the variety is in K^n, whereas the ideal is in $k[\mathbf{x}]$. Similarly, given a subset $V \subset K^n$, we define the ideal $I(V)$ in $k[\mathbf{x}]$, by

$$I(V) := \{f \in k[\mathbf{x}] : f(\mathbf{x}) = 0, \forall \mathbf{x} \in V\}.$$

Ideals and varieties illustrate an interplay between algebra and geometry. With ideals of $k[\mathbf{x}]$ (algebraic objects) are associated varieties in K^n (geometric objects), and conversely, with subsets of K^n are associated ideals of $k[\mathbf{x}]$.

Example A.1. Let $I_1 = \langle f_1 \rangle \subset \mathbb{R}[\mathbf{x}]$ with $f_1(\mathbf{x}) = x_1^2 + x_2^2$, and let $I_2 = \langle f_2, f_3 \rangle \subset \mathbb{R}[\mathbf{x}]$ with $f_2(\mathbf{x}) = x_1 + x_2$, and $f_3(\mathbf{x}) = x_2$. We have $V_{\mathbb{R}}(I_1) = V_{\mathbb{R}}(I_2) = \{0, 0\}$. On the other hand, $V_{\mathbb{C}}(I_2) = \{(0,0)\}$, whereas $V_{\mathbb{C}}(I_1)$ is the union of the lines $x_2 = \pm i x_1$. The example illustrates that I_1 and I_2 are really different ideals, but the set of solutions in $k = \mathbb{R}$ of the corresponding system of equations has too few solutions to distinguish them. Enlarging $k = \mathbb{R}$ to $\mathbb{C}$ shows that the corresponding varieties are indeed different.

A fundamental result in algebraic geometry is the so-called **Hilbert Nullstellensatz.**

Theorem A.3 (Weak Hilbert Nullstellensatz). *Let $I \subset k[\mathbf{x}]$ be an ideal whose associated algebraic variety*

$$V_{\bar{k}}(I) := \left\{\mathbf{x} \in \bar{k}^n \ : \quad g(\mathbf{x}) = 0, \ \forall g \in I\right\},$$

is empty. Then $I = k[\mathbf{x}]$.

In particular, consider a family $\{f_i\} \subset \mathbb{C}[\mathbf{x}]$, and the associated set of polynomials equations

$$f_i(\mathbf{x}) = 0, \quad i = 1, \ldots, m. \tag{A.1}$$

Then, the above system has no solution $\mathbf{x} \in \mathbb{C}^n$ if and only if there exist m polynomials $g_i \in \mathbb{C}[\mathbf{x}]$ such that

$$1 = \sum_{i=1}^{m} f_i \, g_i. \tag{A.2}$$

In other words, the g_i's provide a **certificate** of infeasibility of the system of equations (A.1). Notice that Theorem A.3 also includes the fundamental theorem of algebra. Indeed, if I consists of a single univariate (non constant) polynomial $f \in \mathbb{C}[x]$, then $V_{\mathbb{C}}(I)$ is never empty because we would have that $1 \equiv gf$ for some $g \in \mathbb{C}[x]$, which is clearly impossible.

Example A.2. Let $f_1(\mathbf{x}) = x_1^2 + x_2^2 + 1$ $f_2(\mathbf{x}) = x_1 + x_2$, $f_3(\mathbf{x}) = x_2$ and let $I = \langle f_1, f_2, f_3\rangle$. Then we have $V_{\mathbb{C}}(I) = \emptyset$, which is confirmed by the identity $1 = f_1 - x_1 f_2 + (x_1 - x_2) f_3$.

Theorem A.4 (Strong Hilbert Nullstellensatz). *For every ideal $I \subset k[\mathbf{x}]$, one has $I(V_{\bar{k}}(I)) = \sqrt{I}$.*

Theorem A.4 says that if I is radical, i.e., if $I = \sqrt{I}$, then I is in a sense a minimal algebraic description, without redundances (like squares).

Definition A.6. An ideal I of a commutative ring A is said to be **real** if, for every sequence $a_1, \ldots, a_p$ of elements of A, one has

$$\sum_{i=1}^{p} a_i^2 \in I \quad \Rightarrow \quad a_i \in I, \quad i = 1, \ldots, p.$$

We now state the Real-Nullstellensatz, the analogue of Theorem A.4 for real closed fields.

Theorem A.5 (Real Nullstellensatz). *Let k be a real closed field and I an ideal of $k[\mathbf{x}]$. Then $I = I(V_k(I))$ if and only if I is real.*

For instance, let $k = \mathbb{R}$ and $g \in k[x]$ with $g(x) = x$. Then, the ideal $I = \langle g^2 \rangle$ is not real, because $g^2 \in I$, whereas $g \notin I$. In fact, we have $I(V_k(I)) = \langle g \rangle \neq I$.

Given an ideal $I \subset k[\mathbf{x}]$, the ideal denoted $\sqrt[k]{I}$ and defined by

$$\sqrt[k]{I} := \{ f \in k[\mathbf{x}] \,:\, \exists p \in \mathbb{N},\, (f_i)_{i=1}^m \subset k[\mathbf{x}] \quad \text{s.t.} \quad f^{2p} + \sum_{j=1}^{m} f_i^2 \in I \}$$

is called the *real* radical of I. And one has another form of the Real Nullstellensatz:

Theorem A.6 (Real Nullstellensatz). *Let k be a real closed field and I an ideal of $k[\mathbf{x}]$. Then $\sqrt[k]{I} = I(V_k(I))$.*

A.4 Preordering

The following definitions present concepts that play an essential role in real algebraic geometry.

Definition A.7. A subset T of a commutative ring A, with $1 \in T$, is called:
(a) a **preprime** of A if

$$0 \in T, \quad T+T \subseteq T, \quad T \cdot T \subseteq T, \quad \text{and} \quad -1 \notin T.$$

(b) a **quadratic module** on A if

$$T+T \subseteq T, \quad A^2 \cdot T \subseteq T, \quad \text{and} \quad -1 \notin T.$$

(c) a **preordering** of A if

$$T+T \subseteq T, \quad T \cdot T \subseteq T, \quad A^2 \subseteq T, \quad \text{and} \quad -1 \notin T.$$

where A^2 stands for the set of squared elements of A.

Clearly, every preordering is a preprime and a quadratic module, and since every quadratic module contains ΣA^2, the preordering ΣA^2 is the smallest quadratic module on A, as well as being the smallest preordering of A. A preordering enjoys the basic properties of **positive** elements.

The maximal quadratic modules S on A also satisfy

$$S \cup -S = A \quad \text{and} \quad S \cap -S \text{ is a prime ideal.}$$

Thus, if A is a field, $S \cap -S = \{0\}$. In this latter case, the ordering

$$a \leq b \quad \Leftrightarrow \quad b-a \in S, \qquad a, b \in A,$$

linearly orders A, so that for all $a, b, c \in A$,

$$a \leq b \Rightarrow a+c \leq b+c,$$
$$0 \leq a \Rightarrow 0 \leq a\, b^2.$$

Those linear orderings of fields are called **semiorderings**.

Example A.3. We present three examples of preorderings. With $A := \mathbb{R}$, let $T := \mathbb{R}_+$, and with $A := \mathscr{C}(\mathscr{X}, \mathbb{R})$, the ring of continuous functions $f : \mathscr{X} \to \mathbb{R}$ from a nonempty topological space $\mathscr{X}$ to $\mathbb{R}$, let $T := \{f \in \mathscr{C}(\mathscr{X}, \mathbb{R}) \mid f \geq 0\}$.

Another example of a preordering, extensively used in the sequel, is obtained as follows. Let $\mathbb{R}[\mathbf{x}]$ be the ring of real polynomial in the n variables $x_1, \ldots, x_n$ (also an $\mathbb{R}$-algebra), and let $\Sigma[\mathbf{x}] \subset \mathbb{R}[\mathbf{x}]$ denotes the sums of squares of elements of $\mathbb{R}[\mathbf{x}]$. Given a family $F := \{f_1, \ldots, f_m\} \subset \mathbb{R}[\mathbf{x}]$, and a set $J \subset \{1, \ldots, m\}$ we let by $f_J(\mathbf{x}) = \prod_{j \in J} f_j(\mathbf{x})$, with the convention that $f_\emptyset \equiv 1$. Then, the set

$$P(f_1, \ldots, f_m) := \left\{ \sum_{J \subseteq \{1,\ldots,m\}} q_J\, f_J \quad : \quad q_J \in \Sigma[\mathbf{x}] \right\} \tag{A.3}$$

is a preordering, called the **preordering generated by** F.

Finally, with the constant polynomial $f_0 = 1$, the set

$$Q(f_0, f_1, \ldots, f_m) := \left\{ \sum_{j=0}^{m} q_j\, f_j \quad : \quad q_j \in \Sigma[\mathbf{x}] \right\} \tag{A.4}$$

is called the quadratic module generated by the family F. Observe that both $P(f_1, \ldots, f_m)$ and $Q(f_0, \ldots, f_m)$ are also *convex cones* of $\mathbb{R}[\mathbf{x}]$.

A.5 Algebraic and Semi-algebraic Sets over a Real Closed Field

In this section we introduce algebraic and semi-algebraic sets.

Definition A.8. Let k be a real closed field.
(a) An **algebraic** set A of k^n is the set of zeros of some subset $B \subset k[\mathbf{x}]$, that is,

$$A = V(B) = \{\mathbf{x} \in k^n \, : \, f(\mathbf{x}) = 0, \; \forall f \in B\},$$

for some $B \subset k[\mathbf{x}]$.
(b) A **semi-algebraic** set $A \subset k^n$ is a set of the form

$$A = \bigcup_{i=1}^{s} \bigcap_{j=1}^{r_i} \{\mathbf{x} \in k^n \, : \, f_{ij}(\mathbf{x}) \diamond 0\},$$

for some finite family $(f_{ij}) \subset k(\mathbf{x}]$, and where "$\diamond$" is either "$<$" or "$=$."
(c) A **basic open semi-algebraic** set $A \subset k^n$ is a semi-algebraic set of the form

$$A = \{\mathbf{x} \in k^n \, : \, f_j(\mathbf{x}) > 0, \quad j = 1, \ldots, m\},$$

for some family $(f_j)_{i=1}^m \subset k[\mathbf{x}]$.
(d) A **basic closed semi-algebraic** set $A \subset k^n$ is a semi-algebraic set of the form

$$A = \{\mathbf{x} \in k^n \, : \, f_j(\mathbf{x}) \geq 0, \quad j = 1, \ldots, m\},$$

for some family $(f_j)_{j=1}^m \subset k[\mathbf{x}]$.

From their definition, semi-algebraic sets are invariant under finite union, finite intersection and complementation. They are also invariant under projection, and in fact, every semi-algebraic set of k^n is the projection of an algebraic set of k^{n+1}.

If k is a real closed field and if the coefficients of the polynomials that define a semi-algebraic set belong to a subring $D \subset k$, the semi-algebraic set is said to be defined over D. The following result is the important projection theorem.

Theorem A.7. *Let k be a real closed field and $D \subset k$ be a subring of k. Given a semi-algebraic set of k^{n+1} defined over D, its projection to k^n is a semi-algebraic set defined over D.*

Notice that by Theorem A.7, the projection of a basic semi-algebraic set is semi-algebraic. However, in general it is *not* a basic semi-algebraic set.

Every semi-algebraic set $A \subset k^n$ can be written as a finite union of semi-algebraic sets of the form

$$\{\mathbf{x} \in k^n \,:\, f_i(\mathbf{x}) = 0,\; g_j(\mathbf{x}) > 0,\; i = 1, \ldots, m,\; j = 1, \ldots, p\},$$

for some families $\{f_i\}, \{g_j\} \subset k[\mathbf{x}]$.

We equip k^n with the usual Euclidean topology coming from the ordering of k, so that the open ball $B(\mathbf{x}, r)$ is the usual subset

$$\{\mathbf{y} \in k^n \,:\, \|\mathbf{y} - \mathbf{x}\| < r\}.$$

So, the closure $\overline{S}$ of a semi-algebraic subset $S \subset k^n$ is

$$\overline{S} = \{\mathbf{x} \in k^n \,:\, \forall t \in k,\, \exists \mathbf{y} \in S \text{ s.t. } \|\mathbf{y} - \mathbf{x}\| < t^2 \text{ or } t = 0\}.$$

The closure and the interior of a semi-algebraic set are semi-algebraic.

Warning: Notice that in general, the closure $\overline{S}$ of $S \subset k^n$ is not obtained from S by just relaxing the strict inequalities in the definition of S.

Example A.4. Let $S := \{\mathbf{x} \in \mathbb{R}^2 \,:\, x_1^3 - x_1^2 - x_2^2 > 0\}$. Relaxing $>$ to $\geq$ would imply $\mathbf{x} = \mathbf{0} \in \overline{S}$ which is not true because $\mathbf{x} \in S$ implies $x_1 \geq 1$.

Similarly, in general, the interior $\mathrm{int}(S)$ of the semi-algebraic set $S = \{\mathbf{x} \in \mathbb{R}^n \,:\, g_j(\mathbf{x}) \geq 0,\, j \in I\}$ is not obtained from S by replacing the inequalities $g_j \geq 0$ with the strict inequalities $g_j > 0$.

Example A.5. With $n = 1$, let $S := \{x \in \mathbb{R} : g(x) \geq 0\}$, with $g(x) = (1-x)x^2(1+x)$, so that $S = [-1, 1]$ and $0 \in \mathrm{int}(S) = (-1, 1)$. But the set $\{x \in \mathbb{R} : g(x) > 0\}$ does not contain 0.

Note also that closed and bounded semi-algebraic sets are not necessarily compact for real closed fields other than $\mathbb{R}$.

A.6 Notes and Sources

Sources for the introductory material are from Bochnak *et al.* (1998), Basu *et al.* (2003), and Adams and Loustaunau (1994). Specifically, Example

A.1 if from Adams and Loustaunau (1994). The real Nullstellensatz in the form of Theorem A.5 is due to Risler (1970).

Appendix B

Measures, Weak Convergence and Marginals

B.1 Weak Convergence of Measures

Let $\mathbf{X}$ be a metric space with its usual associated Borel σ-algebra $\mathscr{B}$. Let $\mathscr{M}(\mathbf{X})$ (resp. $C(\mathbf{X})$) be the space of finite signed Borel measures (resp. bounded continuous functions) on $\mathbf{X}$ and let $\mathscr{M}(\mathbf{X})_+$ be its positive cone, the space of finite Borel measures on $\mathbf{X}$. We next introduce the *weak convergence* of measures, the most commonly used in probability.

Definition B.9. Let $\mathscr{M}(\mathbf{X})$ be equipped with the weak topology $\sigma(\mathscr{M}(\mathbf{X}), C(\mathbf{X}))$. A sequence $(\mu_n) \subset \mathscr{M}(\mathbf{X})_+$ converges weakly to $\mu \in \mathscr{M}(\mathbf{X})_+$, denoted $\mu_n \Rightarrow \mu$, if and only if, as $n \to \infty$,

$$\int_{\mathbf{X}} f \, d\mu_n \to \int_{\mathbf{X}} f \, d\mu \qquad \forall f \in C(\mathbf{X}). \tag{B.1}$$

Definition B.10. Let $\Pi \subset \mathscr{M}(\mathbf{X})_+$ be a set of probability measures on $\mathbf{X}$. Then Π is said to be:

(a) **tight** if for every $\epsilon > 0$ there is a compact set $\mathbb{K}_\epsilon \subset \mathbf{X}$ such that $\mu(\mathbb{K}_\epsilon) \geq 1 - \epsilon$ for every $\mu \in \Pi$.

(b) **relatively compact** if every sequence in Π contains a weakly convergent subsequence, that is, for every $(\mu_n) \subset \Pi$ there exists $\mu \in \mathscr{M}(\mathbf{X})_+$ and $(\mu_{n_k}) \subset (\mu_n)$ such that $\mu_{n_k} \Rightarrow \mu$ as $k \to \infty$.

Tightness and relatively compactness are related via the following result:

Theorem B.8 (Prohorov). *Let* $\mathbf{X}$ *be a metric space and let* $\Pi \subset \mathscr{M}(\mathbf{X})_+$ *be a set of probability measures on* $\mathbf{X}$*. Then:*
(a) *If* Π *is tight then* Π *is relatively compact.*
(b) *If* $\mathbf{X}$ *is separable and complete and* Π *is relatively compact, then* Π *is tight.*

Thus if $\mathbf{X}$ is a *Polish* (i.e., separable and complete metric) space, then Prohorov's Theorem yields that Π is tight if and only if it is relatively compact. This is also true if $\mathbf{X}$ is a locally compact separable metric space in which case it can be given a metric under which it is also complete.

Therefore, in view of the above, it is important to detect when a set of probability measures is relatively compact.

Definition B.11. A function $f : \mathbf{X}{\to}\mathbb{R}_+$ is said to be:

(a) a **moment** if there exists a sequence of compact sets $\mathbb{K}_n \uparrow \mathbf{X}$ such that

$$\lim_{n\to\infty} \inf \{f(\mathbf{x}) \,:\, \mathbf{x} \in \mathbf{X} \setminus \mathbb{K}_n\} = +\infty.$$

(b) **inf-compact** if the level set $\mathbf{A}_r := \{\mathbf{x} \in \mathbf{X} \,:\, f(\mathbf{x}) \leq r\}$ is compact for every $r \in \mathbb{N}$.

Of course, f is a moment if it is inf-compact. Conversely, if $f : \mathbf{X}{\to}\mathbb{R}_+$ is a moment, then the closure of $\mathbf{A}_r$ is compact for every $r \in \mathbb{N}$. Moment functions are useful as we have:

Theorem B.9. *Let* $\Pi \subset \mathscr{M}(\mathbf{X})_+$ *be a set of probability measures and let* $f : \mathbf{X}{\to}\mathbb{R}_+$ *be a moment function.*

(a) Π *is tight if*

$$\sup_{\mu\in\Pi} \int_{\mathbf{X}} f \, d\mu < +\infty. \tag{B.2}$$

(b) *Conversely, if* $\mathbf{X}$ *is a locally compact separable metric space and* Π *is tight, then there exists a moment function* $f : \mathbf{X}{\to}\mathbb{R}_+$ *that satisfies (B.2).*

Recall that $\mathbf{X}$ is a metric space. A function $f : \mathbf{X}{\to}\mathbb{R}$ is lower-semicontinuous (l.s.c.) and bounded below if there exists a nondereasing sequence $(v_n) \subset C(\mathbf{X})$ such that $v_n(\mathbf{x}) \uparrow f(\mathbf{x})$ for every $\mathbf{x} \in \mathbf{X}$. Let $L(\mathbf{X})$ denote the space of functions on $\mathbf{X}$ that are l.s.c. and bounded from below. The space $L(\mathbf{X})$ is useful to characterize weak convergence of measures.

Proposition B.10. *Let* $(\mu_n) \subset \mathscr{M}(\mathbf{X})_+$ *be a bounded sequence. Then* $\mu_n \Rightarrow \mu$ *if and only if*

$$\liminf_{n\to\infty} \int_{\mathbf{X}} f\, d\mu_n \geq \int_{\mathbf{X}} f\, d\mu, \qquad \forall f \in L(\mathbf{X}). \tag{B.3}$$

Finally we end up this section with the following two important results. Recall that the *support* of a Borel measure μ on $\mathbb{R}^n$ is a closed Borel set, the complement of the largest open set $O \subset \mathbb{R}^n$ with $\mu(O) = 0$.

Theorem B.11. *Let* $f_1, \ldots, f_m : \mathbf{X} \to \mathbb{R}$ *be Borel measurable on a measurable space* $\mathbf{X}$ *and let* μ *be a probability measure on* $\mathbf{X}$ *such that* f_i *is integrable with respect to* μ *for each* $i = 1, \ldots, m$. *Then there exists a probability measure* φ *with finite support on* $\mathbf{X}$, *such that:*

$$\int_{\mathbf{X}} f_i\, d\varphi = \int_{\mathbf{X}} f_i\, d\mu, \qquad \forall\, i = 1, \ldots, m.$$

Moreover, the support of φ *may consist of at most* $m + 1$ *points.*

A *cubature* formula of degree d for a Borel measure μ on $\mathbb{R}^n$, is given by an integer $m \geq 1$, points $\{\mathbf{x}_i\}_{i=1}^m$ in the support of μ and positive scalars $\{\lambda_i\}_{i=1}^m$ such that

$$\int_{\mathbb{R}^n} f(\mathbf{x})\, d\mu = \sum_{i=1}^m \lambda_i f(\mathbf{x}_i),$$

for all polynomials $f \in \mathbb{R}[\mathbf{x}]$ of degree at most d. The next result is the basis of existence of cubature formulas for numerical integration and is an extension of the celebrated Tchakaloff's theorem.

Theorem B.12. *Let* μ *be a finite Borel measure with support* $\mathbb{K} \subset \mathbb{R}^n$ *and let* $d \geq 1$ *be a fixed positive integer. Then there exist* m *points* $\{\mathbf{x}_i\}_{i=1}^m \subset \mathbb{K}$ *(with* $m \leq \binom{n+d}{n}$*) and positive weights* λ_i, $i = 1, \ldots, m$, *such that*

$$\int_{\mathbb{K}} f\, d\mu = \sum_{i=1}^m \lambda_i\, f(\mathbf{x}_i),$$

for every polynomial $f \in \mathbb{R}[\mathbf{x}]$ *of degree at most* d.

Theorem B.12 states that for an arbitrary finite Borel measure μ supported on $\mathbb{K} \subset \mathbb{R}^n$, one may always find a Borel measure φ finitely supported on $\mathbb{K}$, such that all moments of φ up to degree m, match exactly those of μ. It extends to arbitrary measures supported on $\mathbb{K}$, Tchakaloff's theorem proved for measures supported on a compact set $\mathbb{K}$ and absolutely continuous with respect to the Lebesgue measure on $\mathbb{R}^n$ (see Tchakaloff (1957)).

B.2 Measures with Given Marginals

In this section we address the following issue:

Let $I = \{1, \ldots, n\} = \cup_{i=1}^p I_k$ with $|I_k| =: n_k$, and let $I_{jk} := I_j \cap I_k$ (whenever $I_J \cap I_k \neq \emptyset$), with $|I_{jk}| =: n_{jk}$, $j \neq k$. In general, the I_k's do not form a partition of I as one may have $\sum_k n_k > n$. Let $\mathbf{x} = (x_1, \ldots, x_n)$ and for any $J \subseteq I$ denote by $\mathbf{x}(J)$ the vector $(x_i \,:\, i \in J)$.

For every $i = 1, \ldots, p$, let μ_i be a given probability measure on $\mathbb{R}^{n_i}$, supported on a compact set $\mathbb{K}_i \subset \mathbb{R}^{n_i}$. Whenever $I_j \cap I_k \neq \emptyset$, denote by $\pi_{jk}\mu_j$ the projection of μ_j on $\mathbb{R}^{n_{jk}}$ (its marginal on the variables $\mathbf{x}(I_{jk})$).

Let $\mathbb{K} \subset \mathbb{R}^n$ be the set

$$\mathbb{K} = \{\, \mathbf{x} \in \mathbb{R}^n \,:\, \mathbf{x}(I_i) \in \mathbb{K}_i, \quad i = 1, \ldots, p \,\}, \tag{B.4}$$

and assume that

$$\pi_{jk}\, \mu_j \;=\; \pi_{kj}\, \mu_k, \qquad \forall j, k = 1, \ldots, p;\; I_j \cap I_k \neq \emptyset. \tag{B.5}$$

Can we construct a probability measure μ on $\mathbb{R}^n$ and supported on $\mathbb{K}$, such that its projection $\pi_j \mu$ on $\mathbb{R}^{n_j}$ (i.e. its marginal on the variables $\mathbf{x}(I_j)$) is equal to μ_j, for every $j = 1, \ldots, p$?

Lemma B.13. *Let $I = \{1, \ldots, n\} = \cup_{i=1}^p I_k$ with $|I_k| =: n_k$. Let $\mathbb{K}_i \subset \mathbb{R}^{n_i}$ be given compact sets, and let $\mathbb{K} \subset \mathbb{R}^n$ be the compact set defined in (B.4). Let μ_i be given probability measures supported on $\mathbb{K}_i$, $i = 1, \ldots, p$, and such that (B.5) holds.*

If the collection of sets $\{I_k\}$ satisfies the running intersection property, i.e.,

$$I_k \bigcap \left(\bigcup_{j=1}^{k-1} I_j \right) \subseteq I_s \quad \textit{for some } s \leq k, \quad k = 2, \ldots, p, \tag{B.6}$$

then there exists a probability measure μ *supported on* $\mathbb{K}$, *such that*

$$\pi_j \mu = \mu_j, \qquad j = 1, \ldots, p,$$

that is, μ_j *is the marginal of* μ *on* $\mathbb{R}^{n_j}$ *(i.e., with respect to the variables* $\mathbf{x}(I_j)$, $j = 1, \ldots, p$*).*

Proof. For every $1 \leq m \leq p$, let $n(m) := \sum_{j=1}^{m} n_j$. Let $\mathscr{B}(\mathbb{R}^n)$ denote the Borel σ-algebra associated with $\mathbb{R}^n$. We construct μ by induction on $k = 1, \ldots, p$.
Induction property (H):

With $1 \leq m < p$, *there is a probability measure* ν_m *supported on*

$$\mathbb{K}(m) := \{\mathbf{x} \in \mathbb{R}^{n(m)} \,:\, \mathbf{x}(I_j) \in \mathbb{K}_j, \quad j = 1, \ldots, m\}$$

such that $\pi_k \nu_m = \mu_k$, *i.e., the projection of* ν_m *on* $\mathbb{R}^{n_k}$ *(on the variables* $\mathbf{x}(I_k)$*) is* μ_k, *for every* $k = 1, \ldots, m$.

(H) is true for $m = 1$ (just take $\nu_1 := \mu_1$). We next prove that it is true for $m+1$. Define a measure ν_{m+1} on $\mathbb{R}^{(m+1)}$ as follows: Let $I^m := I_{m+1} \cap (\cup_{k=1}^{m} I_k)$ with cardinal τ.

(a) If $I^m = \emptyset$ then

$$\nu_{m+1}(A \times B) := \nu_m(A)\, \mu_{m+1}(B), \quad \forall A \in \mathscr{B}(\mathbb{R}^{n(m)}),\, B \in \mathscr{B}(\mathbb{R}^{n_{m+1}})$$

And ν_{m+1} satisfies property (H).

(b) If $I^m \neq \emptyset$ then by (B.6) $I^{m+1} \subseteq I_s$ for some $s \leq m$. Let $d\varphi(\mathbf{x}(I^m))$ be the projection of μ_{m+1} on $\mathbb{R}^\tau$ (on the variables $\mathbf{x}(I^m)$). By (B.5), φ is also the projection of ν_m on $\mathbb{R}^\tau$ (with respect to $\mathbf{x}(I^m)$).

Next, the probability measure ν_m can be disintegrated into the product of its associated conditional probability $\nu_m(\bullet \,|\, \mathbf{x}(I^m))$ and its marginal φ on $\mathbb{R}^\tau$ (with respect to the variables $\mathbf{x}(I^m)$). That is

$$\nu_m(A \times B) = \int_B \nu_m(A \,|\, \mathbf{x}(I^m))\, d\varphi(\mathbf{x}(I^m))$$

for all hyper-rectangles $A \times B$ with $A \in \mathscr{B}(\mathbb{R}^{n(m)-\tau})$ and $B \in \mathscr{B}(\mathbb{R}^\tau)$. Similarly, μ_{m+1} can also be disintegrated into the product of its associated conditional probability $\nu_{m+1}(\bullet \,|\, \mathbf{x}(I^m))$ and its marginal φ, that is:

$$\mu_{m+1}(A \times B) = \int_B \mu_{m+1}(A \,|\, \mathbf{x}(I^m))\, d\varphi(\mathbf{x}(I^m))$$

for every hyper-rectangle $A \times B$ with $A \in \mathscr{B}(\mathbb{R}^{n_{m+1}-\tau})$, and $B \in \mathscr{B}(\mathbb{R}^\tau)$.

Therefore, define ν_{m+1} on $\mathbb{R}^{n(m+1)}$ as:

$$\nu_{m+1}(A \times B \times C) := \int_B \nu_m(A \,|\, \mathbf{x}(I^{m+1}))\, \mu_{m+1}(C \,|\, \mathbf{x}(I^m))\, d\varphi(\mathbf{x}(I^m)),$$

for every hyper-rectangle $A \times B \times C$, with $A \in \mathscr{B}(\mathbb{R}^{n(m)-\tau})$, $B \in \mathscr{B}(\mathbb{R}^\tau)$, and $C \in \mathscr{B}(\mathbb{R}^{n_{m+1}-\tau})$.

Then, with $A := \mathbb{R}^{n(m)-\tau}$, $\nu_m(\mathbb{R}^{n(m)-\tau}|\mathbf{x}(I^m)) = 1$ so that

$$\nu_{m+1}(A \times B \times C) := \int_B \mu_{m+1}(C \,|\, \mathbf{x}(I^{m+1}))\, d\varphi(\mathbf{x}(I^m)) \;=\; \mu_{m+1}(B \times C),$$

which shows that $\mu_{m+1} = \pi_{m+1}\nu_{m+1}$, the marginal of ν_{m+1} on $\mathbb{R}^{n_{m+1}}$ (on the variables $\mathbf{x}(I_{m+1})$). Similarly, with $C := \mathbb{R}^\tau$, $\mu_{m+1}(\mathbb{R}^\tau|\mathbf{x}(I^m)) = 1$ so that

$$\nu_{m+1}(A \times B \times C) := \int_B \nu_m(A \,|\, \mathbf{x}(I^{m+1}))\, d\varphi(\mathbf{x}(I^m)) \;=\; \nu_m(A \times B),$$

which shows that ν_m is the marginal of ν_{m+1} on $\mathbb{R}^{n(m)}$ (on the variables $\mathbf{x}(\cup_{i=1}^m I_k)$). But by (H), $\pi_k\nu_m = \mu_k$, for every $k = 1, \ldots, m$ and so $\pi_k\nu_{m+1} = \mu_k$, for all $k = 1, \ldots, m$. Hence

$$\pi_k\nu_{m+1} \;=\; \mu_k, \qquad k = 1, \ldots, m+1.$$

As ν_m is supported on $\mathbb{K}(m)$ and μ_{m+1} on $\mathbb{K}_m$, it follows that ν_{m+1} is supported on

$$\mathbb{K}(m+1) := \{\mathbf{x} \in \mathbb{R}^{n(m+1)} \;:\; \mathbf{x}(I_j) \in \mathbb{K}_j, \quad j = 1, \ldots, m+1\}.$$

Hence (H) propagates to $m+1$. Finally, as $\mathbb{K}(p) \equiv \mathbb{K}$, we obtain the desired result that $\mu\,(:= \nu_p)$ is a probability measure on $\mathbb{K}$, with marginal μ_i on $\mathbb{K}_i$ for every $i = 1, \ldots, p$. □

B.3 Notes and Sources

Most of Section B.1 is from (Hernández-Lerma and Lasserre, 2003, Chap. 1) and (Ash, 1972, A6). Theorem B.11 can be found in e.g. Mulholland and Rogers (1958); see also (Anastassiou, 1993, Theor. 2.1.1, p. 39), and Theorem B.12 is from Bayer and Teichmann (2006) who refined a result of Putinar (2000). Section B.2 is from Lasserre (2006a).

Appendix C

Some Basic Results in Optimization

C.1 Non Linear Programming

Let $O \subset \mathbb{R}^n$ be an open subset and consider the optimization problem **P** defined by:

$$\mathbf{P}: \quad \inf_{\mathbf{x}} \{ f(\mathbf{x}) \,:\, \mathbf{x} \in \mathbb{K} \} \tag{C.1}$$

with

$$\mathbb{K} := \{\mathbf{x} \in \mathbb{R}^n \,:\, g_j(\mathbf{x}) \geq 0,\ j = 1, \ldots, m; \quad h_k(\mathbf{x}) = 0,\ k = 1, \ldots, p\} \tag{C.2}$$

where $f : O \to \mathbb{R}$ is differentiable and $g_j, h_k : \mathbb{R}^n \to \mathbb{R}$ are continuously differentiable functions for every $j = 1, \ldots, m$, and $k = 1, \ldots, p$.

Recall that $\mathbf{x}^* \in O \cap \mathbb{K}$ is a *local* minimizer of f on $\mathbb{K}$ if there exists an open neighborhood C of $\mathbf{x}^*$ such that $f(\mathbf{x}) \geq f(\mathbf{x}^*)$ for all $\mathbf{x} \in C \cap \mathbb{K}$.

Theorem C.14 (Fritz-John optimality conditions). *If* $\mathbf{x}^* \in O \cap \mathbb{K}$ *is a local minimum of* f *on* $\mathbb{K}$*, there exist* $\boldsymbol{\lambda} \in \mathbb{R}^m$ *and* $\boldsymbol{\psi} \in \mathbb{R}^{p+1}$ *such that:*

$$(i)\ \psi_0 \nabla f(\mathbf{x}^*) = \sum_{j=1}^{m} \lambda_j \nabla g_j(\mathbf{x}^*) + \sum_{k=1}^{p} \psi_k \nabla h_k(\mathbf{x}^*) \tag{C.3}$$

$$(ii) \quad \lambda_j \geq 0, \quad j = 1, \ldots, m \tag{C.4}$$

$$(iii)\ \lambda_j\, g_j(\mathbf{x}^*) = 0,\ j = 1, \ldots, m. \tag{C.5}$$

The variables $(\boldsymbol{\lambda}, \boldsymbol{\psi})$ in Theorem C.14 are called the dual *multipliers.* Notice that if $\psi_0 = 0$, the above conditions are not very informative about a local minimizer and so, one would like to find a condition that ensures $\psi_0 \neq 0$ in (C.3). Therefore, given $\mathbf{x} \in \mathbb{K}$, consider the so-called *constraint qualification* condition:

$$\mathrm{QC}(\mathbf{x}) \left\{ \begin{array}{l} \exists \mathbf{u}_0 \in \mathbb{R}^n : \langle \nabla h_k(\mathbf{x}), \mathbf{u}_0 \rangle = 0, \quad \forall k = 1, \ldots, p. \\ \qquad\qquad\quad \text{the vectors} \{\nabla h_k(\mathbf{x})\}_{k=1}^p \text{ are linearly independent} \\ \qquad\qquad\quad \langle \nabla g_j(\mathbf{x}), \mathbf{u}_0 \rangle > 0 \text{ whenever } g_j(\mathbf{x}) = 0. \end{array} \right.$$

Under some additional condition one may obtain the more informative optimality conditions due to Karush, Kuhn and Tucker (and called the KKT optimality conditions) where the dual multiplier ψ_0 in (C.3) can be taken to be 1.

Theorem C.15 (KKT optimality conditions). *If $\mathbf{x}^* \in O \cap \mathbb{K}$ is a local minimum of f on $\mathbb{K}$, and if the constraint qualification* $\mathrm{QC}(\mathbf{x}^*)$ *holds, there exist $\boldsymbol{\lambda} \in \mathbb{R}^m$ and $\boldsymbol{\psi} \in \mathbb{R}^p$ such that:*

$$(i) \quad \nabla f(\mathbf{x}^*) = \sum_{j=1}^{m} \lambda_j \nabla g_j(\mathbf{x}^*) + \sum_{k=1}^{p} \psi_k \nabla h_k(\mathbf{x}^*) \qquad \text{(C.6)}$$

$$(ii) \quad \lambda_j \geq 0, \quad j = 1, \ldots, m \qquad \text{(C.7)}$$

$$(iii) \; \lambda_j \, g_j(\mathbf{x}^*) = 0, \, j = 1, \ldots, m \qquad \text{(C.8)}$$

The dual variables $(\boldsymbol{\lambda}, \boldsymbol{\psi}) \in \mathbb{R}^m \times \mathbb{R}^p$ are called the Lagrange-KKT multipliers. Another constraint qualification, stronger than $\mathrm{QC}(\mathbf{x})$, states that all vectors $\{\nabla h_k(\mathbf{x})\}$ and $\{\nabla g_j(\mathbf{x})\}_{j: g_j(\mathbf{x}^*)=0}$ are linearly independent.

Convexity: In the presence of convexity the KKT-conditions become sufficient, that is,

Theorem C.16. *Let f be convex on the open convex $O \subset \mathbb{R}^n$, $h_k : \mathbb{R}^n \to \mathbb{R}$ be affine for every $k = 1, \ldots, p$, and $g_j : \mathbb{R}^n \to \mathbb{R}$ be concave differentiable for every $j = 1, \ldots, m$. If $\mathbf{x}^* \in O \cap \mathbb{K}$ satisfies (C.6)-(C.8) then $\mathbf{x}^*$ is a local minimizer of f sur $O \cap \mathbb{K}$.*

In addition, because of convexity, every local minimizer $\mathbf{x}^* \in O \cap \mathbb{K}$ is a global minimizer of f on $O \cap \mathbb{K}$.

Finally, under the convexity conditions of Theorem C.16, write the affine function h_k as $\mathbf{x} \mapsto h_k(\mathbf{x}) := \mathbf{a}_k'\mathbf{x} + b_k$ fort some $\mathbf{a}_k \in \mathbb{R}^n, b_k \in \mathbb{R}$, for every $k = 1, \ldots, p$. The following so-called Slater's condition ensures that QC($\mathbf{x}$) holds at every $\mathbf{x} \in \mathbb{K}$.

$$\text{Slater:} \begin{cases} \text{The vectors } \{\mathbf{a}_k\}_{k=1}^p \text{ are linearly independent and} \\ \exists \mathbf{x}_0 \in \mathbb{K} : g_j(\mathbf{x}_0) > 0, \; j = 1, \ldots, m; \quad h_k(\mathbf{x}_0) = 0, \; k = 1, \ldots, p. \end{cases}$$

Hence, under the conditions of Theorem C.16 and Slater's condition, the conditions (C.6)-(C.8) are necessary and sufficient for $\mathbf{x}^* \in O \cap \mathbb{K}$ to be a global minimizer of f on $O \cap \mathbb{K}$.

Next, the *Lagrangian* $L_f : O \to \mathbb{R}$ associated with $\mathbf{x}^* \in O \cap \mathbb{K}$ and $(\boldsymbol{\lambda}, \boldsymbol{\psi}) \in \mathbb{R}^m \times \mathbb{R}^p$, is defined as:

$$\mathbf{x} \mapsto L_f(\mathbf{x}) := f(\mathbf{x}) - f(\mathbf{x}^*) - \sum_{j=1}^m \lambda_j \, g_j(\mathbf{x}) - \sum_{k=1}^p \psi_k \, h_k(\mathbf{x}). \qquad \text{(C.9)}$$

If $\mathbf{x}^* \in O \cap \mathbb{K}$ and $(\boldsymbol{\lambda}, \boldsymbol{\psi}) \in \mathbb{R}^m \times \mathbb{R}^p$ satisfy the KKT-optimality conditions (C.6)-(C.8) then $\mathbf{x}^*$ is a stationary point of L_f. Moreover, in the convex case, i.e., under the conditions of Theorem C.16, the Lagrangian L_f is convex and nonnegative on O with (global) minimum $L_f(\mathbf{x}^*) = 0$ on O attained at $\mathbf{x}^* \in \mathbb{K}$. In particular, if $O = \mathbb{R}^n$ then L_f is convex nonnegative on the whole space $\mathbb{R}^n$ with minimum 0 attained at $\mathbf{x}^* \in \mathbb{K}$.

Most Nonlinear Programming algorithms try to find a pair $(\mathbf{x}^*, (\boldsymbol{\lambda}^*, \boldsymbol{\psi}^*)) \in \mathbb{R}^n \times \mathbb{R}^m \times \mathbb{R}^p$ that satisfy the KKT-optimaliy conditions (C.6)-(C.8).

C.2 Semidefinite Programming

Conic programming is a subarea of convex optimization that refers to linear optimization problems over general convex cones. And semidefinite programming (in short, SDP) is a particular case of conic programming when one considers the convex cone of positive semidefinite matrices, whereas linear programming considers the positive orthant of $\mathbb{R}^n$, a *polyhedral* convex cone.

Let $\mathcal{S}_p \subset \mathbb{R}^{p\times p}$ be the space of real $p\times p$ symmetric matrices. Whenever $\mathbf{A}, \mathbf{B} \in \mathcal{S}_p$, the notation $\mathbf{A} \succeq \mathbf{B}$ (resp. $\mathbf{A} \succ \mathbf{B}$) stands for $\mathbf{A} - \mathbf{B}$ is *positive semidefinite* (resp. positive definite). Also, the notation $\langle \mathbf{A}, \mathbf{B} \rangle$ stands for trace $(\mathbf{AB})$ ($=$ trace $(\mathbf{BA})$). In canonical form, a semidefinite program reads:

$$\mathbf{P}: \quad \inf_{\mathbf{x}} \{\mathbf{c}'\mathbf{x} \,:\, \mathbf{F}_0 + \sum_{i=1}^{n} \mathbf{F}_i\, x_i \succeq 0\,\} \tag{C.10}$$

where $\mathbf{c} \in \mathbb{R}^n$, and $\{\mathbf{F}_i\}_{i=0}^n \subset \mathcal{S}_p$ for some $p \in \mathbb{N}$. Denote by $\inf \mathbf{P}$ the optimal value of $\mathbf{P}$ (possibly $-\infty$ if unbounded or $+\infty$ if $\mathbf{P}$ has no feasible solution $\mathbf{x} \in \mathbb{R}^n$). If the optimal value is attained at some $\mathbf{x}^* \in \mathbb{R}^n$ then write $\inf \mathbf{P} = \min \mathbf{P}$.

The semidefinite constraint $\mathbf{F}_0 + \sum_{i=1}^n \mathbf{F}_i\, x_i \succeq 0$ is also called a *Linear Matrix Inequality* (LMI).

That $\mathbf{P}$ in (C.10) is a convex optimization problem can be seen as follows: Consider the mapping $\mathbf{F} : \mathbb{R}^n \to \mathcal{S}_p$ defined by:

$$\mathbf{x} \mapsto \mathbf{F}(\mathbf{x}) := \mathbf{F}_0 + \sum_{i=1}^{n} \mathbf{F}_i\, x_i, \qquad \mathbf{x} \in \mathbb{R}^n.$$

The constraint $\mathbf{F}(\mathbf{x}) \succeq 0$ is the same as $\lambda_{min}(\mathbf{F}(\mathbf{x})) \geq 0$ where the function $\mathbf{x} \mapsto \lambda_{min}(\mathbf{F}(\mathbf{x}))$ maps $\mathbf{x} \in \mathbb{R}^n$ to the *smallest* eigenvalue of the real symmetric matrix $\mathbf{F}(\mathbf{x}) \in \mathcal{S}_p$. But the smallest eigenvalue of a real symmetric matrix is a concave function of its entries, and the entries of $\mathbf{F}(\mathbf{x})$ are affine functions of $\mathbf{x}$. Hence the set $\{\mathbf{x} \,:\, \mathbf{F}(\mathbf{x}) \succeq 0\}$ is convex and therefore, $\mathbf{P}$ consists of minimizing a linear functional on a convex set, i.e., a convex optimization problem.

Observe that if the matrices $(\mathbf{F}_i)$ are diagonal then $\mathbf{P}$ is just a finite dimensional linear programming problem. Conversely, consider the linear programming problem

$$\inf_{\mathbf{x}\in\mathbb{R}^n} \{\,\mathbf{c}'\mathbf{x} \,:\, \mathbf{a}_j'\mathbf{x} \geq b_j,\ j = 1,\ldots,p; \quad \mathbf{x} \geq 0\,\},$$

and let $\mathbf{F}(\mathbf{x}) \in \mathcal{S}_p$ be the diagonal matrix defined by $\mathbf{F}(\mathbf{x})_{jj} = \mathbf{a}_j'\mathbf{x} - b_j$ for every $j = 1,\ldots,p$. Then the above linear program is also the SDP $\inf_{\mathbf{x}} \{\mathbf{c}'\mathbf{x} \,:\, \mathbf{F}(\mathbf{x}) \succeq 0\,\}$.

Semidefinite programming is a non trivial extension of linear programming. Indeed, while the latter considers the positive orthant $\mathbb{R}^n_+$ (a polyhedral convex cone), the former considers the non polyhedral convex cone of positive semidefinite matrices.

Duality

The *dual* problem associated with $\mathbf{P}$ is the convex optimization problem:

$$\mathbf{P}^*: \quad \sup_{\mathbf{Z}\in\mathcal{S}_p} \{-\langle \mathbf{F}_0, \mathbf{Z}\rangle \,:\, \langle \mathbf{F}_i, \mathbf{Z}\rangle = c_i, \quad i=1,\ldots,n;\, \mathbf{Z}\succeq 0\} \quad (C.11)$$

with optimal value denoted $\sup \mathbf{P}^*$ (possibly $+\infty$ if unbounded or $-\infty$ is there is no feasible solution $\mathbf{Z}\succeq 0$). If the optimal value is attained at some $\mathbf{Z}^*\succeq 0$ then write $\sup \mathbf{P}^* = \max \mathbf{P}^*$.

In fact $\mathbf{P}^*$ is also a semidefinite program. Indeed, the set

$$\mathbf{S} := \{\mathbf{Z}\in\mathcal{S}_p \,:\, \langle \mathbf{F}_i, \mathbf{Z}\rangle = c_i, \quad i=1,\ldots,n\}$$

can be put in the form

$$\mathbf{S} = \{\mathbf{G}(\mathbf{z}) = \mathbf{G}_0 + \sum_{j=1}^{s} \mathbf{G}_j z_j \,:\, \mathbf{z}\in\mathbb{R}^s\}$$

for some appropriate matrices $(\mathbf{G}_j)_{j=0}^{s+1}\subset\mathcal{S}_p$. (For instance, if the $(\mathbf{F}_i)_{i=1}^n$ are linearly independent then $s=p(p+1)/2-m$.) Next, setting $\mathbf{d}=(d_j)\in\mathbb{R}^s$ with $d_j := \langle \mathbf{F}_0, \mathbf{G}_j\rangle$ for every $j=0,\ldots,s$, $\mathbf{P}^*$ reads:

$$\langle -\mathbf{F}_0, \mathbf{G}_0\rangle + \max_{\mathbf{z}} \{-\mathbf{d}'\mathbf{z} \,:\, \mathbf{G}_0 + \sum_{j=1}^{s} \mathbf{G}_j\, z_j \succeq 0\},$$

which is of the same form as $\mathbf{P}$.

Weak duality states that $\inf \mathbf{P} \geq \sup \mathbf{P}^*$ and holds without any assumption. Indeed, let $\mathbf{x}\in\mathbb{R}^n$ and $0\preceq\mathbf{Z}\in\mathcal{S}_p$ be any two feasible solutions of $\mathbf{P}$ and $\mathbf{P}^*$ respectively. Then:

$$\mathbf{c}'\mathbf{x} = \sum_{i=1}^{n} \langle \mathbf{F}_i, \mathbf{Z}\rangle\, x_i = \left\langle \sum_{i=1}^{n} \mathbf{F}_i\, x_i\,,\, \mathbf{Z}\right\rangle \geq \langle -\mathbf{F}_0, \mathbf{Z}\rangle,$$

where we have used that if $\mathbf{Z} \succeq 0$ and $\mathbf{A} \succeq \mathbf{B}$ then $\langle \mathbf{A}, \mathbf{Z} \rangle \geq \langle \mathbf{B}, \mathbf{Z} \rangle$. And so $\inf \mathbf{P} \geq \sup \mathbf{P}^*$. The nonnegative quantity $\inf \mathbf{P} - \sup \mathbf{P}^*$ is called the *duality gap.*

The absence of a duality gap (i.e., $\inf \mathbf{P} = \sup \mathbf{P}^*$) does not hold in general. However, it happens under some strict feasibility conditions.

Theorem C.17 (Strong duality). *Let* $\mathbf{P}$ *and* $\mathbf{P}^*$ *be as in (C.10) and (C.11) respectively.*

(i) If there exists $\mathbf{x} \in \mathbb{R}^n$ *such that* $\mathbf{F}(\mathbf{x}) \succ 0$ *then* $\inf \mathbf{P} = \sup \mathbf{P}^*$ *and* $\inf \mathbf{P} = \max \mathbf{P}^*$ *if the optimal value is finite.*

(ii) If there exists $\mathbf{Z} \succ 0$ *feasible for* $\mathbf{P}^*$ *then* $\inf \mathbf{P} = \sup \mathbf{P}^*$ *and* $\min \mathbf{P} = \sup \mathbf{P}^*$ *if the optimal value is finite.*

(iii) If there exists $\mathbf{Z} \succ 0$ *feasible for* $\mathbf{P}^*$ *and* $\mathbf{x} \in \mathbb{R}^n$ *such that* $\mathbf{F}(\mathbf{x}) \succ 0$, *then* $\min \mathbf{P} = \max \mathbf{P}^*$.

The strict feasiblity condition in Theorem C.17(i) (resp. Theorem C.17(ii)) is a specialization of Slater's condition in convex programming, when applied to the convex problem $\mathbf{P}$ (resp. applied to the convex problem $\mathbf{P}^*$).

Computational complexity

What makes semidefinite programming a powerful technique is its computational complexity when using algorithms based on *interior point* methods. Indeed one may find an approximate solution within prescribed accuracy $\epsilon > 0$, in a time that is polynomial in the input size of the problem. For instance, in using path following algorithms based on the Nesterov-Todd search directions, one only needs $O(\sqrt{n} \ln(1/\epsilon))$ iterations to find such an approximate solution. See e.g. Tuncel (2000).

C.3 Infinite-dimensional Linear Programming

Let $\mathscr{X}, \mathscr{Y}$ be two arbitrary real vector spaces and let $\langle \cdot, \cdot \rangle$ be a bilinear form on $\mathscr{X} \times \mathscr{Y}$. The couple $(\mathscr{X}, \mathscr{Y})$ is called a *dual pair* of vector spaces if it satisfies:

- For each $0 \neq \mathbf{x} \in \mathscr{X}$ there exists $\mathbf{y} \in \mathscr{Y}$ such that $\langle \mathbf{x}, \mathbf{y} \rangle \neq 0$.
- For each $0 \neq \mathbf{y} \in \mathscr{Y}$ there exists $\mathbf{x} \in \mathscr{X}$ such that $\langle \mathbf{x}, \mathbf{y} \rangle \neq 0$.

Given a dual pair $(\mathscr{X}, \mathscr{Y})$, the *weak topology* on $\mathscr{X}$ (denoted $\sigma(\mathscr{X}, \mathscr{Y})$)

is the coarsest topology under which all the elements of $\mathscr{Y}$ are continuous when regarded as linear forms $\langle \cdot, \mathbf{y} \rangle$ on $\mathscr{X}$. Equivalently, in this topology, the base of neighborhoods of the origin is given by

$$\{\mathbf{x} \in \mathscr{X} \, : \, |\langle \mathbf{x}, \mathbf{y} \rangle| \leq 1, \, \forall \mathbf{y} \in I\},$$

where I is a finite subset of $\mathscr{Y}$.

An important particular case is when $\mathscr{Y}$ is a *normed* vector space and $\mathscr{X}$ is the strong dual of $\mathscr{Y}$ (denoted $\mathscr{Y}^*$), i.e., $\mathscr{X}$ is the space of linear forms on $\mathscr{Y}$, continuous for the topology induced by the norm. Then the weak topology $\sigma(\mathscr{X}, \mathscr{Y})$ ($= \sigma(\mathscr{Y}^*, \mathscr{Y})$) is called the weak $\star$ topology on $\mathscr{X}$ and has the following fundamental property.

Theorem C.18 (Banach-Alaoglu Theorem). *The unit ball* $B := \{\mathbf{x} \in \mathscr{Y}^* \, : \, \|\mathbf{x}\| \leq 1\}$ *of* $\mathscr{Y}^*$ *is compact in the weak* $\star$ *topology. In addition, if* $\mathscr{Y}$ *is separable then* B *is weak* $\star$ *sequentially compact.*

An infinite-dimensional linear program requires the following components:

- two dual pairs of vector spaces $(\mathscr{X}, \mathscr{Y})$, $(\mathscr{Z}, \mathscr{W})$;
- the induced associated weak topologies on each vector space;
- a weakly continuous linear map $\mathbf{G} : \mathscr{X} \to \mathscr{Z}$, with adjoint $\mathbf{G}^* : \mathscr{W} \to \mathscr{Y}$;
- a positive convex cone C in $\mathscr{X}$, with dual cone C^* in $\mathscr{Y}$; and
- vectors $\mathbf{b} \in \mathscr{Z}$ and $\mathbf{c} \in \mathscr{Y}$.

Then the primal linear program is

$$\mathbf{P}: \quad \begin{array}{l} \text{minimize } \langle \mathbf{x}, \mathbf{c} \rangle \\ \text{subject to: } \mathbf{G}\mathbf{x} = \mathbf{b}, \quad \mathbf{x} \in C. \end{array} \tag{C.12}$$

The corresponding dual linear program is

$$\mathbf{P}^*: \quad \begin{array}{l} \text{maximize } \langle \mathbf{b}, \mathbf{w} \rangle \\ \text{subject to: } \mathbf{c} - \mathbf{G}^*\mathbf{w} \in C^*, \quad \mathbf{w} \in \mathscr{W}. \end{array} \tag{C.13}$$

Notice that there is a lot of freedom for choosing the vector spaces $\mathscr{X}, \mathscr{Y}, \mathscr{W}, \mathscr{Z}$. For instance, when $\mathscr{X}$ and $\mathscr{Z}$ are normed linear spaces, in general $\mathscr{Y}$ is not the *strong* dual $\mathscr{X}^*$ of $\mathscr{X}$, and similarly, $\mathscr{W}$ is not the strong dual $\mathscr{Z}^*$ of $\mathscr{Z}$. This is because one usually wants to work with spaces whose elements have a simple physical interpretation.

An element $\mathbf{x} \in \mathscr{X}$ is called feasible for $\mathbf{P}$ if $\mathbf{x} \in C$ and $\mathbf{G}\mathbf{x} = \mathbf{b}$; $\mathbf{P}$ is said to be consistent, if it has a feasible solution. If $\mathbf{P}$ is consistent, then its value is defined as

$$\inf \mathbf{P} := \inf \{\langle \mathbf{x}, \mathbf{c} \rangle : \quad \mathbf{x} \text{ is feasible for } \mathbf{P}\};$$

otherwise, $\inf \mathbf{P} = +\infty$. The linear program $\mathbf{P}$ is solvable if there is some feasible solution $\mathbf{x}^* \in \mathscr{X}$ that achieves the value $\inf \mathbf{P}$; then $\mathbf{x}^*$ is an optimal solution of $\mathbf{P}$, and one then writes $\inf \mathbf{P} = \min \mathbf{P}$. The same definitions apply for the dual linear program $\mathbf{P}^*$.

The next weak duality result can be proved as in elementary (finite-dimensional) linear programming.

Proposition C.19. *If* $\mathbf{P}$ *and* $\mathbf{P}^*$ *are both consistent, then their values are finite and satisfy* $\sup \mathbf{P}^* \leq \inf \mathbf{P}$.

There is no duality gap if $\sup \mathbf{P}^* = \inf \mathbf{P}$, and strong duality holds if $\max \mathbf{P}^* = \min \mathbf{P}$, i.e., if there is no duality gap, and both $\mathbf{P}^*$ and $\mathbf{P}$ are solvable.

Theorem C.20. *Let* $\mathbf{D}, \mathbf{H}$ *be the sets in* $\mathscr{Z} \times \mathbb{R}$*, defined as*

$$\mathbf{D} := \{(\mathbf{G}\mathbf{x}, \langle \mathbf{x}, \mathbf{c} \rangle) : \quad \mathbf{x} \in C\} \tag{C.14}$$

$$\mathbf{H} := \{(\mathbf{G}\mathbf{x}, \langle \mathbf{x}, \mathbf{c} \rangle + r) : \quad \mathbf{x} \in C,\ r \in \mathbb{R}_+\} \tag{C.15}$$

$$\tag{C.16}$$

If $\mathbf{P}$ *is consistent with finite value, and* $\mathbf{D}$ *or* $\mathbf{H}$ *is weakly closed (i.e., closed in the weak topology* $\sigma(\mathscr{Z} \times \mathbb{R}, \mathscr{W} \times \mathbb{R})$*), then* $\mathbf{P}$ *is solvable and there is no duality gap, i.e.,* $\sup \mathbf{P}^* = \min \mathbf{P}$.

C.4 Proof of Theorem 1.3

(a) To simplify notation we prove Theorem 1.3 in the case where all constraints are equality constraints. Let $B(\mathbb{K})$ be the space of bounded measurable functions on $\mathbb{K}$. The pairs of vector spaces:

$$(\mathscr{X}, \mathscr{Y}) := (\mathscr{M}(\mathbb{K}), B(\mathbb{K})); \quad (\mathscr{Z}, \mathscr{W}) := (\mathbb{R}^{|\Gamma|}, \mathbb{R}^{|\Gamma|}),$$

form two *dual* pairs with duality bracket $\langle \mu, g \rangle := \int_{\mathbb{K}} g \, d\mu$, and the usual scalar product in $\mathbb{R}^{|\Gamma|}$. Then (1.1) is the infinite-dimensional linear optimization problem (C.12) with $C := \mathcal{M}(\mathbb{K})_+$, and where the linear map $\mathbf{G} : \mathscr{X} \to \mathscr{Z}$ is given by

$$\mu \mapsto \mathbf{G}\mu = \left(\int_{\mathbb{K}} h_j \, d\mu \right) \in \mathbb{R}^{|\Gamma|},$$

with adjoint $\mathbf{G}^* : \lambda \mapsto \mathbf{G}^*\lambda := \sum_j \lambda_j h_j$. The map $\mathbf{G}$ is weakly continuous because $\mathbf{G}^*(\mathbb{R}^{|\Gamma|}) \subset B(\mathbb{K})$ (as the h_j's are bounded on $\mathbb{K}$). Therefore, as we maximize, it suffices to prove that the set $\mathbf{H}$ in (C.15), i.e.,

$$\mathbf{H} := \{(\mathbf{G}\mu, \langle \mu, f \rangle - r) : \quad \mu \in C, \, r \in \mathbb{R}_+\},$$

is weakly closed (i.e., here, closed in the usual topology of $\mathbb{R}^{|\Gamma|+1}$).

Let $(\mu_n, r_n) \in C \times \mathbb{R}_+$ be such that $\mathbf{G}\mu_n \to \mathbf{b}$ and $\int_{\mathbb{K}} f d\mu_n - r_n \to a$ for some $(\mathbf{b}, a) \in \mathbb{R}^{|\Gamma|} \times \mathbb{R}$. As $\mathbb{K} \subset \mathbb{R}^n$ is compact and $h_k > 0$ for some $k \in \Gamma$, from $\int_{\mathbb{K}} h_k d\mu_n \to b_k$, one deduces that the sequence $(\mu_n(\mathbb{K}))_n$ is bounded and therefore, there is a subsequence $\{n_j\}$ and a measure $\mu \in C$ such that $\mu_{n_j} \Rightarrow \mu$ (see Definition B.1 in Appendix B). Hence, as h_i is continuous for every $i \in \Gamma$, we obtain $\mathbf{b} = \lim_{j\to\infty} \mathbf{G}\mu_{n_j} = (\int_{\mathbb{K}} h_j d\mu)$. Next, recall that f is upper semicontinuous (and bounded on $\mathbb{K}$). Therefore by Proposition B.10, $\limsup_j \int_{\mathbb{K}} f d\mu_{n_j} \leq \int_{\mathbb{K}} f d\mu$. Moreover, from $a = \lim_{j\to\infty} \int_{\mathbb{K}} f d\mu_{n_j} - r_{n_j}$ we infer that $r_{n_j} \subset \mathbb{R}_+$ is bounded above and so for another subsequence of $\{n_j\}$ (still denoted $\{n_j\}$ for convenience) we have $r_{n_j} \to r \in \mathbb{R}_+$. But then

$$a = \limsup_{j\to\infty} \int_{\mathbb{K}} f d\mu_{n_j} - r_{n_j} \leq \int_{\mathbb{K}} f d\mu - r$$

so that $a = \int_{\mathbb{K}} f d\mu - r'$ for some $r \leq r' \in \mathbb{R}_+$. Hence we have proved that $(\mathbf{b}, a) = (\int_{\mathbb{K}} h_j d\mu, \int_{\mathbb{K}} f d\mu - r')$ for some $(\mu, r') \in C \times \mathbb{R}_+$, that is, the set H is weakly closed. From Theorem C.20, the moment problem (1.1) is solvable and there is no duality gap, i.e., $\rho_{\text{mom}} = \rho_{\text{pop}}$.

(b) Let μ be an optimal solution of (1.1). As f and the h_j's are all bounded on $\mathbb{K}$, then by Theorem B.11, there exists a measure ν supported on finitely many points of $\mathbb{K}$ and such that $\int_{\mathbb{K}} f d\nu = \int_{\mathbb{K}} f d\mu$, and $\int_{\mathbb{K}} h_j d\nu = \int_{\mathbb{K}} h_j d\mu$, for all $j \in \Gamma$.

C.5 Notes and Sources

Most of Section C.1 is from Hiriart-Urruty (1998) while Section C.2 is inspired from Vandenberghe and Boyd (1996) and most of Section C.3 is taken from Anderson and Nash (1987); in particular, see (Anderson and Nash, 1987, Theorems 3.10 and 3.22.).

Appendix D

The GloptiPoly Software

D.1 Presentation

In this chapter we describe and report on the GloptiPoly software that implements the methodology described in earlier chapters.

GloptiPoly is a Matlab[1] freeware to solve (or at least approximate) the Generalized Moment Problem (GMP) in (1.1) with polynomial data, that is, when the set $\mathbb{K}$ is a basic semi-algebraic set and $f, (h_j)$ are polynomials. It can also solve the several extensions of the GMP described in Chapter 4.

GloptiPoly 3 is an extension of the former version 2 of GloptiPoly described in Henrion and Lasserre (2003) dedicated so solving the global optimization problem (5.2) of Chapter 5. The software automatically generates and solves the hierarchy of semidefinite relaxations (4.5).

It is a user-friendly package that researchers and students can experiment easily. It can also be used as a tutorial material to illustrate the use of semidefinite relaxations for solving the GMP. So far, GloptiPoly is aimed at solving small- and medium-scale problems.

D.2 Installation

GloptiPoly 3 is a freeware subject to the General Public Licence (GPL) policy. It is available for Matlab 7.2 (Release 2006) and later versions. It can be downloaded at

`www.laas.fr/~henrion/software/gloptipoly3`

[1]Matlab is a trademark of The MathWorks, Inc.

The package, available as a compressed archive, consists of several m-files and subdirectories, and it contains no binaries. Extracted files are placed in a `gloptipoly3` directory that should be declared in the Matlab working path, e.g. using the Matlab command

```
>> addpath gloptipoly3
```

To solve the semidefinite relaxation (4.5) GloptiPoly 3 uses by default the semidefinite programming solver SeDuMi of Sturm (1999); so this package should be properly installed. Other semidefinite solvers can also be used provided they are installed and interfaced through YALMIP of Löfberg (2004).

D.3 Getting started

Please type the command

```
>> gloptipolydemo
```

to run interactively the basic example that follows.

Consider the classical problem of minimizing globally the two-dimensional six-hump camel back function Henrion and Lasserre (2003)

$$\min_{\mathbf{x}\in\mathbb{R}^2} g_0(\mathbf{x}) = 4x_1^2 + x_1x_2 - 4x_2^2 - 2.1x_1^4 + 4x_2^4 + \frac{1}{3}x_1^6.$$

The function has six local minima, two of them being global minima. Using GloptiPoly 3, this optimization problem can be modeled as an instance of the moment problem as follows:

```
>> mpol x1 x2
>> g0 = 4*x1^2+x1*x2-4*x2^2-2.1*x1^4+4*x2^4+x1^6/3
Scalar polynomial
4x1^2+x1x2-4x2^2-2.1x1^4+4x2^4+0.33333x1^6
>> P = msdp(min(g0));
GloptiPoly 3.0
Define moment SDP problem
...
(GloptiPoly output suppressed)
...
Generate moment SDP problem
>> P
```

```
Moment SDP problem
  Measure label               = 1
  Relaxation order            = 3
  Decision variables          = 27
  Semidefinite inequalities = 10x10
```

Once the moment problem is modeled, a semidefinite solver (here SeDuMi) can be used to solve it numerically.

```
>> [status,obj] = msol(P)
GloptiPoly 3.0
Solve moment SDP problem
*****************************************************
Calling SeDuMi
SeDuMi 1.1R3 by AdvOL, 2006 and Jos F. Sturm, 1998-2003.
...
(SeDuMi output suppressed)
...
2 globally optimal solutions extracted
>> status
status =
     1
>> obj
obj =
    -1.0316
>> x = double([x1 x2]);
x(:,:,1) =
    0.0898   -0.7127
x(:,:,2) =
   -0.0898    0.7127
```

The flag `status = 1` means that the moment problem is solved successfully and that GloptiPoly 3 can extract two globally optimal solutions reaching the objective function `obj = -1.0316`.

D.4 Description

GloptiPoly 3 uses advanced Matlab features for object-oriented programming and overloaded operators. The user should be familiar with the following basic objects.

D.4.1 *Multivariate polynomials* (`mpol`)

A multivariate polynomial is an affine combination of monomials, each monomial depending on a set of variables. Variables can be declared in the Matlab working space as follows:

```
>> clear
>> mpol x
>> x
Scalar polynomial
x
>> mpol y 2
>> y
2-by-1 polynomial vector
(1,1):y(1)
(2,1):y(2)
>> mpol z 3 2
>> z
3-by-2 polynomial matrix
(1,1):z(1,1)
(2,1):z(2,1)
(3,1):z(3,1)
(1,2):z(1,2)
(2,2):z(2,2)
(3,2):z(3,2)
```

Variables, monomials and polynomials are defined as objects of class `mpol`.

All standard Matlab operators have been uploaded for `mpol` objects:

```
>> y*y'-z'*z+x^3
2-by-2 polynomial matrix
(1,1):y(1)^2-z(1,1)^2-z(2,1)^2-z(3,1)^2+x^3
(2,1):y(1)y(2)-z(1,1)z(1,2)-z(2,1)z(2,2)-z(3,1)z(3,2)+x^3
(1,2):y(1)y(2)-z(1,1)z(1,2)-z(2,1)z(2,2)-z(3,1)z(3,2)+x^3
(2,2):y(2)^2-z(1,2)^2-z(2,2)^2-z(3,2)^2+x^3
```

Use the instruction

```
>> mset clear
```

to delete all existing GloptiPoly variables from the Matlab working space.

D.4.2 *Measures* (meas)

Variables can be associated with real-valued measures, and one variable is associated with only one measure. For GloptiPoly 3, measures are identified with a label, a positive integer. When starting a GloptiPoly session, the default measure has label 1. By default, all created variables are associated with the current measure. Measures can be handled with the class meas as follows:

```
>> mset clear
>> mpol x
>> mpol y 2
>> meas
Measure 1 on 3 variables: x,y(1),y(2)
>> meas(y) % create new measure
Measure 2 on 2 variables: y(1),y(2)
>> m = meas
1-by-2 vector of measures
1:Measure 1 on 1 variable: x
2:Measure 2 on 2 variables: y(1),y(2)
>> m(1)
Measure number 1 on 1 variable: x
```

The above script creates a measure $\mu_1(dx)$ on $\mathbb{R}$ and a measure $\mu_2(d\mathbf{y})$ on $\mathbb{R}^2$. Use the instruction

```
>> mset clearmeas
```

to delete all existing GloptiPoly measures from the working space. Note that this does not delete existing GloptiPoly variables.

D.4.3 *Moments* (mom)

Linear combinations of moments of a given measure can be manipulated with the mom class as follows:

```
>> mom(1+2*x+3*x^2)
Scalar moment
I[1+2x+3x^2]d[1]
>> mom(y*y')
2-by-2 moment matrix
(1,1):I[y(1)^2]d[2]
```

```
(2,1):I[y(1)y(2)]d[2]
(1,2):I[y(1)y(2)]d[2]
(2,2):I[y(2)^2]d[2]
```

The notation `I[p]d[k]` stands for $\int p\, d\mu_k$ where p is a polynomial of the variables associated with the measure μ_k, and k is the measure label.

Note that it makes no sense to define moments over several measures, or nonlinear moment expressions:

```
>> mom(x*y(1))
??? Error using ==> mom.mom
Invalid partitioning of measures in moments
>> mom(x)*mom(y(1))
??? Error using ==> mom.times
Invalid moment product
```

Note also the distinction between a constant term and the mass of a measure:

```
>> 1+mom(x)
Scalar moment
1+I[x]d[1]
>> mom(1+x)
Scalar moment
I[1+x]d[1]
>> mass(x)
Scalar moment
I[1]d[1]
```

Finally, let us mention three equivalent notations to refer to the mass of a measure:

```
>> mass(meas(y))
Scalar moment
I[1]d[2]
>> mass(y)
Scalar moment
I[1]d[2]
>> mass(2)
Scalar moment
I[1]d[2]
```

The first command refers explicitly to the measure, the second command is a handy short-cut to refer to a measure via its variables, and the third command refers to GloptiPoly's labeling of measures.

D.4.4 *Support constraints* (supcon)

By default, a measure on n variables is defined on the whole $\mathbb{R}^n$. We can restrict the support of a mesure to a given semialgebraic set as follows:

```
>> 2*x^2+x^3 == 2+x
Scalar measure support equality
2x^2+x^3 == 2+x
>> disk = (y'*y <= 1)
Scalar measure support inequality
y(1)^2+y(2)^2 <= 1
```

Support constraints are modeled by objects of class supcon. The first command which also reads $x^3 + 2x^2 - x - 2 = (x - 1)(x + 1)(x + 2) = 0$, means that the measure μ_1 must be discrete, a linear combination of three Dirac at 1, -1 and -2. The second command restricts the measure μ_2 to be supported on the unit disk.

Note that it makes no sense to define a support constraint on several measures:

```
>> x+y(1) <= 1
??? Error using ==> supcon.supcon
Invalid reference to several measures
```

D.4.5 *Moment constraints* (momcon)

We can constrain linearly the moments of several measures:

```
>> mom(x^2+2) == 1+mom(y(1)^3*y(2))
Scalar moment equality constraint
I[2+x^2]d[1] == 1+I[y(1)^3y(2)]d[2]
>> mass(x)+mass(y) <= 2
Scalar moment inequality constraint
I[1]d[1]+I[1]d[2] <= 2
```

Moment constraints are modeled by objects of class momcon.

For GloptiPoly an objective function to be minimized or maximized is considered as a particular moment constraint:

```
>> min(mom(x^2+2))
Scalar moment objective function
min I[2+x^2]d[1]
>> max(x^2+2)
Scalar moment objective function
max I[2+x^2]d[1]
```

The latter syntax is a handy short-cut which directly converts an `mpol` object into an momcon object.

D.4.6 *Floating point numbers* (`double`)

Variables in a measure can be assigned numerical values:

```
>> m1 = assign(x,2)
Measure 1 on 1 variable: x
supported on 1 point
```

which is equivalent to enforcing a discrete support for the measure. Here μ_1 is set to the Dirac at the point 2.

The `double` operator converts a measure or its variables into a floating point number:

```
>> double(x)
ans =
     2
>> double(m1)
ans =
     2
```

Polynomials can be evaluated in a similar fashion:

```
>>double(1-2*x+3*x^2)
ans =
     9
```

Discrete measure supports consisting of several points can be specified in an array:

```
>> m2 = assign(y,[-1 2 0;1/3 1/4 -2])
Measure 2 on 2 variables: y(1),y(2)
supported on 3 points
>> double(m2)
```

```
ans(:,:,1) =
   -1.0000
    0.3333
ans(:,:,2) =
    2.0000
    0.2500
ans(:,:,3) =
      0
     -2
```

D.5 Solving Moment Problems (msdp)

Once a moment problem is defined, it can be solved numerically with the instruction `msol`. In the sequel we give several examples of GPMs handled with GloptiPoly 3.

D.5.1 *Unconstrained minimization*

In Section D.3 we already encountered an example of an unconstrained polynomial optimization solved with GloptiPoly 3. Let us revisit this example:

```
>> mset clear
>> mpol x1 x2
>> g0 = 4*x1^2+x1*x2-4*x2^2-2.1*x1^4+4*x2^4+x1^6/3
Scalar polynomial
4x1^2+x1x2-4x2^2-2.1x1^4+4x2^4+0.33333x1^6
>> P = msdp(min(g0));
...
>> msol(P)
...
2 globally optimal solutions extracted
Global optimality certified numerically
```

This indicates that the global minimum is attained with a discrete measure supported on two points. The measure can be constructed from the knowledge of its first moments of degree up to 6:

```
>> meas
```

```
Measure 1 on 2 variables: x1,x2
  with moments of degree up to 6, supported on 2 points
>> double(meas)
ans(:,:,1) =
    0.0898
   -0.7127
ans(:,:,2) =
   -0.0898
    0.7127
>> double(g0)
ans(:,:,1) =
   -1.0316
ans(:,:,2) =
   -1.0316
```

When converting to floating point numbers with the operator `double`, it is essential to make the distinction between `mpol` and `mom` objects:

```
>> v = mmon([x1 x2],2)'
1-by-6 polynomial vector
(1,1):1
(1,2):x1
(1,3):x2
(1,4):x1^2
(1,5):x1x2
(1,6):x2^2
>> double(v)
ans(:,:,1) =
    1.0000    0.0898   -0.7127    0.0081   -0.0640    0.5079
ans(:,:,2) =
    1.0000   -0.0898    0.7127    0.0081   -0.0640    0.5079
>> double(mom(v))
ans =
    1.0000    0.0000   -0.0000    0.0081   -0.0640    0.5079
```

The first instruction mmon generates a vector of monomials v of class `mpol`, so the command `double(v)` calls the convertor `@mpol/double` which evaluates a polynomial expression on the discrete support of a measure (here two points). The last command `double(mom(v))` calls the convertor `@mom/double` which returns the value of the moments obtained after solving the moment problem.

Note that when inputing moment problems on a unique measure whose mass is not constrained, GloptiPoly assumes by default that the measure has mass one, i.e. that we are seeking a probability measure. Therefore, if g0 is the polynomial defined previously, the two instructions

```
>> P = msdp(min(g0));
```

and

```
>> P = msdp(min(g0), mass(meas(g0))==1);
```

are equivalent.

D.5.2 *Constrained minimization*

Consider the non-convex quadratic problem

$$\begin{array}{ll}\min & -2x_1 + x_2 - x_3 \\ \text{s.t.} & 24 - 20x_1 + 9x_2 - 13x_3 + 4x_1^2 - 4x_1x_2 + 4x_1x_3 + 2x_2^2 - 2x_2x_3 + 2x_3^2 \geq 0 \\ & x_1 + x_2 + x_3 \leq 4, \quad 3x_2 + x_3 \leq 6 \\ & 0 \leq x_1 \leq 2, \quad 0 \leq x_2, \quad 0 \leq x_3 \leq 3\end{array}$$

Each constraint in this problem is interpreted by GloptiPoly 3 as a support constraint on the measure associated with $\mathbf{x}$.

```
>> mpol x 3
>> x(1)+x(2)+x(3) <= 4
Scalar measure support inequality
x(1)+x(2)+x(3) <= 4
```

The whole problem can be entered as follows:

```
>> mpol x 3
>> g0 = -2*x(1)+x(2)-x(3);
>> K = [24-20*x(1)+9*x(2)-13*x(3)+4*x(1)^2-4*x(1)*x(2) ...
 +4*x(1)*x(3)+2*x(2)^2-2*x(2)*x(3)+2*x(3)^2 >= 0, ...
 x(1)+x(2)+x(3) <= 4, 3*x(2)+x(3) <= 6, ...
 0 <= x(1), x(1) <= 2, 0 <= x(2), 0 <= x(3), x(3) <= 3];
>> P = msdp(min(g0), K)
...
Moment SDP problem
  Measure label              = 1
  Relaxation order           = 1
  Decision variables         = 9
```

```
  Linear inequalities        = 8
  Semidefinite inequalities = 4x4
```

The moment problem can then be solved:

```
>> [status,obj] = msol(P)
GloptiPoly 3.0
Solve moment SDP problem
...
Global optimality cannot be ensured
status =
     0
obj =
    -6.0000
```

Since `status=0` the moment SDP problem can be solved but it is impossible to detect global optimality. The value `obj=-6.0000` is then a lower bound on the global minimum of the quadratic problem.

The measure associated with the problem variables can be retrieved as follows:

```
>> mu = meas
Measure 1 on 3 variables: x(1),x(2),x(3)
  with moments of degree up to 2
```

Its vector of moments can be built as follows:

```
>> mv = mvec(mu)
10-by-1 moment vector
(1,1):I[1]d[1]
(2,1):I[x(1)]d[1]
(3,1):I[x(2)]d[1]
(4,1):I[x(3)]d[1]
(5,1):I[x(1)^2]d[1]
(6,1):I[x(1)x(2)]d[1]
(7,1):I[x(1)x(3)]d[1]
(8,1):I[x(2)^2]d[1]
(9,1):I[x(2)x(3)]d[1]
(10,1):I[x(3)^2]d[1]
```

These moments are the decision variables of the SDP problem solved with the above `msol` command. Their numerical values can be retrieved as follows:

```
>> double(mv)
ans =
     1.0000
     2.0000
    -0.0000
     2.0000
     7.6106
     1.4671
     2.3363
     4.8335
     0.5008
     8.7247
```

The numerical moment matrix can be obtained using the following commands:

```
>> double(mmat(mu))
ans =
     1.0000     2.0000    -0.0000     2.0000
     2.0000     7.6106     1.4671     2.3363
    -0.0000     1.4671     4.8335     0.5008
     2.0000     2.3363     0.5008     8.7247
```

As explained in Chapter 4, one can build up a hierarchy or relaxations, whose associated monotone sequence optimal values converges to the global optimum, under mild technical assumptions. By default the command msdp builds the relaxation of lowest order, equal to half the degree of the highest degree monomial in the polynomial data. An additional input argument can be specified to build higher order relaxations:

```
>> P = msdp(min(g0), K, 2)
...
Moment SDP problem
  Measure label             = 1
  Relaxation order          = 2
  Decision variables        = 34
  Semidefinite inequalities = 10x10+8x(4x4)
>> [status,obj] = msol(P)
...
```

```
Global optimality cannot be ensured
status =
     0
obj =
   -5.6922
>> P = msdp(min(g0), K, 3)
...
Moment SDP problem
  Measure label                = 1
  Relaxation order             = 3
  Decision variables           = 83
  Semidefinite inequalities = 20x20+8x(10x10)
>> [status,obj] = msol(P)
...
Global optimality cannot be ensured
status =
     0
obj =
    -4.0684
```

Observe that the semidefinite programming problems involve an increasing number of variables and constraints. They generate a monotone non decreasing sequence of lower bounds on the global optimum, which is eventually reached numerically at the fourth relaxation:

```
>> P = msdp(min(g0), K, 4)
...
Moment SDP problem
  Measure label                = 1
  Relaxation order             = 4
  Decision variables           = 164
  Semidefinite inequalities = 35x35+8x(20x20)
>> [status,obj] = msol(P)
...
2 globally optimal solutions extracted
Global optimality certified numerically
status =
     1
obj =
   -4.0000
```

```
>> double(x)
ans(:,:,1) =
    2.0000
    0.0000
    0.0000
ans(:,:,2) =
    0.5000
    0.0000
    3.0000
>> double(g0)
ans(:,:,1) =
   -4.0000
ans(:,:,2) =
   -4.0000
```

D.5.3 *Several measures*

GloptiPoly 3 can handle several measures whose moments are linearly related. For example, consider the GMP arising when solving polynomial optimal control problems as detailed in Chapter 10. We are seeking two occupation measures $\mu(d\mathbf{x}, d\mathbf{u})$ and $\nu(d\mathbf{x})$ of a state vector $\mathbf{x}(t)$ and control vector $\mathbf{u}(t)$ which satisfy the ODE

$$\frac{d\mathbf{x}(t)}{dt} = f(\mathbf{x}, \mathbf{u}), \; \mathbf{x}(0) = \mathbf{x}_0,$$

with $f(\mathbf{x}, \mathbf{u})$ a given polynomial mapping and $\mathbf{x}_0$ a given initial condition. The measure μ is supported on a given semialgebraic set $\mathbb{K}_1$ corresponding to constraints on $\mathbf{x}$ and $\mathbf{u}$. The measure ν is supported on a given semialgebraic set $\mathbb{K}_2$ corresponding to performance requirements. For example $\mathbb{K}_2 = 0$ indicates that state $\mathbf{x}$ must reach the origin.

Given a polynomial test function $g(\mathbf{x})$ we can relax the dynamics constraint with the moment constraint

$$\int_{\mathbb{K}_2} g(\mathbf{x}) d\nu(\mathbf{x}) - g(\mathbf{x}_0) = \int_{\mathbb{K}_1} \langle \nabla g(\mathbf{x}), f(\mathbf{x}, \mathbf{u}) \rangle d\mu(\mathbf{x}, \mathbf{u})$$

linking linearly moments of μ and ν. A lower bound on the minimum time achievable by any feedback control law $\mathbf{u}(\mathbf{x})$ is then obtained by minimizing the mass of μ over all possible measures μ, ν satisfying the support and moment constraints. The gap between the lower bound and the exact minimum time is narrowed by enlarging the class of test functions g.

In the following script we solve this moment problem in the case of a double integrator with state and input constraints:

```
% bounds on minimal achievable time for optimal control of
% double integrator with state and input constraints

x0 = [1; 1]; u0 = 0; % initial conditions
d = 6; % maximum degree of test function


% analytic minimum time
if x0(1) >= -(x0(2)^2-2)/2
 tmin = 1+x0(1)+x0(2)+x0(2)^2/2;
elseif x0(1) >= -x0(2)^2/2*sign(x0(2))
 tmin = 2*sqrt(x0(1)+x0(2)^2/2)+x0(2);
else
 tmin = 2*sqrt(-x0(1)+x0(2)^2/2)-x0(2);
end

% occupation measure for constraints
mpol x1 2
mpol u1
m1 = meas([x1;u1]);

% occupation measure for performance
mpol x2 2
m2 = meas(x2);

% dynamics
scaling = tmin; % time scaling
f = scaling*[x1(2);u1];

% test function
g1 = mmon(x1,d);
g2 = mmon(x2,d);

% initial condition
assign([x1;u1],[x0;u0]);
g0 = double(g1);

% moment problem
P = msdp(min(mass(m1)),...
 u1^2 <= 1,... % input constraint
 x1(2) >= -1,... % state constraint
 x2'*x2 <= 0,... % performance = reach the origin
 mom(g2) - g0 == mom(diff(g1,x1)*f)); % linear moment constraints

% solve
[status,obj] = msol(P);
obj = scaling*obj;
```

```
disp(['Minimum time = ' num2str(tmin)]);
disp(['LMI ' int2str(d) ' lower bound = ' num2str(obj)])
```

For the initial condition $\mathbf{x}_0 = [1\ \ 1]$ the exact minimum time is equal to 3.5. In Table D.1 we report the monotone non decreasing sequence of lower bounds obtained by solving moment problems with test functions of increasing degrees. We used the above script and the semidefinite solver SeDuMi 1.1R3.

Table D.1 Minimum time optimal control for double integrator with state and input constraints: lower bounds on exact minimal time 3.5 achieved by solving moment problems with test functions of increasing degrees.

degree	4	6	8	10	12	14	16
bound	2.3700	2.5640	2.9941	3.3635	3.4813	3.4964	3.4991

D.6 Notes and Sources

The material of this chapter is taken from Henrion *et al.* (2009a). To our knowledge, GloptiPoly 3 is the first software package to solve (or approximate) the generalized moment problem (1.1) and its extensions of Chapter 4.

The software SOSTOOLS of Prajna *et al.* (2002) is dedicated to solving problems involving sums of squares by building up a hierarchy of semidefinite programs in the spirit of (5.15) but allowing products of the polynomials (g_j) as in Schmüdgen's Positivstellensatz in Theorem 2.13.

SparsePOP described in Waki *et al.* (2009) is software that implements the sparse semidefinite relaxations (5.17)-(5.18). In particular it can also build up a sparsity pattern $I = \{1, \ldots, n\} = \cup_{j=1}^{p} I_j$ only from the initial data f, g_j, and with *no* a priori knowledge of any sparsity pattern.

Finally ,YALMIP developed by J. Löfberg, is a Matlab toolbox for rapid prototyping of optimization problems, also implements the moment and s.o.s. approach. See e.g. `http://control.ee.ethz.ch/~joloef/yalmip.php`

Glossary

- $\mathbb{N}$, the set of natural numbers.
- $\mathbb{Z}$, the set of integers.
- $\mathbb{Q}$, the set rational numbers.
- $\mathbb{R}$, the set of real numbers.
- $\mathbb{R}_+$, the set of nonnegative real numbers.
- $\mathbb{C}$, the set of complex numbers.
- $\leq$, less than or equal to
- $\leqq$, inequality "$\leq$" or equality "$=$"
- $\mathbf{A}$, matrix in $\mathbb{R}^{m\times n}$,
- $\mathbf{A}_j$, column j of matrix $\mathbf{A}$.
- $\mathbf{A} \succeq 0$ ($\succ 0$), $\mathbf{A}$ is positive semidefinite (definite)
- x, scalar $x \in \mathbb{R}$
- $\mathbf{x}$, vector $\mathbf{x} = (x_1, \ldots, x_n) \in \mathbb{R}^n$
- $\boldsymbol{\alpha}$, vector $\boldsymbol{\alpha} = (\alpha_1, \ldots, \alpha_n) \in \mathbb{N}^n$
- $|\boldsymbol{\alpha}|$, $= \sum_{i=1}^n \alpha_i$ for $\boldsymbol{\alpha} \in \mathbb{N}^n$.
- $\mathbb{N}_d^n$, $\subset \mathbb{N}^n$, the set $\{\boldsymbol{\alpha} \in \mathbb{N}^n : |\boldsymbol{\alpha}| \leq d\}$

- $\mathbf{x}^{\boldsymbol{\alpha}}$, vector $\mathbf{x}^{\boldsymbol{\alpha}} = (x_1^{\alpha_1} \cdots x_n^{\alpha_n})$, $\mathbf{x} \in \mathbb{C}^n$ or $\mathbf{x} \in \mathbb{R}^n$, $\boldsymbol{\alpha} \in \mathbb{N}^n$.
- $\mathbb{R}[x]$; ring of real univariate polynomials
- $\mathbb{R}[\mathbf{x}]$, $= \mathbb{R}[x_1, \ldots, x_n]$, ring of real multivariate polynomials
- $(\mathbf{x}^{\boldsymbol{\alpha}})$, canonical monomial basis of $\mathbb{R}[\mathbf{x}]$
- $V_{\mathbb{C}}(I) \subset \mathbb{C}^n$, the algebraic variety associated with an ideal $I \subset \mathbb{R}[\mathbf{x}]$
- $\sqrt{I}$, the radical of an ideal $I \subset \mathbb{R}[\mathbf{x}]$
- $\sqrt[\mathbb{R}]{I}$, the real radical of an ideal $I \subset \mathbb{R}[\mathbf{x}]$
- $I(V(I))$, $\subset \mathbb{C}^n$, the vanishing ideal $\{f \in \mathbb{R}[\mathbf{x}] : f(\mathbf{z}) = 0 \,\forall \mathbf{z} \in V_{\mathbb{C}}(I)\}$.
- $V_{\mathbb{R}}(I)$, $\subset \mathbb{R}^n$, the real variety associated with an ideal $I \subset \mathbb{R}[\mathbf{x}]$
- $I(V_{\mathbb{R}}(I))$, $\subset \mathbb{R}[\mathbf{x}]$, the real vanishing ideal $\{f \in \mathbb{R}[\mathbf{x}] : f(\mathbf{x}) = 0 \,\forall \mathbf{x} \in$

$V_{\mathbb{R}}(I)\}$.

- $\mathbb{R}[\mathbf{x}]_t$, $\subset \mathbb{R}[\mathbf{x}]$, real multivariate polynomials of degree at most t
- $(\mathbb{R}[\mathbf{x}])^*$, the vector space of linear forms on $\mathbb{R}[\mathbf{x}]$
- $(\mathbb{R}[\mathbf{x}]_t)^*$, the vector space of linear forms on $\mathbb{R}[\mathbf{x}]_t$
- $\mathbf{y} = (y_{\boldsymbol{\alpha}}) \subset \mathbb{R}$, moment sequence indexed in the canonical basis of $\mathbb{R}[\mathbf{x}]$
- $\mathbf{M}_i(\mathbf{y})$, moment matrix of order i associated with the sequence $\mathbf{y}$
- $\mathbf{M}_i(g\,\mathbf{y})$, localizing matrix of order i associated with the sequence $\mathbf{y}$ and $g \in \mathbb{R}[\mathbf{x}]$
- $P(g)$, $\subset \mathbb{R}[\mathbf{x}]$, preordering generated by the polynomials $(g_j) \subset \mathbb{R}[\mathbf{x}]$
- $Q(g)$, $\subset \mathbb{R}[\mathbf{x}]$, quadratic module generated by the polynomials $(g_j) \subset \mathbb{R}[\mathbf{x}]$

- $\operatorname{co}\mathbf{X}$, convex hull of $\mathbf{X} \subset \mathbb{R}^n$
- $\widehat{f}$, convex envelope of $f : \mathbb{R}^n \to \mathbb{R}$
- $B(\mathbf{X})$, space of bounded measurable functions on $\mathbf{X}$.
- $C(\mathbf{X})$, space of bounded continuous functions on $\mathbf{X}$.
- $\mathcal{M}(\mathbf{X})$, vector space of finite signed Borel measures on $\mathbf{X} \subset \mathbb{R}^n$
- $\mathcal{M}(\mathbf{X})_+$, $\subset \mathcal{M}(\mathbf{X})$, space of finite Borel measures on $\mathbf{X} \subset \mathbb{R}^n$
- $\mathcal{P}(\mathbf{X})$, $\subset \mathcal{M}(\mathbf{X})_+$, space of Borel probability measures on $\mathbf{X} \subset \mathbb{R}^n$
- $L_1(\mathbf{X}, \mu)$, Banach of functions on $\mathbf{X} \subset \mathbb{R}^n$ such that $\int_{\mathbf{X}} |f| d\mu < \infty$.
- $L_\infty(\mathbf{X}, \mu)$, Banach space of measurable functions on $\mathbf{X} \subset \mathbb{R}^n$ such that $\|f\|_\infty := \text{ess sup } |f| < \infty$.
- $\sigma(\mathscr{X}, \mathscr{Y})$, weak topology on $\mathscr{X}$ for a dual pair $(\mathscr{X}, \mathscr{Y})$ of vector spaces.
- $\mu_n \Rightarrow \mu$, weak convergence for a sequence $(\mu_n)_n \subset \mathcal{M}(\mathbf{X})_+$
- $\nu \ll \mu$, ν is absolutely continuous with respect to μ (for measures)
- $\uparrow$, monotone convergence for non decreasing sequences.
- $\downarrow$, monotone convergence for non increasing sequences.

- s.o.s., sum of squares
- LP, linear programming
- SDP, semidefinite programming
- GMP, Generalized Moment Problem
- SDr, semidefinite representation (or semidefinite representable)

Bibliography

Adams, W. and Loustaunau, P. (1994). *An Introduction to Gröbner Bases* (American Mathematical Society, Providence, RI).

Akhiezer, N. I. (1965). *The Classical Moment Problem* (Hafner Publishing Company, New York, NY).

Anastassiou, G. A. (1993). *Moments in Probability and Approximation Theory* (Longman Scientific and Technical, New York, NY).

Anderson, E. and Nash, P. (1987). *Linear Programming in Infinite-Dimensional Spaces* (John Wiley and Sons, Chichester).

Andronov, V. G., Belousov, E. G. and Shironin, V. M. (1982). On solvability of the problem of polynomial programming, *Izvestija Akadem. Nauk SSSR, Teckhnicheskaja Kibernetika* **4**, pp. 194–197.

Anjos, M. (2001). *New Convex Relaxations for the Maximum Cut and VLSI Layout Problems*, Ph.D. thesis, University of Waterloo, Ontario, Canada, orion.math.uwaterloo.ca/~hwolkowi.

Ash, R. B. (1972). *Real Analysis and Probability* (Academic Press, Inc., San Diego).

Balas, E., Ceria, S. and Cornuéjols, G. (1993). A lift-and-project cutting plane algorithm for mixed 0/1 programs, *Math. Program.* **58**, pp. 295–324.

Barvinok, A. (1993). Computing the volume, counting integral points and exponentials sums, *Discr. Comp. Geom.* **10**, pp. 123–141.

Basu, S., Pollack, R. and Roy, M.-F. (2003). *Algorithms in Real Algebraic Geometry* (Springer-Verlag, Berlin).

Bayer, C. and Teichmann, J. (2006). The proof of Tchakaloff's theorem, *Proc. Amer. Soc.* **134**, pp. 3035–3040.

Becker, E. and Schwartz, N. (1983). Zum Darstellungssatz von Kadison-Dubois, *Arch. Math.* **40**, pp. 421–428.

Belisle, C., Romeijn, E. and Smith, R. (1993). Hit-and-run algorithms for generating multivariate distributions, *Math. Oper. Res.* **18**, pp. 255–266.

Belousov, E. G. and Klatte, D. (2002). A Frank-Wolfe type theorem for convex polynomial programs, *Comp. Optim. Appl.* **22**, pp. 37–48.

Ben-Tal, A., Ghaoui, L. E. and Nemirovski, A. (2000). Robustness, in H. Wolkowicz, R. Saigal and L. Vandenberghe (eds.), *Handbook of Semidefinite Programming: Theory, Algorithms, and Applications* (Kluwer Academic Publishers, Boston, MA), pp. 139–162.

Ben-Tal, A., Ghaoui, L. E. and Nemirovski, A. (2006). Foreword: special issue on robust optimization, *Math. Program.* **107**, pp. 1–3.

Ben-Tal, A. and Nemirovski, A. (2001). *Lectures on Modern Convex Optimization* (SIAM, Philadelphia).

Berg, C. (1987). The multidimensional moment problem and semi-groups, in *Moments in Mathematics* (American Mathematical Society, Providence, RI), pp. 110–124.

Bernstein, S. (1921). Sur la représentation des polynômes positifs, *Math. Z.* **9**, pp. 74–109.

Bertsimas, D. and Caramanis, C. (2006). Bounds on linear PDEs via semidefinite optimization, *Math. Program.* **108**, pp. 135–158.

Bertsimas, D., Doan, X. V. and Lasserre, J. B. (2008). Approximating integrals of multivariate exponentials: A moment approach, *Oper. Res. Letters* **36**, pp. 205–210.

Bertsimas, D. and Popescu, I. (2002). On the relation between option and stock prices: a convex optimization approach, *Oper. Res.* **50**, pp. 358–374.

Bertsimas, D. and Popescu, I. (2005). Optimal inequalities in probability theory: A convex optimization approach, *SIAM J. Optim.* **15**, pp. 780–804.

Bertsimas, D. and Sethuraman, J. (2000). Moment problems and semidefinite optimization, in H. Wolkovicz, R. Saigal and L. Vandenberghe (eds.), *Handbook on Semidefinite Programming: Theory, Algorithms, and Applications* (Kluwer Academic Publishers, Boston), pp. 469–509.

Bertsimas, D. and Tsitsiklis, J. (1997). *Introduction to Linear Optimization* (Athena Scientific, Belmont, Massachusetts).

Blekherman, G. (2006). There are significantly more nonnegative polynomials than sums of squares, *Israel J. Math.* **153**, pp. 355–380.

Bochnak, J., Coste, M. and Roy, M.-F. (1998). *Real Algebraic Geometry* (Springer Verlag, New York, NY).

Bollobás, B. (1997). Volume estimates and rapid mixing, in *Flavors of Ge-*

ometry (MSRI Publications, AMS, Providence, RI), pp. 151–180.

Bonnans, F. and Shapiro, A. (2000). *Perturbation Analysis of Optimization Problems* (Springer, New York).

Borkar, V. (2002). Convex analytic methods in Markov Decision processes, in E. A. Feinberg and A. Shwartz (eds.), *Hanbook of Markov Decision Processes* (Kluwer Academic Publishers, Dordrecht), pp. 377–408.

Borwein, J. and Lewis, A. (1991a). Convergence of best entropy estimates, *SIAM J. Optim.* **1**, pp. 191–205.

Borwein, J. and Lewis, A. (1992). Partially finite convex programming, part i: Quasi relative interiors and duality, *Math. Program.* **57**, pp. 15–48.

Borwein, J. and Lewis, A. S. (1991b). On the convergence of moment problems, *Trans. Amer. Math. Soc.* **325**, pp. 249–271.

Brighi, B. and Chipot, M. (1994). Approximated convex envelope of a function, *SIAM J. Num. Anal.* **31**, pp. 128–148.

Büeler, B., Enge, A. and Fukuda, K. (2000). Exact volume computation for polytopes : A practical study, in G. Kalai and G. M. Ziegler (eds.), *Polytopes - Combinatorics and Computation* (Birhäuser Verlag, Basel), pp. 131–154.

Byrnes, C. and Lindquist, A. (2006). The generalized moment problem with complexity constraint, *Int. Equat. Oper. Theory* **56**, pp. 163–180.

Capuzzo-Dolcetta, I. and Lions, P. L. (1990). Hamilton-Jacobi equations with state constraints, *Trans. Amer. Math. Soc.* **318**, pp. 643–683.

Carleman, T. (1926). *Les Fonctions Quasi-Analytiques* (Gauthier-Villars, Paris, France).

Cassier, G. (1984). Problème des moments sur un compact de $\mathbb{R}^n$ et représentation de polynômes à plusieurs variables, *J. Funct. Anal.* **58**, pp. 254–266.

Chebyshev, P. (1874). Sur les valeurs limites des intégrales, *J. Math. Pures. Appl.* **19**, pp. 157–160.

Chen, X. and Deng, X. (2006). Settling the complexity of two-player Nash equilibrium, in *47th Annual IEEE Symposium on Foundations of Computer Science (FOCS'06)*, pp. 261–272.

Chlamtac, E. (2007). Approximation algorithms using hierarchies of semidefinite programming relaxations, in *48th Annual IEEE Symposium on Foundations of Computer Science (FOCS'07)*, pp. 691–701.

Chlamtac, E. and Singh, G. (2008). Improved approximation guarantees through higher levels of SDP hierarchies, in *Approximation Randomization and Combinatorial Optimization Problems, LNCS*, Vol. 5171 (Springer), pp. 49–62.

Choquet, G. (1969). *Lectures on Analysis*, Vol. 1, chap. Integration and Topological Vector Spaces (W. A. Benjamin, Inc.).

Chua, C. B. and Tuncel, L. (2008). Invariance and efficiency of convex representations, *Math. Program.* **111**, pp. 113–140.

Cohen, J. and Hickey, T. (1979). Two algorithms for determining volumes of convex polyhedra, *J. ACM* **26**, pp. 401–414.

Curran, M. (1992). Beyond average intelligence, *Risk* **5**, pp. 50–60.

Curto, R. and Fialkow, L. (1991). Recursiveness, positivity, and truncated moment problems, *Houston Math. J.* **17**, pp. 603–635.

Curto, R. and Fialkow, L. (1996). Solution of the truncated complex moment problem for flat data, *Mem. Amer. Math. Soc.* **119**.

Curto, R. and Fialkow, L. (1998). Flat extensions of positive moment matrices: recursively generated relations, *Mem. Amer. Math. Soc.* **136**.

Curto, R. and Fialkow, L. (2000). The truncated complex $\mathbb{K}$-moment problem, *Trans. Amer. Math. Soc.* **352**, pp. 2825–2855.

Dantzig, G. B. (1963). *Linear Programming and Extensions* (Princeton University Press, Princeton).

Daskalakis, C., Goldberg, P. W. and Papadimitriou, C. H. (2006). The complexity of computing a Nash equilibrium, in *Proceedings of the 38th Annual ACM Symposium on Theory of Computing*, pp. 71–78.

Datta, R. S. (2009). Finding all Nash equilibria of a finite game using polynomial algebra, *Economic Theory* To appear.

Dawson, D. A. (1980). Qualitative behavior of geostochastic systems, *Stoc. Proc. Appl.* **10**, pp. 1–31.

de Klerk, E., Laurent, M. and Parrilo, P. (2006). A PTAS for the minimization of polynomials of fixed degree over the simplex, *Theor. Comp. Sci.* **361**, pp. 210–225.

de Klerk, E., Pasechnik, D. V. and Schrijver, A. (2007). Reduction of symmetric semidefinite programs using the regular $\star$-representation, *Math. Program.* **109**, pp. 613–624.

de Oliveira, M., Helton, J., McCullough, S. and Putinar, M. (2008). Engineering systems and free semi-algebraic geometry, in M. Putinar and S. Sullivant (eds.), *Emerging Applications of Algebraic Geometry*, Vol. 149 (Springer, New York, NY), pp. 17–62.

Deun, J. V., Bultheel, A. and Vera, P. G. (2005). On computing rational Gauss-Chebyshev quadrature formulas, *Math. Comp.* **75**, pp. 307–326.

Diaconis, P. (1987). Application of the method of moments in Probability and Statistics, in *Moments in Mathematics* (American Mathematical Society, Providence, RI), pp. 125–142.

Diaconis, P. and Freedman, D. (2006). The Markov moment problem and de Finetti's Theorem Part I, *Math. Z.* **247**, pp. 183–199.

Dresher, M., Karlin, S. and Shapley, L. S. (1950). Polynomial games, in *Contributions to the Theory of Games, Annals of Mathematics Studies*, Vol. 24 (Princeton University Press), pp. 161–180.

Dyer, M. and Frieze, A. (1988). The complexity of computing the volume of a polyhedron, *SIAM J. Comp.* **17**, pp. 967–974.

Dyer, M., Frieze, A. and Kannan, R. (1991). A random polynomial-time algorithm for approximating the volume of convex bodies, *J. ACM* **38**, pp. 1–17.

Fekete, M. (1935). Proof of three propositions of Paley, *Bull. Amer. Math. Soc.* **41**, pp. 138–144.

Feller, W. (1966). *An Introduction to Probability Theory and Its Applications, 2nd Edition* (John Wiley & Sons, New York).

Fleming, W. H. and Vermes, D. (1989). Convex duality approach to the optimal control of diffusions, *SIAM J. Control Optim.* **27**, pp. 1136–1155.

Fletcher, R. (1980). *Practical Methods of Optimization. Vol. 1. Unconstrained Optimization* (John Wiley and Sons Ltd., Chichester).

Floudas, C. A., Pardalos, P. M., Adjiman, C. S., Esposito, W. R., Gümüs, Z. H., Harding, S. T., Klepeis, J. L., Meyer, C. A. and Schweiger, C. A. (1999). *Handbook of Test Problems in Local and Global Optimization* (Kluwer, Boston, MA), titan.princeton.edu/TestProblems.

Gaitsgory, V. and Rossomakhine, S. (2006). Linear programming approach to deterministic long run average problems of optimal control, *SIAM J. Control Optim.* **44**, pp. 2006–2037.

Gaterman, K. and Parrilo, P. (2004). Symmetry group, semidefinite programs and sums of squares, *J. Pure Appl. Alg.* **192**, pp. 95–128.

Gautschi, W. (1981). A survey of Gauss-Christoffel quadrature formulae, in P. Butzer and F. Féher (eds.), *E.B. Christoffel (Aachen/Monschau, 1979)* (Birkhäuser, Basel), pp. 72–147.

Gautschi, W. (1997). *Numerical Analysis: An Introduction* (Birkhäuser, Boston).

Georgiou, T. (2006). Relative entropy and the multi-variable multi-dimensional moment problem, *IEEE Trans. Inform. Theory* **52**, pp. 1052–1066.

Gill, P. E., Murray, W. and Wright, M. H. (1981). *Practical Optimization* (Academic Press, Inc., London-New York).

Glasserman, P., Heidelberger, P. and Shahabuddin, P. (1999). Asymptotically optimal importance sampling and stratification for pricing path dependent options, *Math. Finance* **9**, pp. 117–152.

Godwin, H. (1955). On generalizations of Tchebycheff's inequality, *J. Amer. Stat. Assoc.* **50**, pp. 923–945.

Godwin, H. (1964). *Inequalities on Distribution Functions* (Charles Griffin and Co., London).

Golub, G. and Loan, C. V. (1996). *Matrix Computations* (The John Hopkins University Press, 3rd edition, New York).

Goursat, E. (1894). Solution, *Intermed. des Math.* **1**, p. 251.

Govindan, S. and Wilson, R. (2003). A global Newton method to compute Nash equilibria, *J. Econ. Theory* **110**, pp. 65–86.

Grötschel, M., Lovász, L. and Schrijver, A. (1988). *Geometric Algorithms and Combinatorial Optimization* (Springer-Verlag, New York, New York).

Gürkan, G. and Pang, J. S. (2009). Approximations of Nash equilibria, *Math. Program.* **117**, pp. 223–253.

Handelman, D. (1988). Representing polynomials by positive linear functions on compact convex polyhedra, *Pac. J. Math.* **132**, pp. 35–62.

Handy, C. R. (2001). Generating converging eigenenergy bounds for the discrete states of the $-ix^3$ non-hermitian potential, *J. Phys. A* **34**, pp. L271–L277.

Handy, C. R. and Bessis, D. (1985). Rapidly convergent lower bounds for the Schrödinger-equation ground-state energy, *Phys. Rev. Letters* **55**, pp. 931–934.

Handy, C. R., Bessis, D. and Morley, T. (1988). Generating quantum energy bounds by the moment method: a linear programming approach, *Phys. Rev. A* **37**, pp. 4557–4569.

Handy, C. R., Mantica, G. and B.Gibbons, J. (1989). Quantization of lattice Schrödinger operators via the trigonometric moment problem, *Phys. Rev. A* **39**, pp. 3256–3259.

Hanzon, B. and Jibetean, D. (2003). Global minimization of a multivariate polynomial using matrix methods, *J. Global Optim.* **27**, pp. 1–23.

Harrison, J. M. and Kreps, D. (1979). Martingales and arbitrage in multiperiod security markets, *J. Econ. Theory* **20**, pp. 381–408.

Hartl, R. F., Sethi, S. P. and Vickson, R. G. (1995). A survey of the maximum principles for optimal control problems with state constraints, *SIAM Rev.* **37**, pp. 181–218.

Hausdorff, F. (1915). Summationsmethoden und momentfolgen i, *Soobshch. Kharkov Matem. ob-va ser. 2* **14**, pp. 227–228.

Haviland, E. K. (1935). On the momentum problem for distributions in more than one dimension, I. *Amer. J. Math.* **57**, pp. 562–568.

Haviland, E. K. (1936). On the momentum problem for distributions in more than one dimension, II. *Amer. J. Math.* **58**, pp. 164–168.

Helmes, K., Röhl, S. and Stockbridge, R. H. (2001). Computing moments of the exit time distribution for Markov processes by linear programming, *Oper. Res.* **49**, pp. 516–530.

Helmes, K. and Stockbridge, R. H. (2000). Numerical comparison of controls and verification of optimality for stochastic control problems, *J. Optim. Theory Appl.* **106**, pp. 107–127.

Helton, J. W., Lasserre, J. B. and Putinar, M. (2008). Measures with zeros in the inverse of their moment matrix, *Ann. Prob.* **36**, pp. 1453–1471.

Helton, J. W. and Nie, J. (2010). Semidefinite representation of convex sets, *Math. Program.* To appear.

Helton, J. W. and Nie, J. (2009a). Sufficient and necessary conditions for semidefinite representability of convex hulls and sets, *SIAM J. Optim.* **20**, pp. 759–791.

Helton, J. W. and Putinar, M. (2007). Positive Polynomials in Scalar and Matrix Variables, the Spectral Theorem and Optimization, in M. Bakonyi, A. Gheondea and M. Putinar (eds.), *Operator Theory, Structured Matrices and Dilations* (Theta, Bucharest), pp. 229–306.

Helton, J. W. and Vinnikov, V. (2007). Linear matrix inequality representation of sets, *Comm. Pure Appl. Math.* **60**, pp. 654–674.

Henrion, D. and Lasserre, J. B. (2003). GloptiPoly: Global optimization over polynomials with Matlab and SeDuMi, *ACM Trans. Math. Software* **29**, pp. 165–194.

Henrion, D., Lasserre, J. B. and Löfberg, J. (2009a). GloptiPoly 3: Moments, optimization and semidefinite programming, *Optim. Meth. Softwares* To appear.

Henrion, D., Lasserre, J. B. and Savorgnan, C. (2008). Nonlinear optimal control synthesis via occupation measures, in *Proceedings of the 47th IEEE Conference on Decision and Control*, pp. 4749–4754.

Henrion, D., Lasserre, J. B. and Savorgnan, C. (2009b). Approximate volume and integration for basic semi-algebraic sets, *SIAM Review* **51**, pp. 722–743.

Herings, P. J.-J. and Peeters, R. (2010). Homotopy methods to compute equilibria in Game Theory, *Econom. Theory* **42**, pp. 119–156.

Herings, P. J.-J. and van den Elzen, A. H. (2002). Computation of the Nash equilibrium selected by the tracing procedure in N-person games, *Games and Economic Behavior* **38**, pp. 89–117.

Hernandez-Hernandez, D., Hernández-Lerma, O. and Taksar, M. (1996). The linear programming approach to deterministic optimal control problems, *Appl. Math. (Warsaw)* **24**, pp. 17–33.

Hernández-Lerma, O. and Lasserre, J. B. (1996). *Discrete-Time Markov Control Processes: Basic Optimality Criteria* (Springer Verlag, New York).

Hernández-Lerma, O. and Lasserre, J. B. (1998a). Approximation schemes for infinite linear programs, *SIAM J. Optim.* **20**, pp. 192–215.

Hernández-Lerma, O. and Lasserre, J. B. (1998b). Linear programming approximations for Markov control processes in metric spaces, *Acta Appl. Math.* **51**, pp. 123–139.

Hernández-Lerma, O. and Lasserre, J. B. (1999). *Further Topics in Discrete-Time Markov Control Processes* (Springer Verlag, New York).

Hernández-Lerma, O. and Lasserre, J. B. (2003). *Markov Chains and Invariant Probabilities* (Birkhäuser Verlag, Basel).

Hiriart-Urruty, J.-B. (1998). *Optimisation et Analyse Convexe* (Presses Universitaires de France, Paris).

Hiriart-Urruty, J.-B. and Lemarechal, C. (1993). *Convex Analysis and Minimization Algorithms II* (Springer-Verlag, Berlin).

Isii, K. (1960). The exrema of probability determined by generalized moments. I. Bounded random variables, *Ann. Inst. Stat. Math.* **12**, pp. 119–133.

Isii, K. (1963). On the sharpness of Chebyshev-type inequalities, *Ann. Inst. Stat. Math.* **14**, pp. 185–197.

Jacobi, T. and Prestel, A. (2001). Distinguished representations of strictly positive polynomials, *J. Reine. Angew. Math.* **532**, pp. 223–235.

Jacobson, D., Lele, M. and Speyer, J. L. (1971). New necessary conditions of optimality for control problems with state-variable inequality constraints, *J. Math. Anal. Appl.* **35**, pp. 255–284.

Jibetean, D. and de Klerk, E. (2006). Global optimization of rational functions: a semidefinite programming approach, *Math. Program.* **106**, pp. 93–109.

Jibetean, D. and Laurent, M. (2005). Semidefinite approximations for global unconstrained polynomial optimization, *SIAM J. Optim.* **16**, pp. 490–514.

Jung, C. M. and Stockbridge, R. H. (2002). Linear programming formulation for optimal stopping problems, *SIAM J. Control Optim.* **40**, pp. 1965–1982.

Karlin, S. and Studden, W. (1966). *Tchebycheff Sytems: With Applications in Analysis and Statistics* (John Wiley and Sons, New York).

Kemperman, J. H. B. (1987). Geometry of the moment problem, in *Moments in Mathematics* (American Mathematical Society, Providence, RI), pp. 110–124.

Khachian, L. G. (1979). A polynomial algorithm for linear programming, *Soviet Mathematics Doklady* **20**, pp. 191–194.

Kim, S., Kojima, M. and Waki, H. (2009). Exploiting sparsity in SDP relaxation for sensor network localization, *SIAM J. Optim.* To appear.

Kojima, M., Kim, S. and Maramatsu, M. (2005). Sparsity in sums of squares of squares of polynomials, *Math. Program.* **103**, pp. 45–62.

Kojima, M. and Maramatsu, M. (2007). An extension of sums of squares relaxations to polynomial optimization problems over symmetric cones, *Math. Program.* **110**, pp. 315–336.

Kojima, M. and Maramatsu, M. (2009). A note on sparse SOS and SDP relaxations for polynomial optimization problems over symmetric cones, *Comput. Optim. Appl.* **42**, pp. 31–41.

Krein, M. G. and Nudel'man, A. (1977). *The Markov Moment Problem and Extremal Problems, Transl. Math. Monographs*, Vol. 50 (American Mathematical Society, Providence, RI).

Krivine, J. L. (1964a). Anneaux préordonnés, *J. Anal. Math.* **12**, pp. 307–326.

Krivine, J. L. (1964b). Quelques propriétés des préordres dans les anneaux commutatifs unitaires, *C.R. Acad. Sci. Paris* **258**, pp. 3417–3418.

Krylov, N. and Bogolioubov, N. (1937). La théorie générale de la mesure dans son application à l'étude des systèmes de la mécanique non linéaires, *Ann. Math.* **38**, pp. 65–113.

Kuhlmann, S., Marshall, M. and Schwartz, N. (2005). Positivity, sums of squares and the multi-dimensional moment problem II, *Adv. Geom.* **5**, pp. 583–606.

Kuhlmann, S. and Putinar, M. (2007). Positive polynomials on fibre products, *C. R. Acad. Sci. Paris, Ser. 1* **1344**, pp. 681–684.

Kuhlmann, S. and Putinar, M. (2009). Positive polynomials on projective limits of real algebraic varieties, *Bull. Sci. Math.* .

Kurtz, T. G. and Stockbridge, R. H. (1998). Existence of Markov controls and characterization of optimal Markov controls, *SIAM J. Control Optim.* **36**, pp. 609–653.

Landau, H. (1987a). Classical background of the moment problem, in *Moments in Mathematics* (American Mathematical Society, Providence, RI), pp. 1–15.

Landau, H. (1987b). *Moments in Mathematics*, Vol. 37 (Proc. Sympos. Appl. Math.).

Laraki, R. and Lasserre, J. B. (2008a). Computing uniform convex approximations for convex envelopes and convex hulls, *J. Convex Anal.* **15**, pp. 635–654.

Laraki, R. and Lasserre, J. B. (2012). Semidefinite programming for min-max problems and games, *Math. Program. Sér A* **131**, pp. 305–332.

Lasserre, J. B. (1983). An analytical expression and an algorithm for the volume of a convex polyhedron in R^n, *J. Optim. Theory Appl.* **39**, pp. 363–377.

Lasserre, J. B. (2000). Optimisation globale et théorie des moments, *C. R. Acad. Sci. Paris, Série I* **331**, pp. 929–934.

Lasserre, J. B. (2001). Global optimization with polynomials and the problem of moments, *SIAM J. Optim.* **11**, pp. 796–817.

Lasserre, J. B. (2002a). Bounds on measures satisfying moment conditions, *Adv. Appl. Prob.* **12**, pp. 1114–1137.

Lasserre, J. B. (2002b). An explicit equivalent positive semidefinite program for nonlinear 0-1 programs, *SIAM J. Optim.* **12**, pp. 756–769.

Lasserre, J. B. (2002c). Polynomials nonnegative on a grid and discrete optimization, *Trans. Amer. Math. Soc.* **354**, pp. 631–649.

Lasserre, J. B. (2002d). Semidefinite programming vs. LP relaxations for polynomial programming, *Math. Oper. Res.* **27**, pp. 347–360.

Lasserre, J. B. (2004). Polynomial programming: LP-relaxations also converge, *SIAM J. Optim.* **15**, pp. 383–393.

Lasserre, J. B. (2005). SOS approximations of polynomials nonnegative on a real algebraic set, *SIAM J. Optim.* **16**, pp. 610–628.

Lasserre, J. B. (2006a). Convergent SDP-relaxations in polynomial optimization with sparsity, *SIAM J. Optim.* **17**, pp. 822–843.

Lasserre, J. B. (2006b). The moment problem with bounded density, Tech. Rep. 06446, LAAS-CNRS, Toulouse, France.

Lasserre, J. B. (2006c). A Positivstellensatz which preserves the coupling pattern of variables, Tech. Rep. 06146, LAAS-CNRS, Toulouse, France.

Lasserre, J. B. (2006d). Robust global optimization with polynomials, *Math. Program.* **107**, pp. 275–293.

Lasserre, J. B. (2006e). A sum of squares approximation of nonnegative polynomials, *SIAM J. Optim* **16**, pp. 751–765, also in SIAM Rev. **49**, pp. 651–669, 2007.

Lasserre, J. B. (2007a). Semidefinite programming for gradient and Hessian computation in maximum entropy estimation, in *46th IEEE Conference on Decision and Control*, pp. 3060–3064.

Lasserre, J. B. (2007b). Sufficient conditions for a real polynomial to a sum of squares, *Arch. Math.* **89**, pp. 390–398.

Lasserre, J. B. (2008a). Representation of nonnegative convex polynomials, *Arch. Math.* **91**, pp. 126–130.

Lasserre, J. B. (2008b). A semidefinite programming approach to the generalized problem of moments, *Math. Program.* **112**, pp. 65–92.

Lasserre, J. B. (2009a). Convex sets with semidefinite representation, *Math. Program.* **120**, pp. 457–477.

Lasserre, J. B. (2009b). Convexity in semi-algebraic geometry and polynomial optimization, *SIAM J. Optim.* **19**, pp. 1995–2014.

Lasserre, J. B. (2009c). The $\mathbb{K}$-moment problem with densities, *Math. Program.* **116**, pp. 321–341.

Lasserre, J. B. (2010). Certificates of convexity for basic semi-algebraic sets, *Applied Math. Letters* **23**, pp. 912–916.

Lasserre, J. B., Henrion, D., Prieur, C. and Trélat, E. (2008a). Nonlinear optimal control via occupation measures and LMI-relaxations, *SIAM J. Control Optim.* **47**, pp. 1649–1666.

Lasserre, J. B., Laurent, M. and Rostalski, P. (2008b). Semidefinite characterization and computation of zero-dimensional real radical ideals, *Found. Comp. Math.* **8**, pp. 607–647.

Lasserre, J. B., Laurent, M. and Rostalski, P. (2008c). A unified approach to computing real and complex zeros of zero-dimensional ideals, in M. Putinar and S. Sullivant (eds.), *Emerging Applications of Algebraic Geometry*, IMA Book Series (Springer), pp. 125–156.

Lasserre, J. B., Laurent, M. and Rostalski, P. (2009). A prolongation-projection algorithm for computing the finite real variety of an ideal, *Theor. Comp. Sci.* **410**, pp. 2685–2700.

Lasserre, J. B. and Netzer, T. (2007). SOS approximations of nonnegative polynomials via simple high degree perturbations, *Math. Z.* **256**, pp. 99–112.

Lasserre, J. B. and Priéto-Rumeau, T. (2004). SDP vs. LP relaxations for the moment approach in some performance evaluation problems, *Stoch. Models* **20**, pp. 439–456.

Lasserre, J. B., Priéto-Rumeau, T. and Zervos, M. (2006). Pricing a class of exotic options via moments and SDP relaxations, *Math. Finance* **16**, pp. 469–494.

Lasserre, J. B. and Zeron, E. (2001). A Laplace transform algorithm for the volume of a convex polytope, *J. ACM* **48**, pp. 1126–1140.

Laurent, M. (2003). A comparison of the Sherali-Adams, Lovász-Schrijver and Lasserre relaxations for 0-1 programming, *Math. Oper. Res.* **28**, pp. 470–496.

Laurent, M. (2005). Revisiting two theorems of Curto and Fialkow on moment matrices, *Proc. Amer. Math. Soc.* **133**, pp. 2965–2976.

Laurent, M. (2007a). Semidefinite representations for finite varieties, *Math. Program.* **109**, pp. 1–26.

Laurent, M. (2007b). Strengthened semidefinite programming bounds for codes, *Math. Program.* **109**, pp. 239–261.

Laurent, M. (2008). Sums of squares, moment matrices and optimization over polynomials, in M. Putinar and S. Sullivant (eds.), *Emerging Applications of Algebraic Geometry*, Vol. 149 (Springer, New York, NY), pp. 157–270.

Lawrence, J. (1991). Polytope volume computation, *Math. Comp.* **57**, pp. 259–271.

Lemke, C. E. and Howson, J. T. (1964). Equilibrium points of bimatrix games, *J. SIAM* **12**, pp. 413–423.

Lewis, A. S., Parrilo, P. and Ramana, M. V. (2005). The Lax conjecture is true, *Proc. Amer. Math. Soc.* **133**, pp. 2495–2499.

Lipton, R. and Markakis, E. (2004). Nash equilibria via polynomial equations, in *Proceedings of the Latin American Symposium on Theoretical Informatics, Buenos Aires, Argentina*, Lecture Notes in Computer Sciences (Springer Verlag), pp. 413–422.

Löfberg, J. (2004). Yalmip : a toolbox for modeling and optimization in Matlab, in *Proceedings of the IEEE International Symposium on Computer-Aided Control System Design (CACSD), Taipei, Taiwan*, pp. 284–290.

Lovász, L. and Schrijver, A. (1991). Cones of matrices and set-functions and 0-1 optimization, *SIAM J. Optim.* **1**, pp. 166–190.

Markov, A. (1884). *On Certain Applications of Algebraic Continued Fractions*, Ph.D. thesis, University of St Petersburg, St Petersburg, Russia.

Marshall, M. (2008). *Positive Polynomials and Sums of Squares, AMS Math. Surveys and Monographs*, Vol. 146 (American Mathematical Society, Providence, RI).

Marshall, M. (2009). Representation of non-negative polynomials, degree bounds and applications to optimization, *Canad. J. Math.* **61**, pp. 205–221.

Marshall, M. (2010). Polynomials non-negative on a strip, *Proc. Amer. Math. Soc.* **138**, pp. 1559–1567.

Maurer, H. (1977). On optimal control problems with bounded state variables and control appearing linearly, *SIAM J. Control Optim.* **15**, pp. 345–362.

Mead, L. R. and Papanicolaou, N. (1984). Maximum entropy in the problem of moments, *J. Math. Phys.* **25**, pp. 2404–2417.

Mourrain, B. and Trébuchet, P. (2005). Generalized normal forms and polynomials system solving, in *Proceedings of the 2005 International Symposium on Symbolic and Algebraic Computation (ISSAC 2005)*, pp. 253–260, distinguished paper award.

Mulholland, H. P. and Rogers, C. A. (1958). Representation theorems for distribution functions, *Proc. London Math. Soc.* **8**, pp. 177–223.

Musiela, M. and Rutkowski, M. (1997). *Martingale Methods in Financial Modelling* (Springer, New-York).

Navascués, M., Prironio, S. and Acin, A. (2008). A convergent hierarchy of semidefinite programs characterizing the set of quantum correlations, *New J. Physics* **10**, pp. 1–28.

Nesterov, Y. (2000). Squared functional systems and optimization problems, in H. Frenk, K. Roos, T. Terlaky and S. Zhang (eds.), *High Performance Optimization* (Kluwer, Dordrecht), pp. 405–440.

Nie, J., Demmel, J. and Sturmfels, B. (2006). Semidefinite approximations for global unconstrained polynomial optimization, *Math. Program.* **106**, pp. 587–606.

Nie, J. and Schweighofer, M. (2007). On the complexity of Putinar's Positivstellensatz, *J. Complexity* **23**, pp. 135–150.

Niederreiter, H. (1992). *Random Number Generation and Quasi-Monte Carlo Methods* (SIAM, Philadelphia).

Nussbaum, A. E. (1966). Quasi-analytic vectors, *Ark. Mat.* **5**, pp. 179–191.

Papadimitriou, C. H. (1994). On the complexity of the parity argument and other inefficient proofs of existence, *J. Comp. Syst. Sci.* **48-3**, pp. 498–532.

Parrilo, P. (2002). An explicit construction of distinguished representations of polynomials nonnegative over finite sets, Tech. rep., IfA, ETH, Zurich, Switzerland, technical report AUTO-02.

Parrilo, P. (2006). Polynomial games and sum of squares optimization, in *Proceedings of the 45th IEEE Conference on Decision and Control*, pp. 2855–2860.

Parrilo, P. (2007). Exact semidefinite representations for genus zero curves,

Tech. rep., MIT, Cambridge, Massachusetts, USA.

Parrilo, P. A. (2000). *Structured Semidefinite Programs and Semialgebraic Geometry Methods in Robustness and Optimization*, Ph.D. thesis, California Institute of Technology, Pasadena, CA.

Parrilo, P. A. (2003). Semidefinite programming relaxations for semialgebraic problems, *Math. Program.* **96**, pp. 293–320.

Pesch, H. J. (1994). A practical guide to the solution of real-life optimal control problems, *Control Cybernet.* **23**, pp. 7–60.

Pironio, S., Navascués, M. and Acin, A. (2009). Convergent relaxations of polynomial optimization problems with non commuting variables, Tech. rep., Group of Applied Physics, University of Geneveva, Geneva, Switzerland.

Pólya, G. (1974). *Collected Papers. Vol. II* (MIT Press, Cambridge, Mass-London).

Pólya, G. and Szegö, G. (1976). *Problems and Theorems in Analysis II* (Springer-Verlag).

Powers, V. and Reznick, B. (2000). Polynomials that are positive on an interval, *Trans. Amer. Soc.* **352**, pp. 4677–4692.

Prajna, S., Papachristodoulou, A. and Parrilo, P. A. (2002). Introducing SOSTOOLS: a general purpose sum of squares programming solver, in *Proceedings of the 41st IEEE Conference on Decison and Control* (Las Vegas, USA), pp. 741–746.

Prajna, S. and Rantzer, A. (2007). Convex programs for temporal verification of nonlinear dynamical systems, *SIAM J. Control Optim.* **46**, pp. 999–1021.

Prestel, A. and Delzell, C. N. (2001). *Positive Polynomials* (Springer Verlag, Berlin).

Prieur, C. and Trélat, E. (2005). Robust optimal stabilization of the brockett integrator via a hybrid feedback, *Math. Control Sign. Syst.* **17**, pp. 201–216.

Putinar, M. (1993). Positive polynomials on compact semi-algebraic sets, *Ind. Univ. Math. J.* **42**, pp. 969–984.

Putinar, M. (2000). A note on Tchakaloff's theorem, *Proc. Amer. Soc.* **125**, pp. 2409–2414.

Reznick, B. (2000). Some concrete aspects of Hilbert's 17th problem, in *Real Algebraic Geometry and Ordered Structures; Contemporary Mathematics, 253* (American Mathematical Society, Providence, RI), pp. 251–272.

Risler, J.-J. (1970). Une caractérisation des idéaux des variétés algébriques réelles, *C.R. Acad. Sci. Paris* **271**, pp. 1171–1173.

Rockafellar, R. T. (1970). *Convex Analysis* (Princeton University Press, Princeton, New Jersey).

Rogers, L. C. and Shi, Z. (1995). The value of an asian option, *J. Appl. Prob.* **32**, pp. 1077–1088.

Rosenmüller, J. (1971). On a generalization of the Lemke-Howson algorithm to non cooperative N-person games, *SIAM J. Appl. Math.* **21**, pp. 73–79.

Roughgarden, T. (2010). Computing equilibria: a computational complexity perspective, *Econom. Theory* **42**, pp. 193–236.

Rouillier, F. (1999). Solving zero-dimensional systems through the rational univariate representation, *J. Appl. Alg. Eng. Com. Comp.* **9**, 5, pp. 433–461.

Rouillier, F. and Zimmermann, P. (2003). Efficient isolation of polynomial real roots, *J. Comp. Appl. Math.* **162**, 1, pp. 33–50.

Savani, R. and von Stengel, B. (2006). Hard-to-solve bimatrix games, *Econometrica* **74**, pp. 397–429.

Scheiderer, C. (2008). Positivity and sums of squares: A guide to some recent results, in M. Putinar and S. Sullivant (eds.), *Emerging Applications of Algebraic Geometry*, IMA Book Series (Springer), pp. 271–324.

Scherer, C. W. and Hol, C. W. J. (2004). Sum of qsquares relaxation for polynomial semidefinite programming, in *Proceeding of the 16th International Symposium on Mathematical Theory of Newtworks and Systems, Leuven*, pp. 1–10.

Schmid, J. (1998). *On the degree complexity of Hilbert's 17th problem and the Real Nullstellensatz*, Ph.D. thesis, University of Dortmund, habilitationsschrift zur Erlangug der Lehrbefignis für das Fach Mathematik an der Universität Dortmund.

Schmüdgen, K. (1991). The K-moment problem for compact semi-algebraic sets, *Math. Ann.* **289**, pp. 203–206.

Schmüdgen, K. (2008). Noncommutative real algebraic geometry: some basic concepts and first ideas, in M. Putinar and S. Sullivant (eds.), *Emerging Applications of Algebraic Geometry*, Vol. 149 (Springer, New York, NY), pp. 325–350.

Schneider, R. (1994). *Convex Bodies: The Brunn–Minkowski Theory* (Cambridge University Press, Cambridge, United Kingdom).

Schoenebeck, G. (2008). Linear level Lasserre lower bounds for certain k-CSPs, in *49th Annual IEEE Symposium on Foundations of Computer Science (FOCS'08)*, pp. 593–602.

Schrijver, A. (2005). New codes upper bounds from the Terwilliger algebra and semidefinite programming, *IEEE Trans. Inform. Theory* **51**, pp.

2859–2866.

Schweighofer, M. (2005a). On the complexity of Schmüdgen's Positivstellensatz, *J. Complexity* **20**, pp. 529–543.

Schweighofer, M. (2005b). Optimization of polynomials on compact semialgebraic sets, *SIAM J. Optim.* **15**, pp. 805–825.

Schweighofer, M. (2006). Global optimization of polynomials using gradient tentacles and sums of squares, *SIAM J. Optim.* **17**, pp. 920–942.

Schwerer, E. (2001). A linear programming approach to the steady-state analysis of reflected brownian motion, *Stoch. Models* **17**, pp. 341–368.

Shah, P. and Parrilo, P. (2007). Polynomial stochastic games via sum of squares optimization, in *Proceedings of the 46th IEEE Conference on Decision and Control*, pp. 745–750.

Shapiro, A. (2001). On duality theory of conic linear problems, in M. A. Goberna and M. A. Lopez (eds.), *Semi-Infinite Programming: Recent Advances* (Kluwer Academic Publishers, Dordrecht), pp. 135–165.

Sherali, H. D. and Adams, W. P. (1990). A hierarchy of relaxations between the continuous and convex hull representations for zero-one programming problems, *SIAM J. Discr. Math.* **3**, pp. 411–430.

Sherali, H. D. and Adams, W. P. (1999). *A Reformulation-Linearization Technique for Solving Discrete and Continuous Nonconvex Problems* (Kluwer, Dordrecht, MA).

Sherali, H. D., Adams, W. P. and Tuncbilek, C. H. (1992). A global optimization algorithm for polynomial programming problems using a reformulation-linearization technique, *J. Global Optim.* **2**, pp. 101–112.

Shohat, J. A. and Tamarkin, J. D. (1943). *The Problem of Moments*, Vol. 1 (Amer. Math. Soc., Providence, RI), Mathematical Surveys.

Shor, N. Z. (1970). Utilization of the operation of space dilation in the minimizaton of convex functions, *Cybernetics* **6**, pp. 7–15.

Shor, N. Z. (1987). Quadratic optimization problems, *Tekhnicheskaya Kibernetika* **1**, pp. 128–139.

Shor, N. Z. (1998). *Nondifferentiable Optimization and Polynomial Problems* (Kluwer, Dordrecht).

Simon, B. (1998). The classical moment problem as a self-adjoint finite difference operator, *Adv. Math.* **137**, pp. 82–203.

Smith, J. (1995). Generalized Chebyshev inequalities: Theory and applications in decision analysis, *Oper. Res.* **43**, pp. 807–825.

Sommese, A. and Wampler, C. (2005). *The Numerical Solution of Systems of Polynomials Arising in Engineering and Science.* (World Scientific Press, Singapore).

Soner, M. H. (1986). Optimal control with state-space constraints. I. *SIAM J. Control Optim.* **24**, pp. 552–561.

Stengle, G. (1974). A Nullstellensatz and a Positivstellensatz in semialgebraic geometry, *Math. Ann.* **207**, pp. 87–97.

Stetter, H. J. (2004). *Numerical Polynomial Algebra* (Society for Industrial and Applied Mathematics, Philadelphia, PA, USA).

Stoer, J. and Bulirsch, R. (2002). *Introduction to Numerical Analysis* (Springer-Verlag, New York).

Stoyanov, J. (2001). Moment problems related to the solutions of stochastic differential equations, in *Stochastic Theory and Control*, Lecture Notes in Control and Information (Springer, Berlin), pp. 459–469.

Sturm, J. F. (1999). Using SeDuMi 1.02, a Matlab toolbox for optimizing over symmetric cones, *Opt. Meth. Softwares* **11-12**, pp. 625–653.

Sturmfels, B. (2002). *Solving Systems of Polynomial Equations* (American Mathematical Society, Providence, RI).

Tagliani, A. (2002a). Entropy estimate of probability densities having assigned moments: Hausdorff case, *Appl. Math. Letters* **15**, pp. 309–314.

Tagliani, A. (2002b). Entropy estimate of probability densities having assigned moments: Stieltjes case, *Appl. Math. Comp.* **130**, pp. 201–211.

Tchakaloff, V. (1957). Formules de cubature mécanique à coefficients non négatifs, *Bull. Sci. Math.* **81**, pp. 123–134.

Tong, Y. (1980). *Probability Inequalities in Multivariate Distributions* (Academic Press, New York).

Tuncel, L. (2000). Potential reduction and primal-dual methods, in H. Wolkowicz, R. Saigal and L. Vandenberghe (eds.), *Handbook of Semidefinite Programming: Theory, Algorithms, and Applications* (Kluwer Academic Publishers, Boston, MA), pp. 235–265.

Turnbull, S. M. and Wakeman, L. M. (1991). A quick algorithm for pricing european average options, *J. Fin. Quant. Anal.* **26**, pp. 377–389.

Vallentin, F. (2009). Symmetry in semidefinite programs, *Linear Alg. Appl.* **430**, pp. 360–369.

van den Elzen, A. H. and Talman, A. J. J. (1991). Procedure for finding Nash equilibria in bi-matrix games, *ZOR - Meth. Mod. Oper. Res.* **35**, pp. 27–43.

Vandenberghe, L. and Boyd, S. (1996). Semidefinite programming, *SIAM Rev.* **38**, pp. 49–95.

Vasilescu, F.-H. (2003). Spectral measures and moment problems, in *Spectral Theory and Its Applications*, Theta Ser. Adv. Math. 2 (Theta, Bucharest), pp. 173–215.

Verschelde, J. (1999). PHCPACK: A general-purpose solver for polynomial systems by homotopy continuation, *ACM Trans. Math. Software* **25**, pp. 251–276.

Vinter, R. (1993). Convex duality and nonlinear optimal control, *SIAM J. Control Optim.* **31**, pp. 518–538.

von Stryk, O. and Bulirsch, R. (1992). Direct and indirect methods for trajectory optimization, *Ann. Oper. Res.* **37**, pp. 357–373.

Vui, H. H. and Son, P. T. (2009). Global optimization of polynomials using the truncated tangency variety and sums of squares, *SIAM J. Optim.* **19**, pp. 941–951.

Waki, H., Kim, S., Kojima, M., Muramatsu, M. and Sugimoto, H. (2009). Algorithm 883: SparsePOP–a sparse semidefinite programming relaxation of polynomial optimization problems, *ACM Trans. Math. Software* **35**, pp. 90–104.

Waki, S., Kim, S., Kojima, M. and Maramatsu, M. (2006). Sums of squares and semidefinite programming relaxations for polynomial optimization problems witth structured sparsity, *SIAM J. Optim.* **17**, pp. 218–242.

Wilson, R. (1971). Computing equilibria of N-person games, *SIAM J. Appl. Math.* **21**, pp. 80–87.

Yudin, D. B. and Nemirovski, A. (1977). Informational complexity and efficient methods for the solution of convex extremal problems, *Matekon* **13**, pp. 25–45.

Zhi, L. and Reid, G. (2004). Solving nonlinear polynomial system via symbolic-numeric elimination method, in J. Faugère and F. Rouillier (eds.), *Proceedings of the International Conference on Polynomial System Solving*, pp. 50–53.

Index

algebraic set, 300
algorithm
  extraction, 80, 118, 158, 165
  global optimization, 118
  local optimization, 128
  moment-matrix, 155
arbitrage, 5, 193
Artin, 15

Banach-Alaoglu Theorem, 317
barrier call option, 197
Brownian motion, 197

Carleman, 60, 72
cone, 293, 294
constraint qualification, 312
coset, 156
Cox-Ingersoll-Ross model, 197
cubature, 307
Curto-Fialkow, 72

dual pair, 182, 316, 317
duality
  conic, 7
  gap, 316
  strong, 8, 179, 316
  weak, 8, 315, 318
duality gap, 112, 183, 318, 319
duality theory, 112

eigenvalue method, 158
equations
  polynomial, 148, 293
  stochastic differential, 197
ergodic criteria, 187
exact arithmetic, 161
extension
  flat, 57
  positive, 57

field, 293
  algebraically closed, 293
  ordered, 293
  real, 294
  real closed, 300
field extension
  algebraic, 293
  transcendental, 293

games
  N-player, 273
  zero-sum polynomial, 276
generic element, 155
GloptiPoly, 117, 120, 122, 321
GMP, 3
Goursat Lemma, 48
Gröbner basis, 161

Hilbert, 15
Hilbert basis theorem, 295

ideal, 26, 295
  radical, 38, 134, 149, 296
  zero-dimensional, 37, 38, 134, 149,

156
inf-compact, 306
infinitesimal generator, 198
integrality gap, 145

Jacobi-Prestel, 72

Karush-Kuhn-Tucker (KKT), 127

Lagrange multipliers
KKT, 143
s.o.s. polynomial, 129, 143
Lagrangian, 313
Linear Matrix Inequality, 314
LMI, 314

marginal, 308
Markov chain, 181
martingale, 196, 198
matrix
Hankel moment, 56
localizing, 60
moment, 58
multiplication, 157
measure
atomic, 62
determinate, 52, 201
Dirac, 62
ergodic, 188
exit location, 199
invariant, 182
Lebesgue, 163
marginal, 175, 309
occupation, 199
representing, 52
weak convergence of, 9, 93, 305
min-max optimization problem, 265
moment function, 306
moment problem
full, 51
generalized, 3, 6, 15
Hamburger, 54
Hausdorff, 54
one-dimensional, 52
Stieltjes, 54
truncated, 51

Monte Carlo, 191

Nash equilibria, 273
NullStellenSatz
real, 298
strong Hilbert, 297
weak Hilbert, 297
Nussbaum, 72

optimality conditions
Fritz-John, 311
global, 127, 129
local KKT, 127, 130, 312
option pricing, 193
Ornstein-Uhlenbeck process, 197

Polyá, 25
positive extension
flat, 62
Positivstellensatz
Putinar, 29
Schmüdgen, 29
Stengle, 28
preordering, 26, 236, 298
probability space, 197
problem
mass-transfer, 163, 175
Monge-Kantorovich, 163, 176
program
dual linear, 317
linear, 317
primal linear, 317
semidefinite, 314
programming
conic, 313
linear, 313, 314
semidefinite, 313
Prohorov Theorem, 306

quadratic module, 39, 299, 300
quotient algebra, 156
quotient space, 156

rank sufficient condition, 119
relaxations
linear, 86, 111, 125

semidefinite, 75, 111, 148, 165, 200
sparse semidefinite, 94
representations
sparse, 39
Riesz-Haviland, 53
ring, 295
quotient, 38, 295
running intersection property, 42, 93, 308

s.o.s.-convex, 231
SDP, 313
semi-algebraic set, 26, 300
basic, 15, 28, 66, 73
semidefinite constraints, 75, 314
semidefinite optimization, 110
separation problem, 11
Slater's condition, 8, 128, 313
SOOSTOOLS, 337
SparsePOP, 337
spectral decomposition, 77
Stengle, 26
Stickelberger, 156
stochastic kernel, 181
sum of squares, 15, 17, 22
decomposition, 17
support, 307

Tchakaloff, 307
tightness, 305
topology
weak, 305, 316
weak $\star$, 220, 317
transition probability function, 186

variety, 37, 296
algebraic, 147
finite, 38
real, 37, 147

weak-Feller, 182

www.ingramcontent.com/pod-product-compliance
Lightning Source LLC
Chambersburg PA
CBHW061618220326
41598CB00026BA/3802
9781911299738